IET ENERGY ENGINEERING 128

Characterization of Wide Bandgap Power Semiconductor Devices

Other volumes in this series:

Volume 1	**Power Circuit Breaker Theory and Design** C.H. Flurscheim (Editor)
Volume 4	**Industrial Microwave Heating** A.C. Metaxas and R.J. Meredith
Volume 7	**Insulators for High Voltages** J.S.T. Looms
Volume 8	**Variable Frequency AC Motor Drive Systems** D. Finney
Volume 10	**SF_6 Switchgear** H.M. Ryan and G.R. Jones
Volume 11	**Conduction and Induction Heating** E.J. Davies
Volume 13	**Statistical Techniques for High Voltage Engineering** W. Hauschild and W. Mosch
Volume 14	**Uninterruptible Power Supplies** J. Platts and J.D. St Aubyn (Editors)
Volume 15	**Digital Protection for Power Systems** A.T. Johns and S.K. Salman
Volume 16	**Electricity Economics and Planning** T.W. Berrie
Volume 18	**Vacuum Switchgear** A. Greenwood
Volume 19	**Electrical Safety: A guide to causes and prevention of hazards** J. Maxwell Adams
Volume 21	**Electricity Distribution Network Design, 2nd Edition** E. Lakervi and E.J. Holmes
Volume 22	**Artificial Intelligence Techniques in Power Systems** K. Warwick, A.O. Ekwue and R. Aggarwal (Editors)
Volume 24	**Power System Commissioning and Maintenance Practice** K. Harker
Volume 25	**Engineers' Handbook of Industrial Microwave Heating** R.J. Meredith
Volume 26	**Small Electric Motors** H. Moczala *et al.*
Volume 27	**AC–DC Power System Analysis** J. Arrillaga and B.C. Smith
Volume 29	**High Voltage Direct Current Transmission, 2nd Edition** J. Arrillaga
Volume 30	**Flexible AC Transmission Systems (FACTS)** Y.-H. Song (Editor)
Volume 31	**Embedded generation** N. Jenkins *et al.*
Volume 32	**High Voltage Engineering and Testing, 2nd Edition** H.M. Ryan (Editor)
Volume 33	**Overvoltage Protection of Low-Voltage Systems, Revised Edition** P. Hasse
Volume 36	**Voltage Quality in Electrical Power Systems** J. Schlabbach *et al.*
Volume 37	**Electrical Steels for Rotating Machines** P. Beckley
Volume 38	**The Electric Car: Development and future of battery, hybrid and fuel-cell cars** M. Westbrook
Volume 39	**Power Systems Electromagnetic Transients Simulation** J. Arrillaga and N. Watson
Volume 40	**Advances in High Voltage Engineering** M. Haddad and D. Warne
Volume 41	**Electrical Operation of Electrostatic Precipitators** K. Parker
Volume 43	**Thermal Power Plant Simulation and Control** D. Flynn
Volume 44	**Economic Evaluation of Projects in the Electricity Supply Industry** H. Khatib
Volume 45	**Propulsion Systems for Hybrid Vehicles** J. Miller
Volume 46	**Distribution Switchgear** S. Stewart
Volume 47	**Protection of Electricity Distribution Networks, 2nd Edition** J. Gers and E. Holmes
Volume 48	**Wood Pole Overhead Lines** B. Wareing
Volume 49	**Electric Fuses, 3rd Edition** A. Wright and G. Newbery
Volume 50	**Wind Power Integration: Connection and system operational aspects** B. Fox *et al.*
Volume 51	**Short Circuit Currents** J. Schlabbach
Volume 52	**Nuclear Power** J. Wood
Volume 53	**Condition Assessment of High Voltage Insulation in Power System Equipment** R.E. James and Q. Su
Volume 55	**Local Energy: Distributed generation of heat and power** J. Wood
Volume 56	**Condition Monitoring of Rotating Electrical Machines** P. Tavner, L. Ran, J. Penman and H. Sedding
Volume 57	**The Control Techniques Drives and Controls Handbook, 2nd Edition** B. Drury
Volume 58	**Lightning Protection** V. Cooray (Editor)
Volume 59	**Ultracapacitor Applications** J.M. Miller
Volume 62	**Lightning Electromagnetics** V. Cooray
Volume 63	**Energy Storage for Power Systems, 2nd Edition** A. Ter-Gazarian
Volume 65	**Protection of Electricity Distribution Networks, 3rd Edition** J. Gers
Volume 66	**High Voltage Engineering Testing, 3rd Edition** H. Ryan (Editor)
Volume 67	**Multicore Simulation of Power System Transients** F.M. Uriate
Volume 68	**Distribution System Analysis and Automation** J. Gers
Volume 69	**The Lightening Flash, 2nd Edition** V. Cooray (Editor)
Volume 70	**Economic Evaluation of Projects in the Electricity Supply Industry, 3rd Edition** H. Khatib
Volume 72	**Control Circuits in Power Electronics: Practical issues in design and implementation** M. Castilla (Editor)
Volume 73	**Wide Area Monitoring, Protection and Control Systems: The enabler for smarter grids** A. Vaccaro and A. Zobaa (Editors)
Volume 74	**Power Electronic Converters and Systems: Frontiers and applications** A.M. Trzynadlowski (Editor)
Volume 75	**Power Distribution Automation** B. Das (Editor)
Volume 76	**Power System Stability: Modelling, analysis and control** B. Om P. Malik
Volume 78	**Numerical Analysis of Power System Transients and Dynamics** A. Ametani (Editor)
Volume 79	**Vehicle-to-Grid: Linking electric vehicles to the smart grid** J. Lu and J. Hossain (Editors)
Volume 81	**Cyber-Physical-Social Systems and Constructs in Electric Power Engineering** S. Suryanarayanan, R. Roche and T.M. Hansen (Editors)
Volume 82	**Periodic Control of Power Electronic Converters** F. Blaabjerg, K. Zhou, D. Wang and Y. Yang
Volume 86	**Advances in Power System Modelling, Control and Stability Analysis** F. Milano (Editor)
Volume 87	**Cogeneration: Technologies, optimisation and implementation** C.A. Frangopoulos (Editor)
Volume 88	**Smarter Energy: From smart metering to the smart grid** H. Sun, N. Hatziargyriou, H.V. Poor, L. Carpanini and M.A. Sánchez Fornié (Editors)
Volume 89	**Hydrogen Production, Separation and Purification for Energy** A. Basile, F. Dalena, J. Tong and T.N. Veziroğlu (Editors)
Volume 90	**Clean Energy Microgrids** S. Obara and J. Morel (Editors)
Volume 91	**Fuzzy Logic Control in Energy Systems with Design Applications in MATLAB®/Simulink®** İ.H. Altaş
Volume 92	**Power Quality in Future Electrical Power Systems** A.F. Zobaa and S.H.E.A. Aleem (Editors)
Volume 93	**Cogeneration and District Energy Systems: Modelling, analysis and optimization** M.A. Rosen and S. Koohi-Fayegh
Volume 94	**Introduction to the Smart Grid: Concepts, technologies and evolution** S.K. Salman
Volume 95	**Communication, Control and Security Challenges for the Smart Grid** S.M. Muyeen and S. Rahman (Editors)
Volume 97	**Synchronized Phasor Measurements for Smart Grids** M.J.B. Reddy and D.K. Mohanta (Editors)
Volume 98	**Large Scale Grid Integration of Renewable Energy Sources** A. Moreno-Munoz (Editor)
Volume 100	**Modeling and Dynamic Behaviour of Hydropower Plants** N. Kishor and J. Fraile-Ardanuy (Editors)
Volume 101	**Methane and Hydrogen for Energy Storage** R. Carriveau and D.S.-K. Ting
Volume 104	**Power Transformer Condition Monitoring and Diagnosis** A. Abu-Siada (Editor)
Volume 108	**Fault Diagnosis of Induction Motors** J. Faiz, V. Ghorbanian and G. Joksimović
Volume 110	**High Voltage Power Network Construction** K. Harker
Volume 112	**Wireless Power Transfer: Theory, technology and application** N. Shinohara
Volume 119	**Thermal Power Plant Control and Instrumentation: The control of boilers and HRSGs, 2nd Edition** D. Lindsley, J. Grist and D. Parker
Volume 123	**Power Systems Electromagnetic Transients Simulation, 2nd Edition** N. Watson and J. Arrillaga
Volume 124	**Power Market Transformation** B. Murray
Volume 130	**Wind and Solar Based Energy Systems for Communities** R. Carriveau and D.S.-K. Ting (Editors)
Volume 131	**Metaheuristic Optimization in Power Engineering** J. Radosavljević
Volume 905	**Power System Protection, 4 volumes**

Characterization of Wide Bandgap Power Semiconductor Devices

Fei (Fred) Wang, Zheyu Zhang, and Edward A. Jones

The Institution of Engineering and Technology

Published by The Institution of Engineering and Technology, London, United Kingdom

The Institution of Engineering and Technology is registered as a Charity in England & Wales (no. 211014) and Scotland (no. SC038698).

First published 2018

The Institution of Engineering and Technology
Michael Faraday House
Six Hills Way, Stevenage
Herts, SG1 2AY, United Kingdom

www.theiet.org

British Library Cataloguing in Publication Data
A catalogue record for this product is available from the British Library

ISBN 978-1-78561-491-0 (hardback)
ISBN 978-1-78561-492-7 (PDF)

Typeset in India by MPS Limited
Printed in the UK by CPI Group (UK) Ltd, Croydon

Contents

Biographies

Fei (Fred) Wang is Professor and Condra Chair of Excellence in Power Electronics, and Technical Director of US NSF-DOE Engineering Research Center for Ultra-wide-area Resilient Electric Energy Transmission Networks (CURENT) at the Department of Electrical Engineering and Computer Science at The University of Tennessee, Knoxville, USA. He also holds a joint appointment with Oak Ridge National Lab. Prof. Wang has published over 400 journal and conference papers, authored 3 book chapters, and holds 15 US patents. His achievements have resulted in IEEE IAS Gerald Kliman Innovation Award, seven IEEE prize paper awards, the Dushman Award – GE's highest award for technical team contributions, and three University of Tennessee Faculty Research Achievements Awards. He is a fellow of IEEE and a fellow of US National Academy of Inventors. His research mainly focuses on wide bandgap device-based power electronics, and power electronics applications in transportation, motor drives, and renewable energy systems and power grids.

Zheyu Zhang is a Lead Power Electronics Engineer with General Electric Global Research, Niskayuna, NY, USA. He received the B.S. and M.S. degrees from Huazhong University of Science and Technology, Wuhan, China, and the Ph.D. degree from The University of Tennessee, Knoxville, TN, in 2008, 2011, and 2015, respectively, all in electrical engineering. After his graduation, Zheyu stayed as a Research Assistant Professor in the Department of Electrical Engineering and Computer Science at the University of Tennessee, Knoxville from 2015 to 2018. Also, he was a research and development intern at General Electric Global Research, Shanghai, China in 2011. He has published over 60 papers in most prestigious journals and conference proceedings, four patent applications with one licensed, and two IEEE tutorial seminars. He was the recipient of two prize paper awards from the IEEE Industry Applications Society and IEEE Power Electronics Society. His research interests include wide band-gap semiconductors and application; gate driver technology; motor drive design, operation, integration and reliability; and cryogenic power electronics.

Edward A. Jones is a Senior Applications Engineer with Efficient Power Conversion Corporation. He completed his Ph.D. at The University of Tennessee in 2017, where he was a Chancellor's Fellow, a CURENT Fellow, and a Bredesen Energy Sciences and Engineering Fellow. He earned his B.S.E.E. in 2007 at Virginia Tech, where he was awarded the prestigious Bradley Scholarship and co-founded the Virginia Tech chapter of Engineers Without Borders. He has published over 20 peer-reviewed IEEE papers, an IEEE tutorial seminar, and a patent. His current research interests include wide bandgap device characterization and modeling, and GaN-based converter design, including thermal improvement and integrated overcurrent protection.

Acknowledgments

This book made use of Engineering Research Center Shared Facilities supported by the Engineering Research Center Program of the National Science Foundation and DOE under NSF Award Number EEC-1041877 and the Center for Ultra-wide-area Resilient Electric Energy Transmission Networks (CURENT) Industry Partnership Program. The authors would like to acknowledge the following people for their direct or indirect help: Bo Liu, Paige Williford, Kyle Goodrick, Ren Ren, and Shiqi Ji of CURENT at UTK , Zhiqiang (Jack) Wang of Oak Ridge National Laboratory, Ben Guo of United Technology Research Center, Weimin Zhang of Tesla, Zhuxian Xu and Fan Xu of Ford, Eric Persson, Tim McDonald, and Deepak Veereddy of Infineon, Julian Styles and Peter Di Maso of GaN Systems, Teja Kothamasu, Michael Alimov of Keysight, Adam Barkley, Jeff Cassidy, and Zheng (Henry) Chen of Wolfspeed, Andrew Lemmon of University of Alabama, and Raghav Khanna of University of Toledo. In particular, the authors are very grateful for Shiqi Ji's contribution to the high-voltage device protection and safety consideration content in Chapter 7.

Chapter 1

Introduction

This book focuses on how to characterize the emerging wide bandgap (WBG) power semiconductor switching devices, including silicon carbide (SiC) and gallium nitride (GaN) devices. Power semiconductor switching devices are at the heart of modern power electronics converters. Today's commercial power semiconductor devices are still dominated by the mature and well-established silicon (Si) technology. Since the advent of Si thyristors in the 1950s, Si power semiconductor devices have gone through many generations of development in the last 60 years and are approaching material theoretical limitations in terms of blocking voltage, operation temperature, and conduction and switching characteristics. These intrinsic physical limits become a barrier to achieving higher performance power conversion.

The emergence of WBG semiconductor devices promises to revolutionize next-generation power electronics converters. Compared with Si devices, WBG devices feature high breakdown electric field, low specific on-resistance, fast switching speed, and high junction temperature capability. All of these characteristics are beneficial for the efficiency, power density, specific power, and/or reliability of power electronics converters. The WBG devices under rapid development and commercialization include SiC and GaN devices, with SiC mainly targeting high-voltage high-power (600 V, kW or above) applications, and GaN for low-voltage low-power (600 V, kW or below) applications [1, 2]. With high breakdown electric field, WBG devices have higher voltage capability, lower loss, and smaller packages; with low specific on-resistance and fast switching speed, WBG devices have lower conduction and switching loss, and, together with high-temperature capability, the cooling requirements can be reduced; with fast switching speed and low switching loss, high switching frequency can be utilized, which can lead to better dynamics and reduced filter and passive component needs in general.

The extremely fast switching and other superior characteristics of WBG devices have nonetheless also posed severe challenges to their applications. The challenges include dv/dt slew rate of greater than 100 V/ns and extremely high corresponding di/dt slew rate pervasive throughout the converters, augmented electromagnetic interference (EMI) emissions, single-device blocking voltages as high as tens of kilovolts with corresponding insulation requirements, switching frequencies in the hundreds of kilohertz to megahertz range, and junction temperatures surpassing 200 °C. All of these have called for a comprehensive reformulation of design procedures developed for Si-based power electronics. Addressing these design and application issues is critical to the adoption and success of WBG power electronics.

One key aspect of power converter design is to obtain the power semiconductor device characteristics. Compared to Si counterparts, special considerations are often needed in characterizing WBG devices due to their distinct features, such as higher voltage, faster switching speed, and smaller size, as mentioned earlier. Many issues specific to WBG device characterization have been identified and new methods have been developed in recent years. This book intends to provide readers with these new developments and updated comprehensive procedures in WBG device characterization.

1.1 Overview of WBG devices

1.1.1 WBG devices in comparison to Si devices

WBG refers to the electronic energy band gaps significantly larger than one electron-volt (eV). WBG materials have several characteristics that make them attractive compared to narrow bandgap Si for power electronics converters. Figure 1.1 highlights some key material properties of WBG semiconductor candidates as compared to traditional Si [3]. Generally speaking, for WBG materials, their energy gap, breakdown electric field, thermal conductivity, melting point, and electron velocity are all significantly higher. These characteristics allow WBG semiconductor-based power devices to operate at much higher voltage, switching frequency, and temperature than Si [4–6].

For example, with the breakdown field higher than that of Si, a thinner drift layer with a higher doping concentration can be used for WBG power devices at the same blocking voltage. For unipolar device such as Schottky diodes and metal

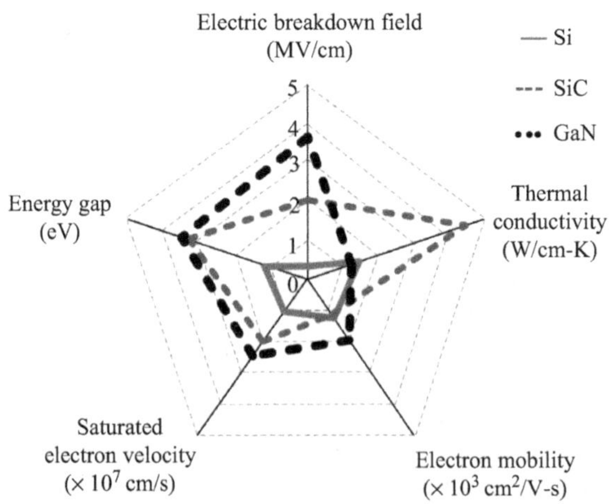

Figure 1.1 Summary of Si, SiC, and GaN relevant material properties [3]

oxide semiconductor field-effect transistors (MOSFETs), the combination of thinner blocking layer and higher doping concentration yields a lower specific on-resistance compared with Si majority carrier devices.

The fast switching-speed capability of WBG devices can be expected due to higher breakdown field and electron velocity. First, with lower on-resistance at the same breakdown voltage, a smaller chip size is achieved in WBG unipolar devices such as MOSFETs. Therefore, the junction capacitance is smaller due to the reduced size, resulting in a fast switching speed. Second, the charges are swept into and out of the junction capacitance at the saturated drift velocity during switching transients. The electron saturated drift velocity of WBG is higher than that of Si, leading to an increased switching speed of WBG devices.

Additionally, the excellent thermal conductivity allows WBG devices' dissipated heat to be readily extracted from the device. Hence, a larger power can be handled by the device at a given junction temperature. Also, higher thermal conductivity, together with the wider bandgap, makes it possible for WBG devices to work at high temperature.

In summary, WBG-based power devices offer low specific on-state resistance, fast switching speed, and high operating temperature and high-voltage capabilities.

1.1.2 WBG device status

There are many III–V and II–VI compound semiconductors (i.e., semiconductor compound composed of elements from periodic table groups III, V and II, VI) with wide bandgaps. Among the possible WBG semiconductor material candidates, SiC and GaN show the best tradeoff among theoretical characteristics (e.g., high blocking voltage capability, high-temperature operation, and high switching speed), real commercial availability of the starting material (e.g., wafers and epitaxial layers), and maturity of their technological processes [5]. Consequently, almost all of the WBG semiconductor power devices on today's market or in research and development phase are SiC and GaN based. This subsection summarizes the available information on SiC and GaN power devices, including device types, voltage/current ratings, as well as status of commercialization in 2017.

1.1.2.1 SiC devices

The availability of high-quality SiC wafers allows a reasonable yield of large-area SiC power devices. Currently, 150-mm or 6-in. SiC wafers are commercially available [7]. Figure 1.2 summarizes the status of SiC-based power devices, including Schottky diodes, PIN diodes, MOSFETs, junction gate FETs (JFETs), insulated gate bipolar transistors (IGBTs), bipolar junction transistors (BJTs), and thyristors with the voltage range from 400 V to 22.6 kV. It is observed that the low-voltage (from 400 V to 1,700 V) SiC devices are becoming commercially available. Among them, the current rating per die approaches up to 100 A, and with multiple dies in parallel, state-of-the-art SiC power modules on market can deliver hundreds of amperes of current. On the other hand, high-voltage SiC devices (referred to here as 3.3 kV and above) are generally in developmental stages with limited commercial availability and small current rating per die [8].

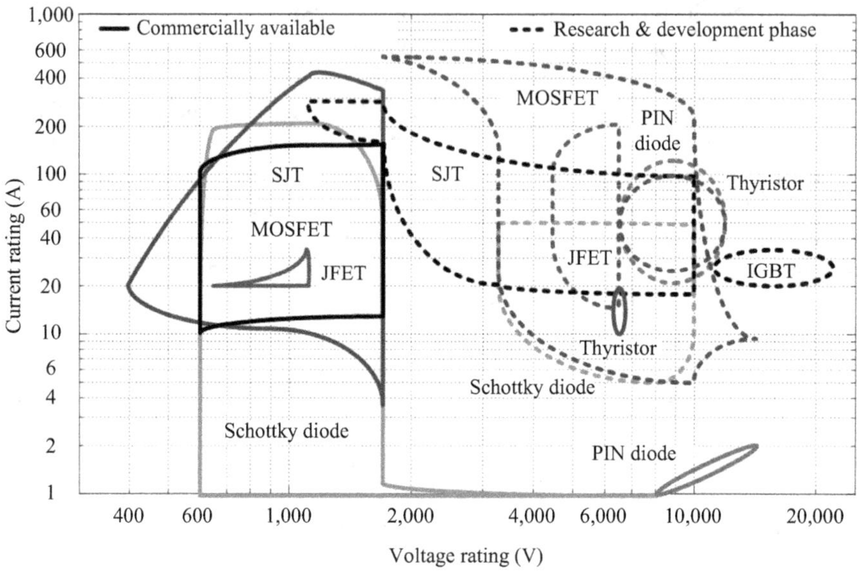

Figure 1.2 Summary of status of SiC power devices (updated till January 2017) [9]

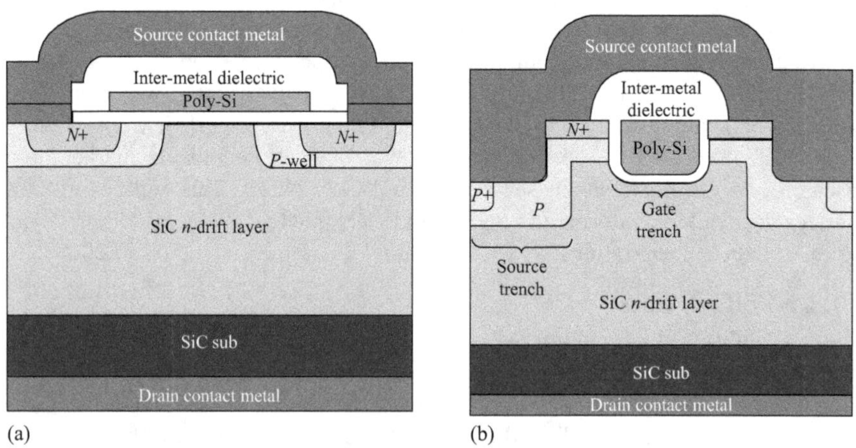

Figure 1.3 Structure of SiC MOSFETs: (a) planar DMOS structure; (b) trench structure

Currently SiC MOFETs are the most developed active switches, with some JFETs, IGBTs, BJTs, and thyristors also available. The commercial SiC MOSFETs are primarily based on two structures: planar double-diffused metal-oxide-semiconductor (DMOS) structure and trench structure, as shown in Figure 1.3.

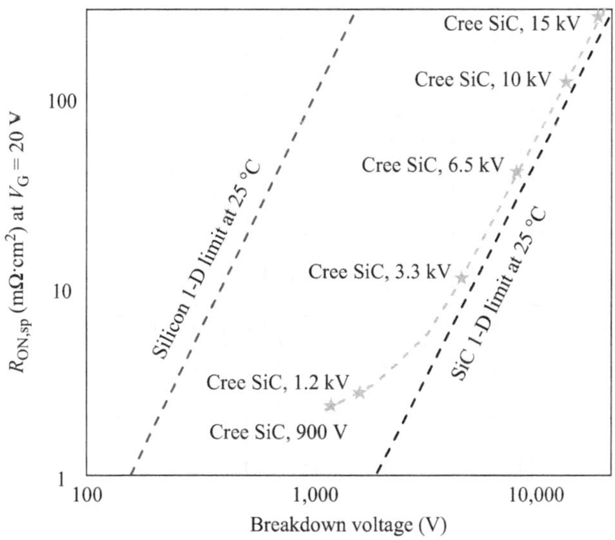

Figure 1.4 Specific on-resistance as function breakdown voltage

With the planar DMOS structure, Figure 1.4 shows the relationship between specific on-resistance and the device breakdown voltage for various SiC MOSFETs from Cree (Wolfspeed) [8]. It can be seen that the SiC MOSFET characteristics closely follow the SiC material one-dimensional (1-D) limit, indicating that the on-resistance is dominated by the drift layer resistance for these devices. Since the specific on-state resistance $R_{ON,sp}$ is proportional to the square of the breakdown voltage for the 1-D limit, it means that the lower voltage devices will have lower normalized resistance. For example, $R_{ON,sp}$ of 6.5 kV device is roughly four times that of a 3.3 kV device. In other words, with today's technology, lower voltage devices (e.g., 3.3 kV) have lower $R_{ON,sp}$ or normalized forward voltage drop.

To overcome this "high-voltage penalty," trench structure is employed, as shown in Figure 1.3(b). It is reported that the trench structure can reduce on-state resistance by half throughout the entire temperature range. Also, the total switching loss reduces by 30% [10]. At present, the "high-voltage penalty" issue is compensated by de-rating the current for high-voltage devices.

For SiC diodes, at low breakdown voltage (<1,700 V), Schottky barrier diodes (SBDs) are popular since they show extremely high switching speed and low on-state loss. But high leakage current and low blocking voltage limit their utilization in high-voltage applications. PIN diodes and junction barrier Schottky (JBS) diodes are preferred in high-voltage applications. Compared with PIN diodes, JBS diodes have excellent reverse recovery characteristic but poor static performance, making them suitable as antiparallel diodes of active switches.

SiC power devices are developing rapidly with several trends highlighted as follows.

Increased current ratings and more available voltage ratings: Current ratings have been significantly improved at both die level and power module level under wide voltage ratings. For example, 650 V/ 17 mΩ discrete SiC MOSFET has been released with the current rating of 118 A at room temperature [10], which is comparable with the current rating of the state-of-the-art 650-V C7 Si CoolMOS on the market. Also, at medium voltage level (e.g., >1.7 kV), several large current power modules have been developed and demonstrated, including but not limited to 1.7 kV/550 A, 3.3 kV/180 A, 10 kV/240 A SiC MOSFETs-based phase-leg power modules, and 6.5 kV/200 A SiC JFETs-based phase-leg power modules [11, 12].

More voltage ratings are available mainly due to the application-orientated consideration. One example is that a 1,000 V SiC MOSFET has been released recently targeting the electric vehicle application. Compared with 900 V and 1,200 V voltage ratings, which are typical values for traditional Si devices, a dedicated 1,000 V SiC MOSFET is developed to achieve better tradeoff between performance (e.g., switching and conduction loss) and reliability (e.g., adequate voltage margin) [13]. As expected, with the acceleration of acceptance and adoption of SiC devices in industrial products, SiC manufacturers may be willing to develop more dedicated power semiconductors to best serve specific markets.

Improved packaging techniques: Advanced packaging techniques have been adopted for SiC devices for parasitic minimization, weight and size reduction, and high-temperature operation.

For example, TO-247-4 pin package with separated Kelvin source has been utilized for discrete SiC MOSFETs at 1,000 and 1,200 V levels [13, 14]. Thanks to the Kelvin source connection and its resultant lower common source inductance, the switching performance can be greatly improved with fast switching and low loss. Also, power modules with <5 nH parasitic inductance are developed for 1,200–1,700 V power modules with more than 400 A current capability [13, 15]. Additionally, ultra-light SiC power modules based on Easy1B PressFIT package are available for high-density power conversion systems [13]. Furthermore, high-temperature packaging techniques are employed so as to allow commercially available SiC devices (e.g., Schottky diodes and supper junction transistors) to operate up to 210 °C [16].

Availability of more power module configuration: In addition to the phase-leg power module, more options with respect to the configuration of power modules are available, such as boost chopper, buck chopper, full bridge, three-level neutral point clamped (NPC), and three-level T-type, triple phase-leg [17]. With this, users are able to more conveniently develop high-performance power electronics circuits based on different topologies for various applications.

Enhanced short-circuit withstand capability: Short-circuit withstand capability is challenging for small and fast SiC devices [18–21]. Compared with traditional Si devices with >10 μs short-circuit withstand time, the typical short-circuit withstand time of SiC MOSFETs is on the order of 1 μs. Recently, an enhanced short-circuit capability SiC MOSFETs has been developed at medium voltage level (>3.3 kV). It is demonstrated that these new devices are capable of sustaining short-circuit current up to 13 μs, which significantly benefits the reliable operation of SiC MOSFETs for voltage source-based high-power conversion system [11].

Better SiC MOSFET's body diode performance and trend of eliminating SBDs in power modules: The body diode of the SiC MOSFET is structurally similar to the *p–n* junction diode formed in the body of a Si MOSFET. The lifetime of minority carriers in SiC is shorter than Si, so the reverse recovery charge is reduced.

To mitigate the reverse recovery induced by MOSFET's body diode, a dedicated SiC SBD is generally added. However, considering the charging of SBD junction capacitance, employing an SBD in parallel does not necessarily result in lower total switching energy loss. Specifically, at room temperature, the charging of SBD junction capacitance is greater than reverse recovery charge introduced by MOSFET's body diode [22]. Under elevated temperature, reverse recovery charge was observed to increase significantly; as a result, the switching energy loss may be higher as compared to the SBD case. Recently, it was demonstrated that for the latest generation Wolfspeed SiC MOSFETs, at 150 °C, the total switching energy loss without SBD does not exceed the loss with employing SBD [22]. Therefore, under wide operating range, the excellent switching performance can be achieved by SiC MOSFET without the extra SBD. Regarding the conduction loss, the channel of SiC MOSFETs, instead of its body diode, can efficiently conduct reverse current. In the end, the penalty without antiparallel SBD with respect to efficiency is limited.

Furthermore, reduced size and cost from eliminating SBDs in a power module can improve the power density and cost of overall power conversion system. Accordingly, it is observed that there is a trend of eliminating SBDs in SiC-based power modules [15].

SiC intelligent power module (IPM): IPMs are advanced power conversion units that combine power semiconductor chips with optimized gate drive and protection circuitry, such as over-current, short-circuit, over-temperature, and under voltage lock out (UVLO). The compact, easily assembled IPMs can be beneficial to reducing system size/weight, cost, and time to market, and have been successfully applied for Si-based high-power conversion system [23].

A similar concept is being leveraged to SiC power semiconductors by different approaches. By replacing Si devices with SiC, SiC-based IPMs are starting to be commercially available [24]. Also, several gate drives optimized for the commercial-off-the-shelf SiC modules were developed and released recently [25, 26]. Although a gate drive is not packaged with the power module in an integrated fashion, the basic function of the IPMs defined earlier can be achieved. More versions of SiC-based power modules integrated with gate drives have been investigated and demonstrated [27–30]. They are generally in developmental stages.

1.1.2.2 GaN transistors

Although GaN theoretically offers better high-frequency and high-voltage performances as compared to SiC due to its higher breakdown electric field and electron velocity (see Figure 1.1), GaN wafer fabrication technology is still a major challenge for vertical devices. Currently, lateral devices called heterojunction field-effect transistors (HFETs), also known as high electron mobility transistors (HEMTs), offer an effective alternative that takes advantage of the high-mobility

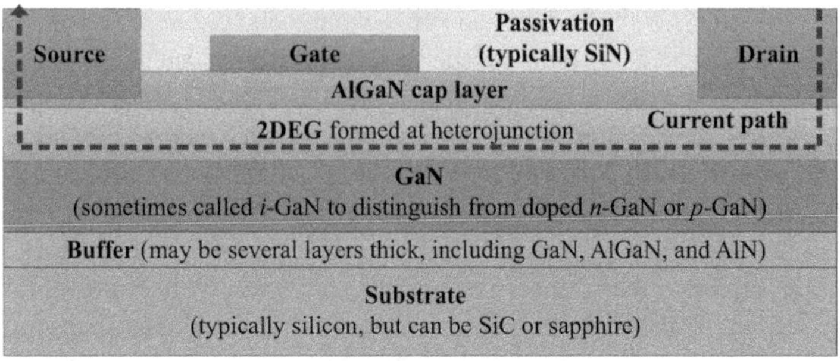

Figure 1.5 GaN HFET basic device structure

layer formed at the heterojunction between GaN and AlGaN. HFETs have been fabricated on substrates of Si, SiC, and sapphire. Si is considerably cheaper than other alternatives, and its conductive properties offer unique advantages as a substrate [31].

Unlike a MOSFET, there is no intentional doping or *p–n* junction in an HFET due to the two-dimensional electron gas (2DEG) generated by the piezoelectric polarization at the heterojunction between GaN and AlGaN. Figure 1.5 shows the basic structure. First, a multilayer buffer is deposited on the substrate wafer, providing crystal lattice strain relief for the thick GaN layer that is next epitaxially deposited. Then, the AlGaN cap layer is deposited, and the 2DEG forms between these two top layers. The 2DEG is a dense native layer of electrons that creates a normally-on low-resistance current path between the drain and source. To turn the device off, a Schottky gate is deposited on the AlGaN cap layer that depletes the 2DEG when a negative voltage is applied. Therefore, the HFET is a fundamentally depletion-mode (normally-on) device. In power conversion systems, due to reliability concerns, depletion-mode devices are less desirable. Three distinct methods are widely used to achieve normally-off HFETs: non-insulated gate enhancement mode, insulated gate enhancement mode, and cascode [32].

Enhancement-mode devices require a modification to the gate structure of the HFET, usually some etching or doping of the AlGaN cap layer. But because there is no source-connected body region or *p–n* junction in an enhancement-mode GaN device, there is no body diode as in a Si or SiC MOSFET. Instead, the symmetry of the drain and source allows for self-commutated reverse conduction with no reverse recovery, but with a high voltage drop that can cause heavy losses during dead time. Additionally, this enhancement-mode gate usually has a very low gate threshold and can easily rupture with gate overvoltage. Alternatively, cascode GaN involves packaging of the device with a low-voltage Si MOSFET to control the gate of a depletion-mode GaN device. The result is a normally-off device with the same rugged gate as a conventional Si power MOSFET, but the fast switching and low on-resistance of a GaN device. Cascode devices utilize the body diode of the low-voltage Si MOSFET for reverse conduction. Unlike reverse conduction in an enhancement-mode GaN device, the cascode body diode has a very low forward voltage drop and a nonzero reverse recovery current [31, 33, 34].

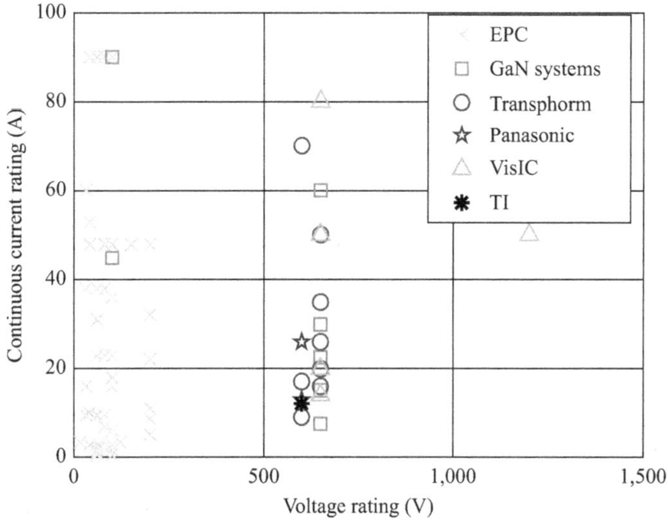

Figure 1.6 Summary of status of GaN transistors (updated June 2017) [35]

Figure 1.6 summarizes the status of GaN transistors based on the latest published data from several leading vendors. For the reasons mentioned previously, commercially available GaN power transistors on today's market use Si as substrate (GaN-on-Si). Due to the lateral structure, their voltage rating is limited to below 650 V with up to about 100 A current rating. Several manufacturers and researchers are exploring other GaN devices, such as vertical GaN-on-GaN JFETs and new HFET gate structures, to increase the breakdown voltage. But these are all in the research and development phase, and no GaN devices have been released yet commercially above 650 V at this time. In this book, we will focus on commercially available enhancement-mode GaN HFET, and will use the terms GaN HFET, GaN HEMT, and GaN FET interchangeably.

1.2 Motivation for WBG device characterization

For the mature Si devices, their characteristics are readily available from the datasheet. For WBG devices, the designers often wish to characterize these devices themselves for various reasons, which can be categorized as follows:

1. Even though the WBG device manufacturers generally provide a datasheet for their devices, these devices are new and relatively immature. As a result, WBG device characteristics often show high variability. It is prudent to characterize the devices to know their real behavior, especially during the design and development stage.
2. The datasheet may not be complete for some new WBG devices or the provided data may not cover the conditions needed for a particular design.

For example, the WBG device datasheet often only provide switching characteristics at limited voltage and temperature points, which may not be sufficient for some converter design.

3. WBG device characterization requires careful testing setup design. For example, a circuit layout with low parasitics is needed for accurate switching characteristics measurement. The process of designing the characterization circuit can directly help the converter design.
4. Characterization of WBG devices is an excellent way to learn about these new emerging devices. In fact, the authors have gained much of their hands-on experience with WBG devices through characterizing various SiC and GaN devices in the laboratory. Significant parts of materials in this book are based on those works. Many industry, academic, and other device users have adopted the approach to learn about a new WBG device.

In principle, similar general procedures in characterizing Si devices can be applied to WBG device characterization. For example, characterizing Si and WBG devices both involves static and dynamic characterization. The dynamic or switching characterization for both Si and WBG devices is generally performed through the so-called double pulse test (DPT). However, due to the significantly different properties of Si and WBG devices, the details in carrying out the WBG device characterization can be very different from those in Si device characterization. The differences can be summarized in several aspects:

1. Device parameters to characterize: Certain parameters matter only or are more important for some WBG devices. For example, one type of GaN HFET, gate injection transistor (GIT), requires a steady-state gate current to operate due to its unique structure, and therefore needs special characterization. The junction capacitances generally play a more important role in determining the switching characteristics of the faster WBG devices compared with Si devices.
2. Testing setup: For dynamic characterization, the WBG device gate drive, layout, current and voltage measurement, temperature control, protection, etc., all have more stringent requirements than those used for Si devices due to the fast switching speed and small size of the WBG devices.
3. Measurement procedure and data processing: Again due to fast switching speed, low loss, and small size, better controlled measurement procedures and more precise data processing are needed for WBG devices. Examples include the pulse control during the static characterization using a curve tracer, and *I–V* alignment technique for dynamic characterization.

Figure 1.7 shows an example of measured switching energy as the function of drain current for a Wolfspeed SiC MOSFET. The datasheet is compared to the measured data obtained by the authors at the University of Tennessee, Knoxville (UTK). The results are quite different for a number of reasons: (1) the UTK gate drive adopts a 0 Ω external gate resistance versus 2.5 Ω in datasheet; (2) the UTK gate drive is an intelligent gate drive with a four-level gate voltage logic compared with the simple two-level voltage used for datasheet to deal with the

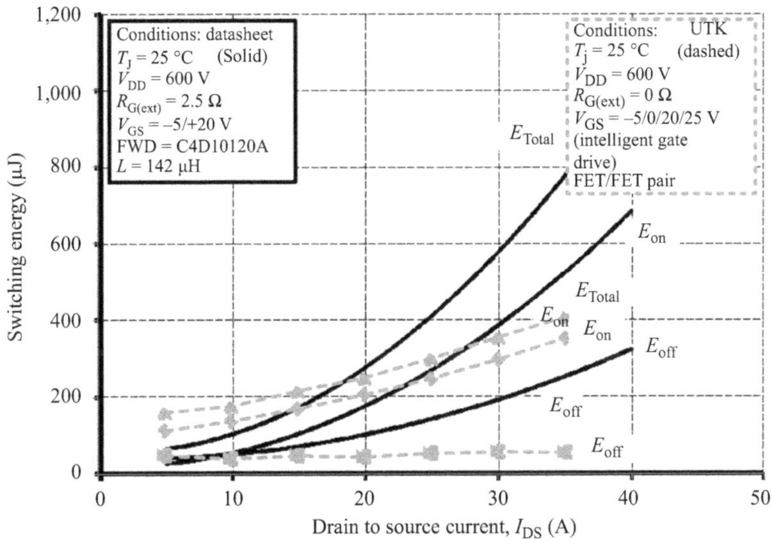

Figure 1.7 Comparison of the switching energy: datasheet data versus custom data

cross-talk issues, which will be discussed later in this book; and (3) the UTK setup for the DPT is a MOSFET/MOSFET pair, while the datasheet corresponds to a MOSFET/diode pair. It is interesting to note that the UTK data shows lower switching energy in general due to faster switching with lower gate resistance and higher gate driving voltage; however, at low current, the UTK data exhibits a higher switching energy due to the higher junction capacitance of the MOSFET/MOSFET pair setup. The conditions and setup used by UTK represent a more realistic setup with a more desirable gate drive, which are not covered by the datasheet. Clearly, in this case, it is necessary to carry out the device characterization.

1.3 About this book

This book is written from the perspective of the WBG device users. It aims at providing a practical step-by-step guide on how to characterize a SiC or GaN WBG device, while also presenting the theory and explanation for design, testing, measurement, and data processing needed for the characterization. The intended readers include university professors and graduate students, practicing industry engineers, and other researchers working on WBG-based power electronics.

The rest of the book has ten more chapters, organized as follows:

Chapter 2, "Pulsed static characterization," covers the characterization of various static characteristics, except the junction capacitances, of a WBG device using the curve tracer type of equipment.

Chapter 3, "Junction capacitance characterization," is dedicated to the characterization of different junction capacitances for a WBG device as they play very important roles in WBG performance.

Chapter 4, "Fundamentals of dynamic characterization," introduces the DPT and its basic design methodology, including main components and control.

Chapter 5, "Gate drive for dynamic characterization," focuses on the gate drive design needed for the switching test, including major component functions and design selection.

Chapter 6, "Layout design and parasitic management," focuses on the layout of the DPT circuit, including performance analysis and layout of gate drive and power stage.

Chapter 7, "Protection design for double pulse test," presents the protection scheme effective in the case of over-current during WBG dynamic characterization, which is very important since the SiC and GaN devices have limited fault current-handling capability compared with Si counterparts. Safety measures are also covered, including for high-voltage device dynamic characterization.

Chapter 8, "Measurement and data processing for dynamic characterization," covers high-speed measurement techniques, grounding, V–I alignment, switching energy calculation, the impact of parasitic ringing, etc.

Chapter 9, "Cross-talk consideration," together with Chapters 10 and 11, discusses several topics important to WBG device performance but not covered in standard characterization in other chapters. This chapter covers cross-talk mechanism, suppression methods including mitigation schemes using intelligent gate drives, and the impact on WBG device characterization.

Chapter 10, "Impact of three-phase system," discusses how the WBG device characteristics will be impacted by a three-phase converter configuration, including the alternate current (AC) load, heat sink, and phase-leg interactions. The approach to minimize such impacts is also discussed.

Chapter 11, "Topology consideration," considers the applicability of DPT for topologies other than half-bridge phase-leg. The topologies covered include current source converters and three-level voltage source converters.

Two appendices are provided, with Appendix A on DPT equipment and components list, and Appendix B on MATLAB® code for dynamic characterization data process.

In addition to the methodology description, examples and case studies are presented in each chapter. Both SiC MOSFET and GaN HFET examples are given throughout the book. It should be noted that these two types of devices are the focus of this book, due to their overwhelming dominance in availability, research, and application. Other types of WBG devices are also being developed, including SiC BJT, SiC IGBT, SiC thyristor, and vertical GaN FET. While the basic principle of device characterization introduced in this book can be applied to these devices, they are beyond the scope.

It should also be noted that most of the examples in the book are centered on low-voltage devices, that is, 1,200 V or 1,700 SiC and 600 V GaN devices. For high-voltage devices, for example, devices rated at 3.3 kV or higher, special attention will be required in their device characterization. Chapter 7 discusses the special protection and safety consideration needed for high-voltage WBG device characterization.

Many of the contents in the book are based on the authors' own experience. In most cases, our own results in SiC and GaN characterization have been used as examples. On the other hand, the methodologies presented are generalized including referencing other work.

References

[1] J. Millan, P. Godignon, X. Perpina, A. Perez-Tomas, and J. Rebollo, "A survey of wide bandgap power semiconductor devices," *IEEE Transactions on Power Electronics*, vol. 29, no. 5, pp. 2155–2163, 2014.

[2] F. Wang, Z. Zhang, T. Ericsen, R. Raju, R. Burgos, and D. Boroyevich, "Advances in power conversion and drives for shipboard systems," *Proceedings of the IEEE*, vol. 103, no. 12, pp. 2285–2311, 2015.

[3] T. P. Chow, "Wide bandgap semiconductor power devices for energy efficient systems," in *Proc. IEEE Workshop on Wide Bandgap Power Devices and Applications (WiPDA)*, 2015, pp. 402–405.

[4] A. Hefner, R. Sei-Hyung, B. Hull, *et al.*, "Recent advances in high-voltage, high-frequency silicon-carbide power devices," in *Industry Applications Conference, 2006. 41st IAS Annual Meeting. Conference Record of the 2006 IEEE*, Tampa, FL, USA, 2006, pp. 330–337.

[5] J. Millan, P. Godignon, X. Perpina, A. Perez-Tomas, and J. Rebollo, "A survey of wide bandgap power semiconductor devices," *Power Electronics, IEEE Transactions on*, vol. 29, pp. 2155–2163, 2014.

[6] J. Wang, "Design, characterization, modeling and analysis of high voltage silicon carbide power devices," Ph.D., North Carolina State University, 2010.

[7] J. W. Palmour, "Silicon carbide power device development for industrial markets," in Electron Devices Meeting (IEDM), 2014 IEEE International, 2014, pp. 1.1.1–1.1.8.

[8] J. W. Palmour, L. Cheng, V. Pala, *et al.*, "Silicon carbide power MOSFETs: breakthrough performance from 900 V up to 15 kV," in *Power Semiconductor Devices & IC's (ISPSD), 2014 IEEE 26th International Symposium on*, 2014, pp. 79–82.

[9] F. Wang and Z. Zhang, "Overview of silicon carbide technology: device, converter, system, and application," *CPSS Transactions on Power Electronics and Applications*, vol. 1, no. 1, pp. 13–32, 2016.

[10] Rohm [Online]. Available: http://www.rohm.com/web/global/groups/-/group/groupname/SiC%20Power%20Devices. Accessed in May 2018.

[11] J. B. Casady, T. McNutt, D. Girder, and J. Palmour, "Medium voltage SiC R&D update," Wolfspeed, 2016.

[12] J. Hostetler, X. Li, P. Alexandrov, A. Bhalla, M. Becher, and J. Sherbondy, "6.5 kV silicon carbide JFETs switch module for high density power conversion system," in *EESAT Technical Conference*, Blacksburg, VA, USA, September 2015.

[13] Wolfspeed [Online]. Available: http://www.wolfspeed.com/power/products. Accessed in May 2018.

[14] Infineon [Online]. Available: https://www.infineon.com/cms/en/product/power/sicarbide-sic/channel.html?channel=ff80808112ab681d0112ab6a50b304a0. Accessed in May 2018.

[15] L. Stevanovic. Industrial readiness of SiC power devices [Online]. Available: https://www.rpi.edu/dept/cfes/AnnualConference/b2%20Stevanoic%20GE.pdf. Accessed in February 2015.

[16] GeneSiC [Online]. Available: http://www.genesicsemi.com/commercial-sic/. Accessed in May 2018.

[17] Microsemi Power Products Group [Online]. Available: http://www.microsemi.com/product-directory/discretes/3613-silicon-carbide-sic. Accessed in May 2018.

[18] M. B. D. Othman, S. Lefebvre, A. Ibrahim, Z. Khatir, and A. Bouzourene, "Comparison study on performances and robustness between SiC MOSFET & JFET devices – abilities for aeronautics application," *European Symposium on the Reliability of Electron Devices, Failure Physics and Analysis*, vol. 52, pp. 1859–1964, 2012.

[19] Rohm, SiC power devices and modules [Online]. Available: http://rohmfs.rohm.com. Accessed in May 2018.

[20] M. Treu, R. Rupp, P. Blaschitz, *et al.*, "Strategic considerations for unipolar SiC switch options: JFET vs. MOSFET," in *Industry Applications Conference, 2007. 42nd IAS Annual Meeting. Conference Record of the 2007 IEEE*, New Orleans, LA, USA, 2007, pp. 324–330.

[21] X. Huang, G. Wang, Y. Li, A. Q. Huang, and B. J. Baliga, "Short-circuit capability of 1200V SiC MOSFET and JFET for fault protection," in *Applied Power Electronics Conference and Exposition (APEC), 2013 Twenty-Eighth Annual IEEE*, Long Beach, CA, USA, 2013, pp. 197–200.

[22] C. DiMarino, W. Zhang, N. Haryani, Q. Wang, R. Burgos, and D. Boroyevich, "A high-density, high-efficiency 1.2 kV SiC MOSFET module and gate drive circuit," in *IEEE Workshop on Wide Bandgap Power Devices and Applications*, 2016.

[23] Powerex. Introduction to IPM [Online]. Available: www.pwrx.com/pwrx/app/IntellimodIntellPwrMods.pdf. Accessed in May 2018.

[24] Powerex. SiC module PMF75CL1A120 [Online]. Available: http://www.pwrx.com/Product/PMF75CL1A120.

[25] Cree. Dual channel differential isolated gate driver CGD15HB62LP [Online]. Available: www.wolfspeed.com/media/downloads/872/CGD15HB62LP.pdf.

[26] AgileSwitch. Gate driver EDEM3 [Online]. Available: http://www.agileswitch.com/sic-products.html.

[27] Z. Wang, X. Shi, L. M. Tolbert, *et al.*, "A High temperature silicon carbide MOSFET power module with integrated silicon-on-insulator-based gate drive," *IEEE Transactions on Power Electronics*, vol. 30, no. 3, pp. 1432–1445, 2015.

[28] Z. Chen, Y. Yao, D. Boroyevich, K. D. T. Ngo, P. Mattavelli, and K. Rajashekara, "A 1200-V, 60-A SiC MOSFET multichip phase-leg module for high-temperature, high-frequency applications," *IEEE Transactions on Power Electronics*, vol. 29, no. 5, pp. 2307–2320, 2014.

[29] L. Zhang, S. Guo, X. Li, Y. Lei, W. Yu, and A. Q. Huang, "Integrated SiC MOSFET module with ultra low parasitic inductance for noise free ultra high speed switching," in *2015 IEEE 3rd Workshop on Wide Bandgap Power Devices and Applications (WiPDA)*, 2015, pp. 224–229.

[30] O. Kreutzer, B. Eckardt, and M. Maerz, "Optimum gate driver design to reach SiC-MOSFET's full potential Speeding up to 200 kV/us," in *2015 IEEE 3rd Workshop on Wide Bandgap Power Devices and Applications (WiPDA)*, 2015, pp. 41–46.

[31] A. Lidow, J. Strydom, M. de Rooij, D. Reusch, *GaN Transistors for Efficient Power Conversion*, 2nd ed., Wiley, 2014.

[32] E. A. Jones, F. Wang, and B. Ozpineci, "Application-based review of GaN HFETs," in *Wide Bandgap Power Devices and Applications (WiPDA), 2014 IEEE Workshop on*, 2014, pp. 24–29.

[33] H. L. J. Roberts and C. McKnight-Macneil, "Advanced SPICE models applied to high power GaN devices and integrated GaN drive circuits," presented at the *Applied Power Electronics Conference and Exposition (APEC)*, Fort Worth, TX, 2014.

[34] C. C. M. Su and S. Rajan, "Prospects for the application of GaN power devices in hybrid electric vehicle drive systems," *Semiconductors Science and Technology*, vol. 28, 2013, pp. 1–9.

[35] B. Liu, Z. Zhang, and F. Wang, "Application of GaN in hard-switching converter: challenges and potential solutions," *Chinese Journal of Power Electronics*, vol. 51, no. 9, pp. 3–13, 2017.

Chapter 2

Pulsed static characterization

The static I–V characteristics of a power device describe its behavior in steady-state. Pulsed I–V characterization is usually performed with a curve tracer to establish the on-resistance, maximum current capability, and maximum voltage capability of the device. Some of the static characteristics are also used to interpret later dynamic test results. While most of this procedure can be found in the user manual for the test equipment (i.e., curve tracer), the unique properties of WBG devices require special care and considerations that may be overlooked in such general instructions. This chapter will describe these considerations, the desired deliverables of static characterization, and the tests required to produce them.

2.1 Fundamentals of pulsed *I–V* testing

Static characterization aims to capture and describe the steady-state behavior of a device under specific operating conditions, such as gate voltage (v_{gs}), drain voltage (v_{ds}), and drain current (i_d). These I–V relationships also have a strong junction temperature (T_j) dependency. Maintaining a device in an on-state produces conduction loss, which may impact that junction temperature, and change the test conditions in an uncontrolled manner. This poses a challenge. How can static on-state behavior be characterized without keeping the device turned on for an extended period of time?

The solution is a pulsed I–V test. Pulsing the device in a temporary on-state allows for precise control of the operating conditions without significant increase of junction temperature. Unlike the dynamic characterization to be described later in this book, the transients at the beginning of the pulse are only a means to an end. The area of interest is the middle of each pulse, neglecting any dynamic effects caused by the turn-on or turn-off transients. It is therefore desirable to avoid hard switching, which injects resonant energy into the circuit and causes ringing.

Many of the techniques described in this chapter are equally applicable to conventional Si switching devices such as MOSFETs and IGBTs. In many cases, the difference is the sensitivity of the measurement accuracy. Most Si power devices have enough thermal capacitance that a relatively long pulse may have a negligible effect on junction temperature. The relatively small die size of a WBG device means a much smaller thermal capacitance, especially as packages become less bulky or even chipscale. Meanwhile, slower Si devices do not achieve the high voltage and current slew rates of a WBG device, so accidental hard switching may not introduce enough resonant energy to observably affect pulsed static measurements. These key differences make proper test setup and procedure more essential for static characterization of WBG devices. Using conventional techniques developed for Si devices can produce inaccurate results or false trends between variables that are actually unrelated.

2.2 Test equipment description

The most commonly used equipment for pulsed static characterization is a parametric curve tracer. This equipment can be purchased from several manufacturers either as a packaged system or as individual modules for sourcing and measuring the different parameters under consideration. One early example of a power device curve tracer is the 370 series from Tektronix, although this product line is no longer manufactured. More recent examples include the B1500 series from Keysight, the 2600 series from Keithley, and the CS series from Iwatsu. Keysight's B1505a will be used for the case studies in this book, although similar methodology can be implemented with other equipment. The hardware and software configuration may differ between manufacturers, including the names of certain settings or the capability of adjusting them. But the underlying principles remain the same.

Figure 2.1 shows the hardware configuration for the Keysight B1505a parametric curve tracer used in the case studies of this chapter as well as Chapter 3.

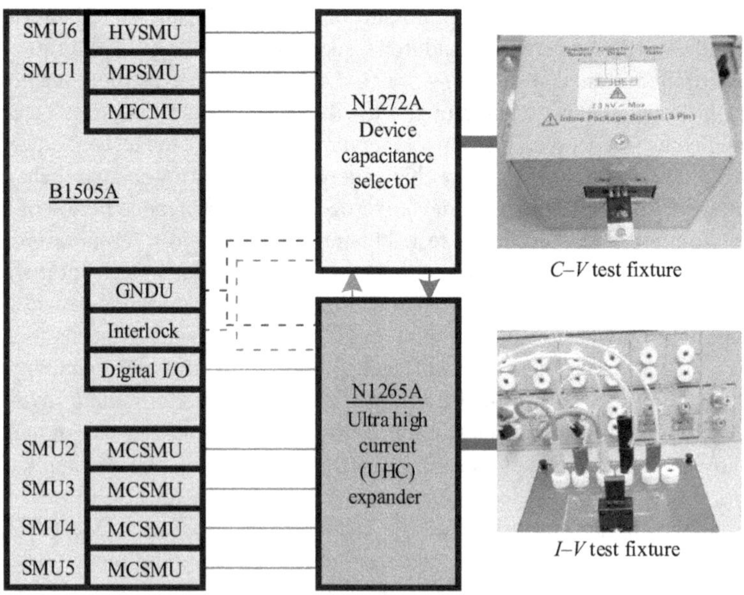

Figure 2.1 Hardware configuration for the Keysight B1505a parametric curve tracer used in this chapter, showing fixtures for a TO-220 packaged device as an example [1]

This particular B1505a is made up of a mainframe, several source/measurement units (SMUs), a capacitance selector unit, current expander unit, and separate test fixtures for *C–V* and *I–V* measurements. Other configurations are possible, such as an additional voltage expander to increase the maximum drain voltage testing capability, or without the current expander if high drain current capability is not required. Each SMU has the capability to either supply or measure voltage, while in either case also measuring current. Some SMUs have a special purpose, such as generating particularly high voltage, measuring capacitance, or maintaining a stable and accurate ground reference.

An important feature of a curve tracer is the Kelvin connection for both drain/collector and source/emitter terminals. Because the equipment provides the drain current, the same test leads cannot be used to accurately measure the voltage as well. The "force" connections (which provide the current) and "sense" connections (which measure the voltage) must be entirely isolated up to the device terminals. Similarly, the source/emitter test lead used as a reference for the gate voltage must be isolated from both the "force" and "sense" source leads. Implementation of this test setup is further explained in the next section.

The Kelvin connection is especially important when a current limiting resistor is included in the "force" loop, which curve tracers often use to protect the equipment against short-circuit conditions. The impact of this resistor, combined with the series resistance of test leads, causes a significant difference between the applied "force" voltage and the actual drain voltage. Figure 2.2 shows a simplified circuit representation of this phenomenon. Because the voltage discrepancy is current-dependent, consideration of this series resistance helps explain the unusual shape of some output characteristic curves, which will be discussed in further detail later in this chapter.

If a curve tracer is not available, it is possible to design a static characterization test setup with more common laboratory equipment (e.g., benchtop power supplies, oscilloscopes, function generator) similar to the dynamic test setup described later in Chapter 4. Running a full sweep of static characteristic curves in this manner

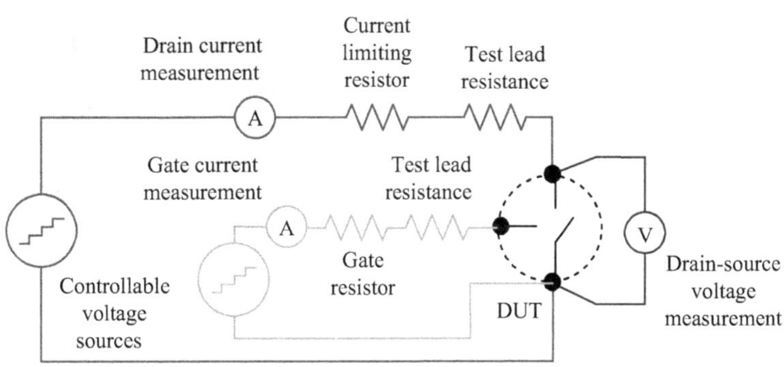

Figure 2.2 Simplified circuit representation of "force" loop, "sense" loop, and gate control loop of a typical curve tracer setup

may be very time consuming compared to a curve tracer. The design considerations of such a system are beyond the scope of this book, but they can be easily inferred from the material in this chapter.

2.3 Test fixture selection/design

Most curve tracers include standard test fixtures for common device packages, namely TO-type packages with pins in a gate–drain–source configuration (in that order). However, modern WBG devices rarely use this package. Many use a four-terminal configuration with a second "source sense" or "Kelvin source" terminal, and may even use a surface-mount or near-chipscale package style. In the opposite extreme, the device may be packaged as part of a large power module, with only screw terminal connections available to access the gate, source, and drain.

Proper selection of a test fixture is critical to the accurate measurement of static device parameters. For example, simply soldering gate, drain, and source leads to a surface-mount package allows it to fit into a TO-style fixture. But the resulting characterization would be invalid, because the measurement would include the resistance of these leads. On the other hand, soldering six leads (two drain, three source, and one gate) directly to the surface-mount package would allow for proper Kelvin connection according to Figure 2.2. But this is not feasible for many package styles and sizes. The most effective solution is designing a custom printed circuit board (PCB) fixture for the device.

Figure 2.3 shows an example of a custom fixture for a surface-mount SiC MOSFET. The key feature is the separation of "force" and "sense" terminals up to the point of common coupling, which is as close as possible to the terminals of the

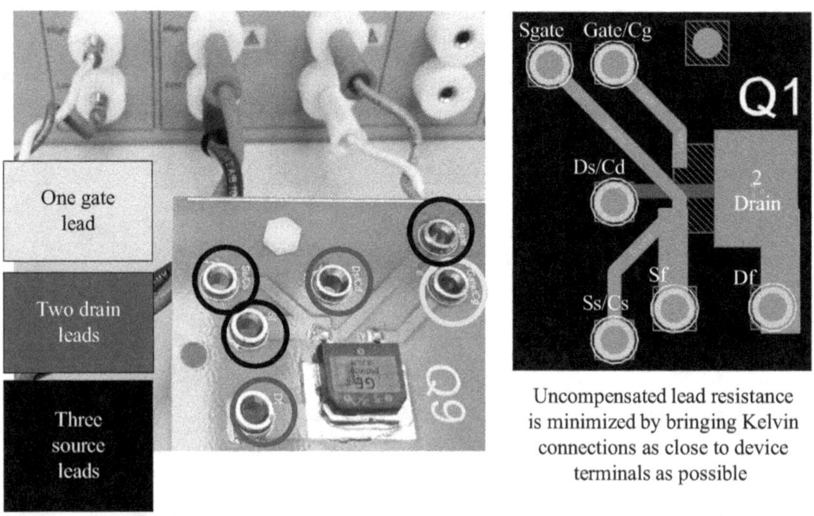

Figure 2.3 Example curve tracer test fixtures for surface-mount SiC MOSFET

packaged device. Additionally, a third source terminal is included here for the gate voltage reference, which can be important in cases where the steady-state gate current is nonzero. This fixture is also designed to meet the requirements of a *C–V* test described in the next chapter, which impacts the positioning of the six terminals on the PCB.

Of course, a "fixture" may not be necessary at all for a large power module with screw terminals. But in this case, the choice of test leads is equally important. Alligator clips add an undesirable contact resistance, so connections should be made in the same manner that the module will be connected in its intended application (e.g., screwed down, crimped, or soldered). Alligator clips may be acceptable in some cases where no other option is available. In either case, the "force" leads should be a larger wire gauge than the "sense" leads, to avoid adding excessive series resistance and reducing the high-current testing capability of the curve tracer.

2.4 Junction temperature control

Most aspects of device behavior are temperature-dependent. It is common to perform static characterization over a range of temperatures, and it can be challenging to control the junction temperature of a device accurately. The externally measurable temperature of the device is not always a good representation of the internal temperature of the device junction, which is the hottest point in a device when it is operating in a converter. Figure 2.4 shows the two most conventional methods to elevate the temperature of a device. The first is an oven, or thermal chamber. A chamber can be used to either heat or cool the device by controlling the ambient temperature of its surrounding environment. An effective thermal chamber must have a means of heating or cooling the air, a fan to ensure thermal equilibrium within the chamber, and accurate measurement of the temperature inside. Digital closed-loop control is also included for some commercial products. Cryogenic applications may also require submerging the device in a medium other than air, such as liquid nitrogen.

The second method for controlling junction temperature is with a hot plate, as pictured in Figure 2.4(b). This method is typically only used for elevated temperature tests, but a similar scheme could be used to cool the device as well. Mounting the device on a hot plate enables manipulation of the case temperature, but it is not as accurate as a chamber.

The source of potential error can be explained with thermal equivalent circuits, which use electrical circuit components to analyze and understand the flow and distribution of thermal energy. In such a circuit, the temperature of any point is analogous to a node voltage, the flow of thermal energy (i.e., thermal power) is analogous to electrical current, and the thermal resistance between any two points is analogous to electrical resistance. The thermal equivalent circuit for a device when operating in a converter is pictured in Figure 2.5(a). This circuit neglects thermal capacitances in order to focus on the steady-state behavior.

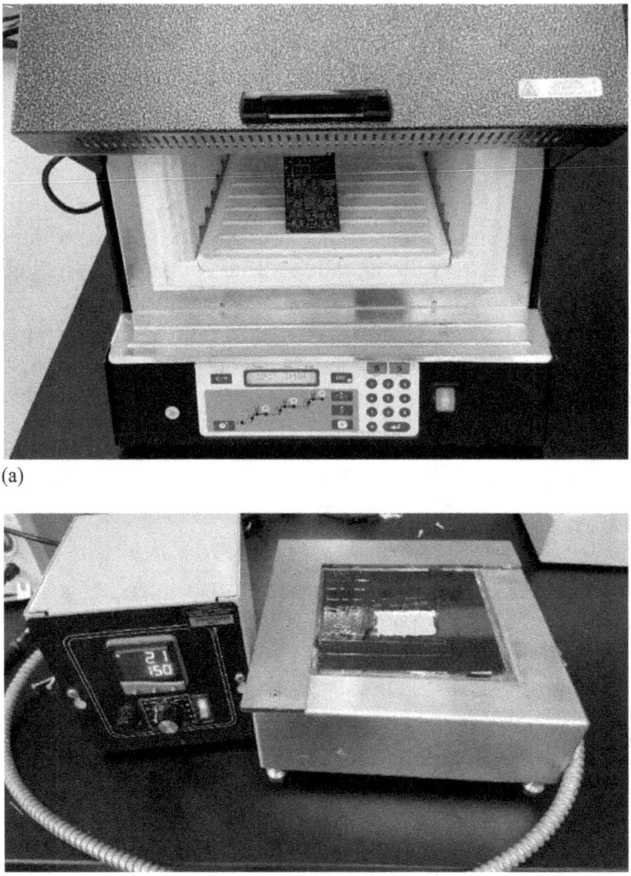

(a)

(b)

Figure 2.4 Two conventional options for junction temperature control for static device characterization: (a) oven/thermal chamber; (b) hot plate

The source of all thermal energy in this circuit is the electrical power loss (P_{loss}) occurring at the junction of the device (T_j), which must eventually escape to the ambient environment (T_A). There are several paths through which this thermal energy can reach the ambient air. It can flow through the top of the device package ($R_{\theta,td\text{-}a}$) or through the bottom of the device package ($R_{\theta,j\text{-}c}$) to the thermal pad of the device (T_{case}). There is an important distinction between the "case" of the device (which refers to the thermally conductive frame intended for heatsink connection) and the top of the device package, which is often constructed from plastic or other poor thermal conductors. In this circuit, neither the case temperature nor the top cover temperature is an accurate representation of the junction temperature because an unknown temperature differential exists due to the flow of P_{loss} through these two thermal resistances. The case temperature can be measured, but there will also be a thermal interface resistance between the case and the sensor ($R_{\theta,c\text{-}sens}$),

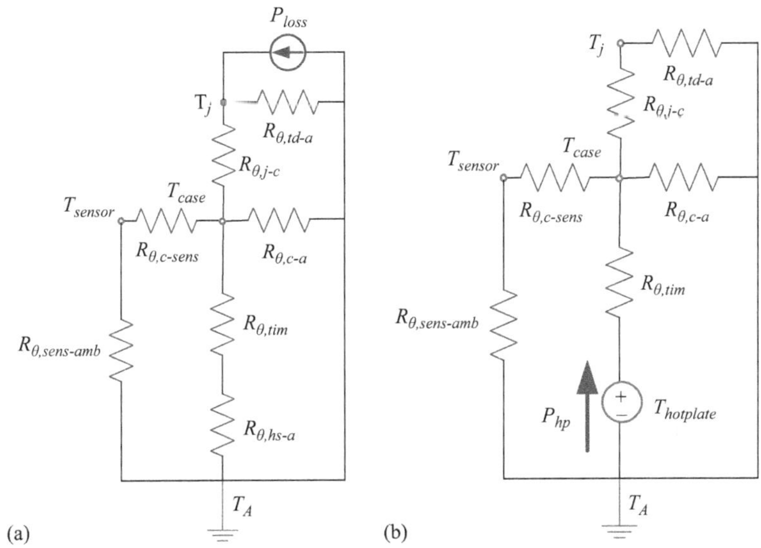

Figure 2.5 Thermal equivalent circuit for (a) power device operating in a converter with a heatsink and (b) power device undergoing characterization on a hot plate

as well as between the exposed side of the sensor and the ambient air ($R_{\theta,sens\text{-}a}$). Meanwhile, the case itself has two paths for heat to flow: a typically weak path directly to ambient ($R_{\theta,c\text{-}a}$), and the much stronger path through the heatsink. The heatsink path includes two primary components: the thermal interface material (including a galvanically isolated pad and sometimes thermally conductive grease) represented by $R_{\theta,tim}$; and the thermal resistance between the heatsink and ambient ($R_{\theta,hs\text{-}a}$). The relative magnitudes of these resistances will determine how effectively the junction temperature and case temperature can be estimated based on measurement near the case or top cover of the device package.

Now, Figure 2.5(b) shows a modified version of this circuit to represent a device mounted on a heatsink during characterization. It is assumed that self-heating is not allowed, so the P_{loss} component is removed. Instead, the source of thermal energy is the surface of the hot plate. Even if this temperature is well-controlled, the thermal power being conducted to the device (P_{hp}) is not easily measured. The hot plate must provide a continuous flow of thermal energy to compensate for cooling of all components in the circuit by the ambient air at all exposed surfaces. Therefore, there are temperature drops across the various elements of this thermal circuit just as there would be in an operating converter. Again, the magnitude of these differentials depends on the relative magnitudes of the thermal resistances involved. In some cases, it may be a reasonable assumption that $T_j \approx T_{case} \approx T_{sensor}$. But without supporting analysis, making this assumption can lead to erroneous conclusions about the temperature dependency of device characteristics, particularly if the goal is to estimate on-resistance for later use in heatsink design.

So, if a hot plate can be such a poor method of junction temperature control, why is it commonly adopted for static characterization? One reason is ease of use. Connecting a device to a curve tracer from inside an oven can be difficult, requiring longer test leads rated for higher temperature. But while an oven has superior accuracy in static characterization, this method may not be practical in later dynamic characterization. Placing an entire DPT setup in an oven may jeopardize the other components on the board or potentially impact their operation in ways that are not relevant to the test. Oscilloscope probes may also be susceptible to damage or inaccuracy when exposed to high temperatures. The hot plate method does not typically present these challenges for a dynamic test, although accurate junction temperature control is equally important.

One solution to this problem is to use the junction of the device itself as a sensor to measure and control its temperature. The full sweep of static *I–V* characterization curves can first be captured with the device in an oven. Once the on-resistance curve is calculated as a function of junction temperature, this on-resistance can be used to calibrate a hot plate test setup. Once the DPT board is well-fixed to the hot plate, the on-resistance can be measured *in situ* by disconnecting the gate drive circuit and connecting test leads to the terminals of the device. A calibration curve is then produced by measuring the on-resistance as a function of hot plate temperature and comparing it with the previously captured $R_{ds,on}$–T_j curve, as pictured in Figure 2.6. The calibration may be sensitive to any variations in the test setup, so it should be repeated if the hot plate connection must be modified in any way. Secondary temperature measurement using a sensor near the device case can improve this calibration as well. After accurate calibration, this hot plate test setup can be used for dynamic testing as well as the capacitance characterization described in Chapter 3.

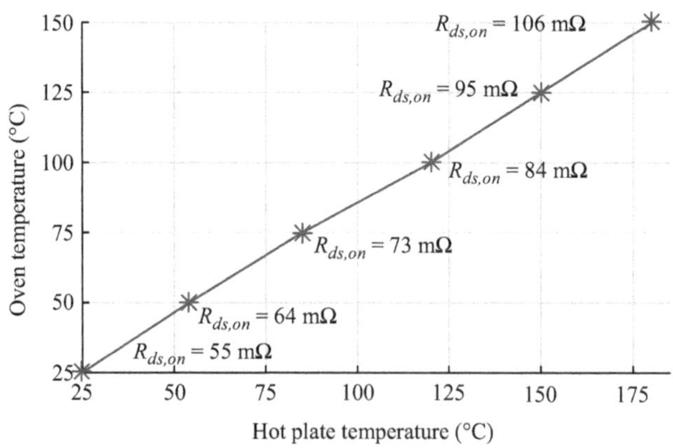

Figure 2.6 Example calibration curve for hot plate control of a device's junction temperature, based on on-resistance measurement in an oven and on a hot plate

2.5 Cryogenic device testing

In addition to testing at elevated temperature, it is sometimes necessary to characterize power device performance at decreased temperature. A special case of particular interest is cryogenic testing, which involves cooling the semiconductor below 0 °C with a liquefied gas such as nitrogen or natural gas. Cryogenic device characterization is important for applications where extremely low-temperature liquid cooling will be employed in the converter design. To characterize power

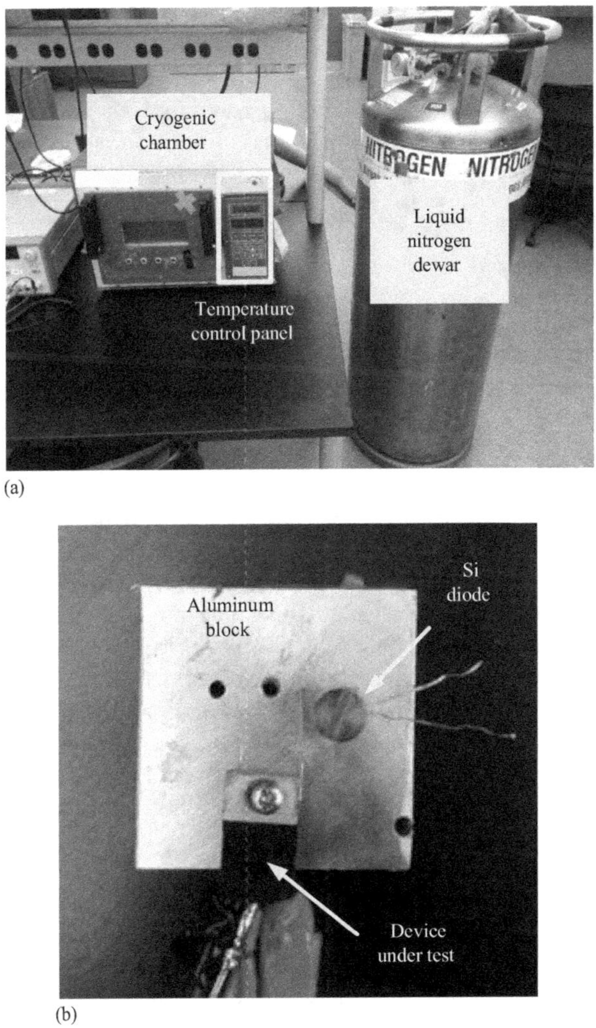

Figure 2.7 Equipment for cryogenic testing: (a) cryogenic chamber and liquid nitrogen dewar; (b) aluminum block for temperature stabilization and measurement setup

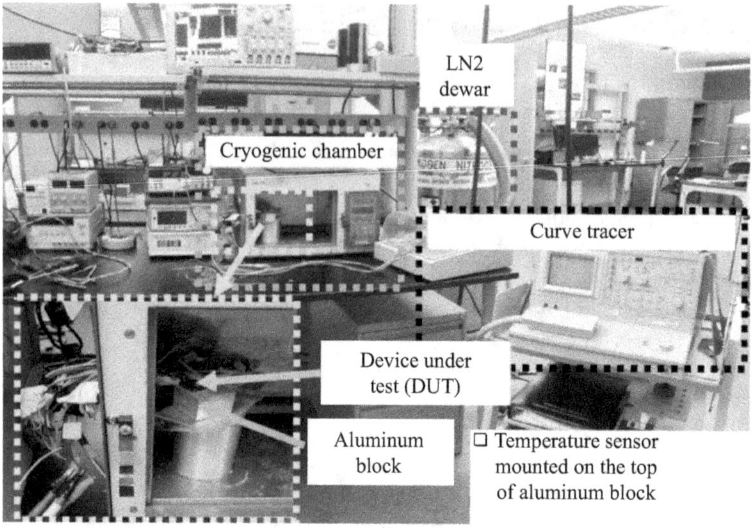

Figure 2.8 Test setup for cryogenic static characterization of power devices

semiconductors in cryogenic conditions, a temperature-controlled environment and monitoring system is required. This section will describe one particular test setup using liquid nitrogen (LN2).

The cryogenic chamber with an attached liquid nitrogen dewar was employed to establish the cryogenic testing platform as shown in Figure 2.7(a). The temperature inside the chamber was regulated by the temperature control panel on the front of the chamber. During the test, the liquid nitrogen from the dewar was injected through an insulated aluminum duct, entering the chamber as vapor. Due to the mechanics of the temperature chamber, the controllable temperature range of the chamber for this setup was limited to a threshold above 93 K. To verify the testing temperature of the device, a Si diode was used as the cryogenic temperature sensor as shown in Figure 2.7(b). The device under test (DUT) and sensor were both attached to an aluminum block to maintain a constant temperature during testing.

A curve tracer was connected to the DUT to measure the device's static characteristics. The Kelvin connection was utilized to compensate for the measurement error introduced by long connecting wires. Insulating tape was used to wrap the device's leads and bare threads inside the chamber to avoid potential short-circuits caused by condensation and icing. The test configuration is shown in Figure 2.8.

A similar cryogenic test setup was also adopted for dynamic characterization using the DPT methodology that will be described later in the book. To maximize the switching performance of the power semiconductor device, the DPT circuitry (such as the gate driver, complementary switch, DC link capacitor bank, and DPT PCB) were located close to the DUT inside the cryogenic chamber as shown in Figure 2.9. This significantly minimizes the parasitics in the gate loop in order to evaluate the true capability of a fast switching power semiconductor device.

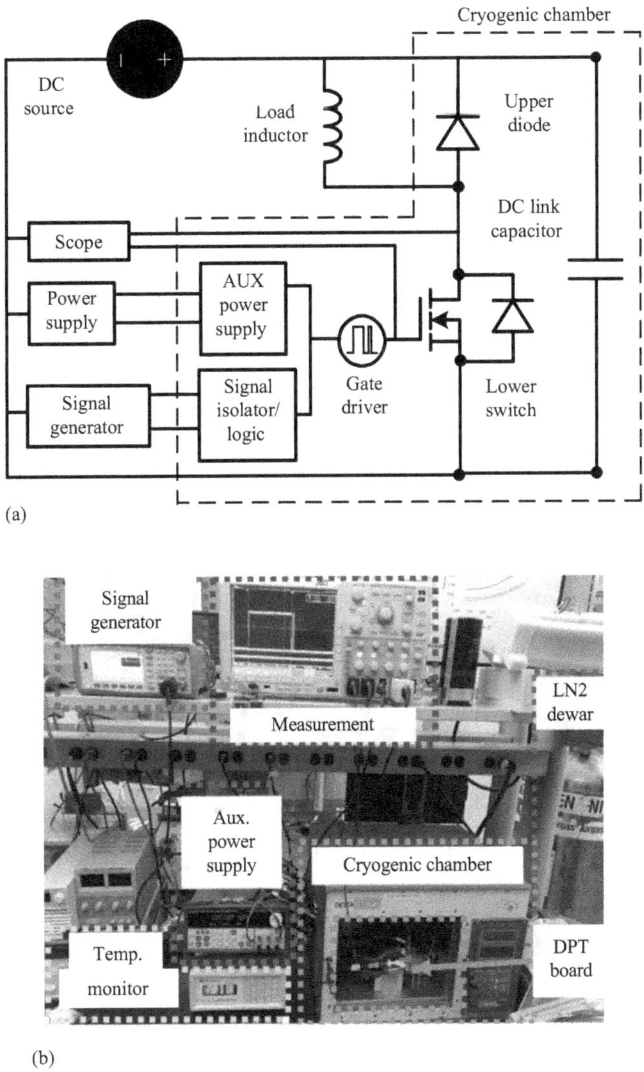

Figure 2.9 Testing for cryogenic dynamic characterization of power devices: (a) equivalent circuit; (b) test bench setup

To ensure all the components inside the chamber could operate properly at cryogenic temperature, the component selection is critical, and all passive components were selected based on datasheet ratings to operate at the desired temperature. As a core element of DPT, the gate drive circuit was designed and independently tested at cryogenic temperature to ensure proper switching waveforms could be achieved. To cope with the noise interference between the signal stage and power stage, the location and interconnection among components inside the cryogenic chamber are critical. Figure 2.10 shows how the DPT was set up

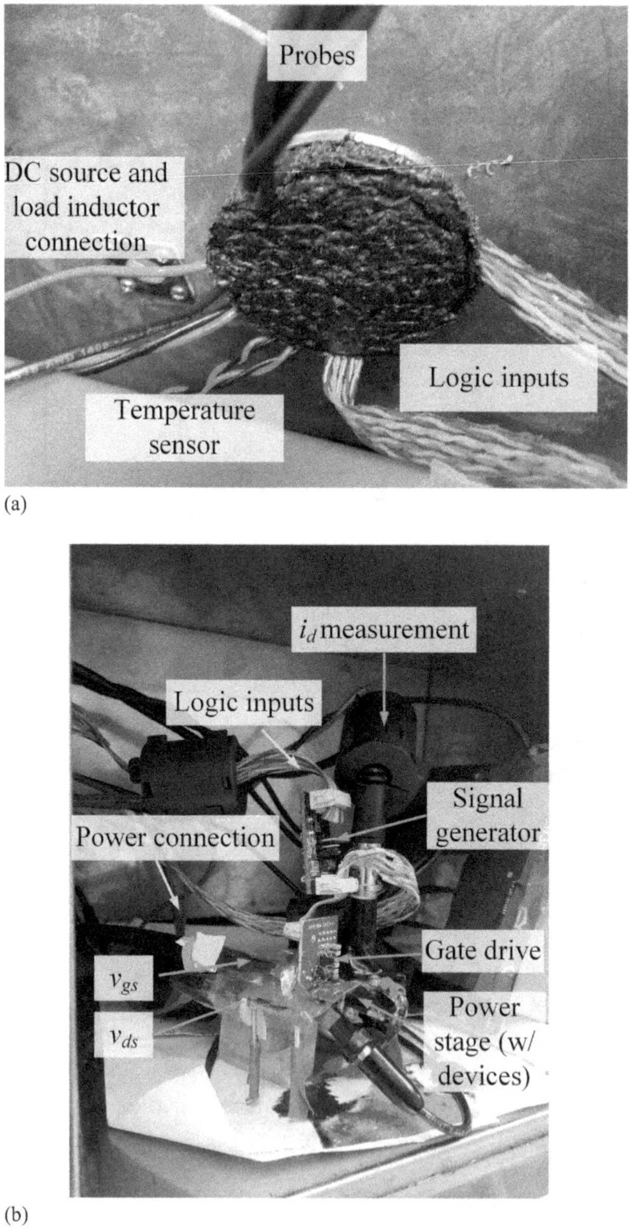

Figure 2.10 Cryogenic chamber setup for dynamic characterization of power devices: (a) test leads inserted into chamber; (b) test setup inside chamber

within the chamber. The high-voltage carrying wires, the oscilloscope probes, and the signal lines were separated as much as possible at the access hole to the chamber. Inside the chamber, all the probes and voltage lines ran below the power stage board and away from the signal generator circuit and wires. The 5 V maximum signal lines are the most sensitive to any noise or interference. With respect to these sensitivities, it is best to keep the signal generator circuit and vulnerable IC components like signal isolators far from the power stage while also perpendicular to avoid dissonant EMI. The signal was carried with twisted pair wire to further minimize any effect of EMI.

2.6 Pulse waveform timing

With a curve tracer or any other test equipment setup, the timing of the gate and drain voltage pulses must be precisely tuned to accurately capture steady-state behavior in a pulsed *I–V* test. The drain voltage pulse must begin after the gate voltage pulse, and only after the gate has had sufficient time to turn fully on. The drain voltage and current measurements can then be captured after a short delay. There are five independent settings that should be considered in the pulse waveform timing: gate pulse width, drain pulse width, drain pulse delay, measurement delay, and pulse frequency. Figure 2.11 shows the definitions of these terms.

The option to adjust these settings was not available in some older curve tracer models, because the consequences of poor pulse timing are much more significant in newer WBG devices than they were in larger-die Si MOSFETs or IGBTs. But this can cause severe measurement issues for devices with relatively low thermal capacitance (i.e., small die area) and high current capability such as SiC or GaN FETs.

The gate pulse width, drain pulse width, and drain pulse delay must be selected to avoid hard switching, which could impact the measurement or even cause damage to the device or equipment. Pulse frequency should be selected based on the potential for self-heating between pulses. There is no minimum frequency constraint, other than the experimenter's patience in waiting for a long sweep to finish. However, setting the frequency too high can cause thermal energy to gradually build up within the device, causing pulses later in the sweep to occur at higher temperature than earlier pulses. This risk can be minimized by individually retesting the last point in any sweep, after giving the device several seconds to cool down, and comparing it to the end of the sweep.

The final setting, measurement delay, can be more challenging to select. An example is shown in Figure 2.12 for a GaN FET, including waveform captures from a curve tracer as well as the resulting output characteristics. Three possible measurement delays are shown. The drain has only reached a stable voltage after ~45 ns, so this is the minimum measurement delay time. However, it is evident from the falling I_d and rising V_{ds} that a true "steady state" is not achieved even after 45 ns. For points in the saturation region such as this, significant self-heating can occur within an individual pulse, so the measurement delay time must be as

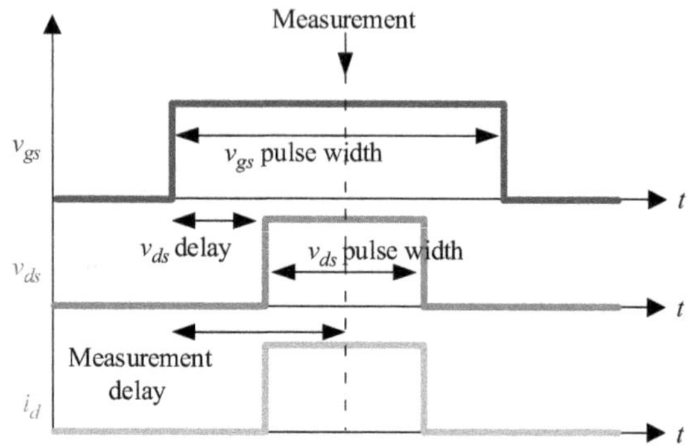

Figure 2.11 Waveform timing settings for pulsed I–V testing

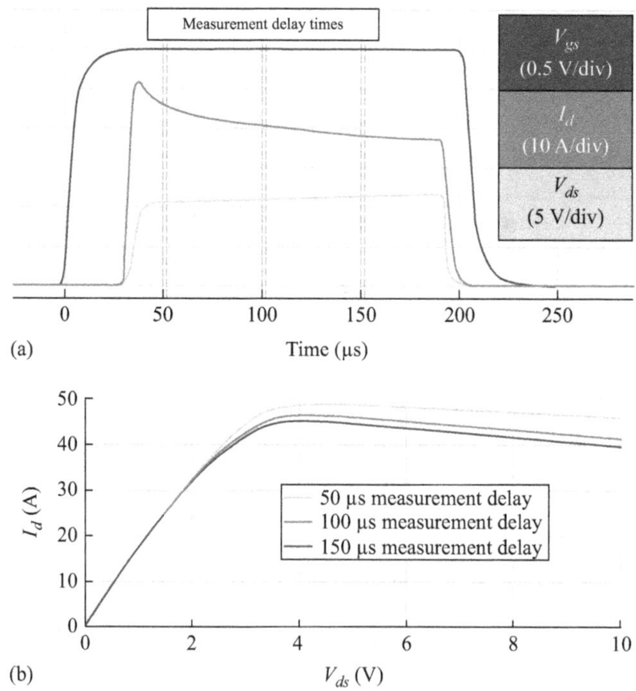

Figure 2.12 Impact of measurement delay selection on output characteristic for GaN FET

short as possible to minimize its effect. The self-heating impact increases further as the device moves into saturation, leading to the three different output characteristic curves shown in Figure 2.12b, which vary with changes in measurement delay.

2.7 Output (I_d–V_{ds}) characteristic

The most fundamental pulsed *I–V* test is the output characteristic curve, which measures the drain current resulting from a sweep of drain voltages, while the gate voltage is held fixed. Figure 2.13(a) and (b) shows example output characteristics for a GaN FET and a SiC MOSFET, respectively. The output characteristic shows the ohmic, saturation, and body diode operation of the device, with both positive and negative drain currents. When v_{gs} is above the threshold voltage (V_{th}) and v_{ds} is below $v_{gs} - V_{th}$, this curve represents the ohmic region. When v_{ds} is above this limit, this curve represents the saturation region.

For gate voltages below the threshold, and with negative drain current, this curve represents the body diode region, or "diode-like behavior" (DLB) in the case of the lateral GaN FET. Both SiC and GaN FETs have reverse conduction

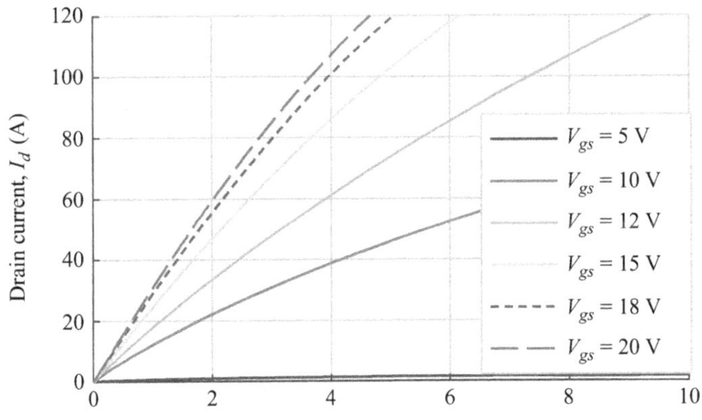

(a)

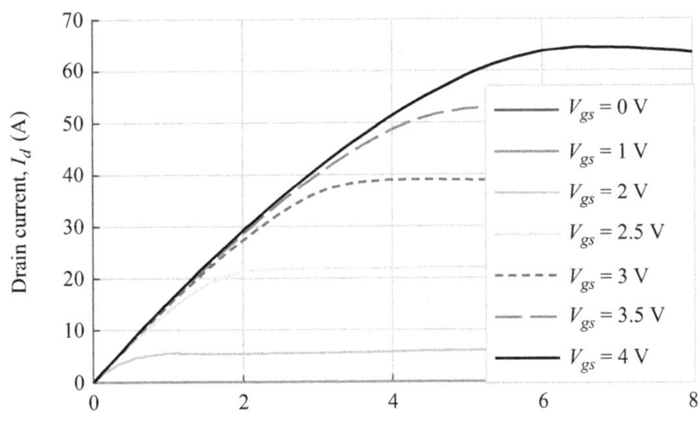

(b)

Figure 2.13 Example first-quadrant output characteristics at 25 °C for (a) SiC MOSFET and (b) GaN FET mechanisms that allow self-commutation with negative drain current when v_{gs} is below the threshold voltage. Likewise, both devices can more efficiently conduct negative drain current through its channel when v_{gs} is above the threshold.

One notable difference is the physical mechanism responsible for the self-commutation. In a vertical MOSFET (including SiC), self-commutated reverse conduction occurs in its body diode, which is a distinctly different current path that does not include the channel. Enhancement-mode GaN FETs do not include body diodes as part of their physical structure, but they do offer an alternative DLB mechanism. In DLB, the negative drain current forward biases the gate-to-drain voltage v_{gd} and turns on the channel of the device. Further description and comparison of these two mechanisms is beyond the scope of this book, but more information can be found in [2]. Reverse conduction should be fully characterized over a range of positive and negative gate voltages, because both mechanisms (body diode and DLB) are variable over v_{gs}. Figure 2.14 shows an example third-quadrant characteristic for a SiC MOSFET and GaN FET at room temperature.

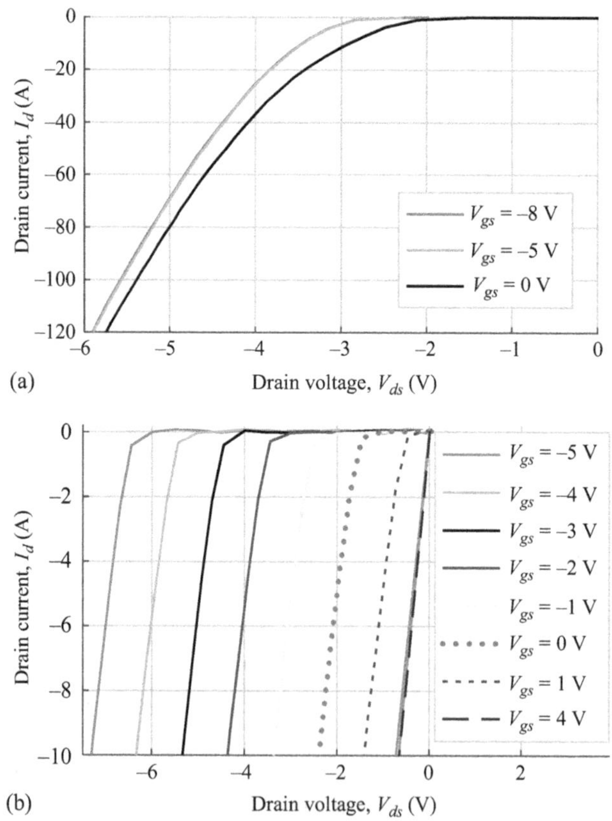

Figure 2.14 Example third-quadrant output characteristics at 25 °C for (a) SiC MOSFET and (b) GaN FET

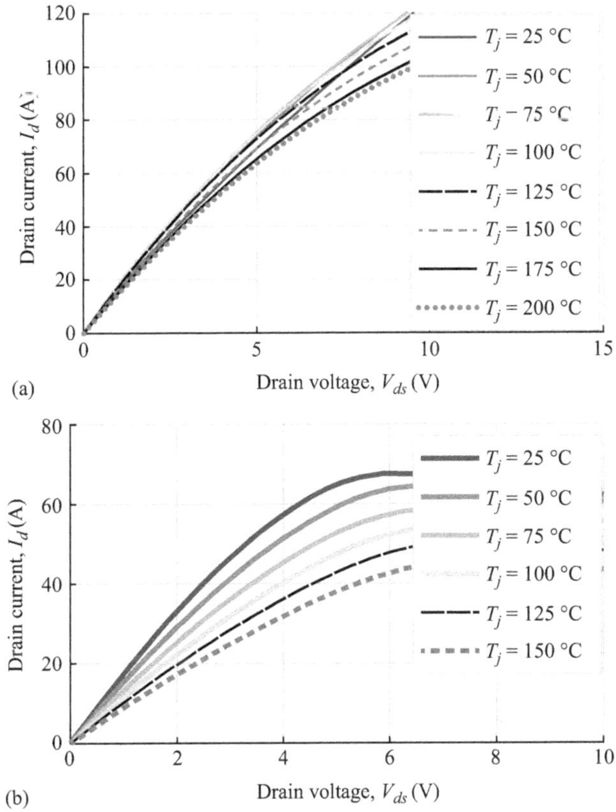

Figure 2.15 Example first-quadrant output characteristics over temperature with a fixed gate voltage for (a) SiC MOSFET and (b) GaN FET

Forward and reverse conduction behavior should also be characterized over a range of junction temperatures, in addition to gate voltage. Figures 2.15 and 2.16 show example output curves for SiC and GaN FETs over temperature, in the first quadrant and third quadrants, respectively.

Perhaps the most useful output of this sweep is the on-resistance, $R_{ds,on}$, which typically decreases with higher gate voltage and increases with higher junction temperature. One notable exception is the SiC MOSFET, which in some cases has a near-zero or even negative temperature coefficient. Another special case is the GaN GIT, which is driven primarily by a gate current ($I_{g,ss}$) rather than a gate voltage in steady state. These devices therefore require on-resistance characterization as a function of gate current, which may not be a directly controllable variable. The $R_{ds,on}$–$I_{g,ss}$ characteristic must therefore be extrapolated from a high-resolution sweep of v_{gs}, combined with measurement of gate current for each test.

Example on-resistance characteristics for a SiC MOSFET and GaN FET as a function of temperature and gate voltage/current is shown in Figure 2.17(a)

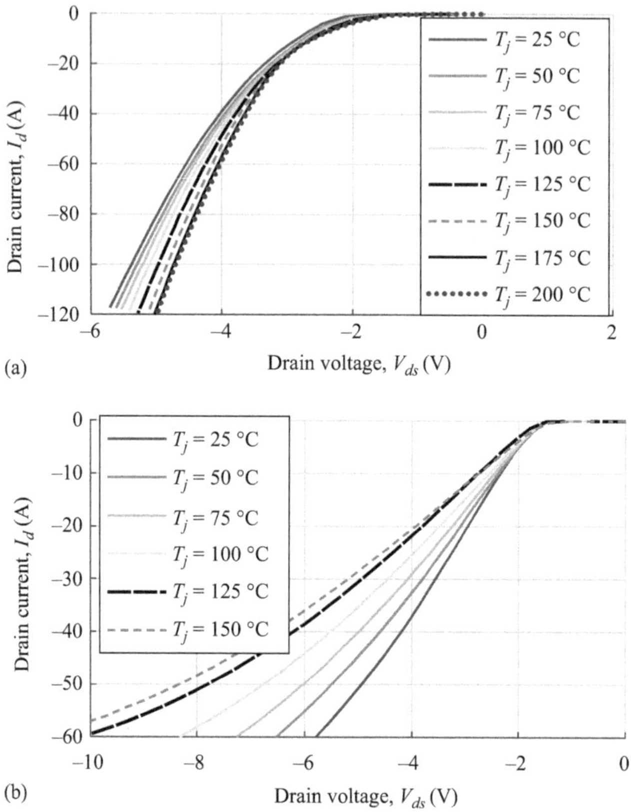

Figure 2.16 Example third-quadrant off-state output characteristics over temperature with $V_{gs} = 0$ V for (a) SiC MOSFET body diode and (b) GaN FET diode-like behavior

and (b), respectively. $R_{ds,on}$ is calculated as V_{ds}/I_d at a particular point in the output characteristic. A common mistake is to calculate the slope of the I–V curve, but this slope is actually the small-signal channel resistance (not typically included in a static characterization). A fixed drain current is typically selected for an on-resistance characteristic, because there is usually some relationship between these two variables. However, it is important to select a drain current that is within the ohmic region for the desired sweep of gate voltages and junction temperatures. $R_{ds,on}$ is not a meaningful calculation in the saturation region. Including saturation behavior in this characteristic implies that the channel resistance becomes large in an operating condition where the channel can no longer be accurately modeled as a resistance. For example, at high temperature or low gate voltage, $R_{ds,on}$ may appear to suddenly "jump" to a higher resistance that is strongly proportional to drain current. Proper selection of parameter ranges can avoid unintentionally portraying false or misleading trends in the characterization.

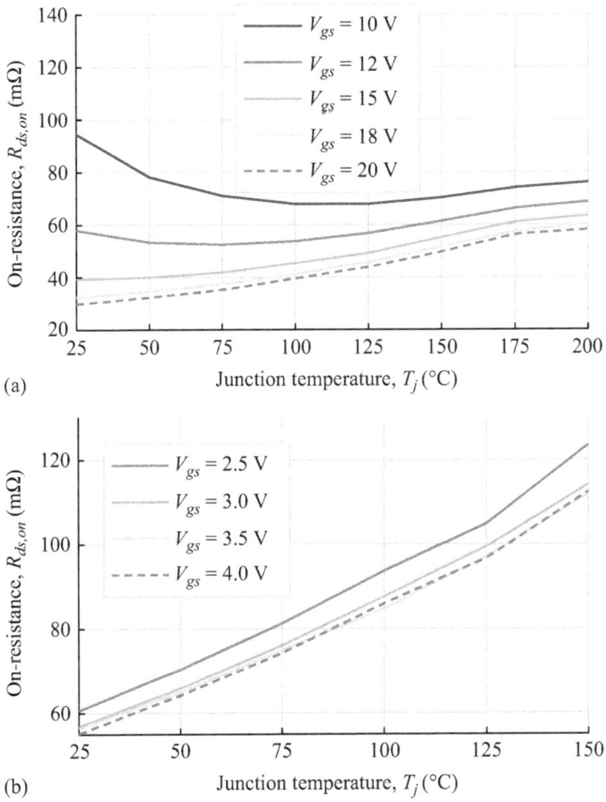

Figure 2.17 Example on-resistance characteristics as a function of temperature and gate voltage (a) a SiC MOSFET and (b) a GaN FET

2.8 Transfer (I_d–V_{gs}) characteristic

The individual points of a transfer characteristic can be extracted from an output characteristic, but typically only a few gate voltage points are chosen. This motivates a separate static test procedure for the transfer characteristic, which depicts the drain current for a high-resolution sweep of gate voltages with a fixed drain voltage. The discrepancy between the applied "force" voltage and measured "sense" voltage on the drain makes this test particularly challenging to implement. For a transfer characteristic, it is necessary to hold the drain voltage at a fixed value, which must be high enough to keep the device in the saturation region up to the maximum tested gate voltage, as given by

$$V_{ds} > V_{gs,max} - V_{th} \tag{2.1}$$

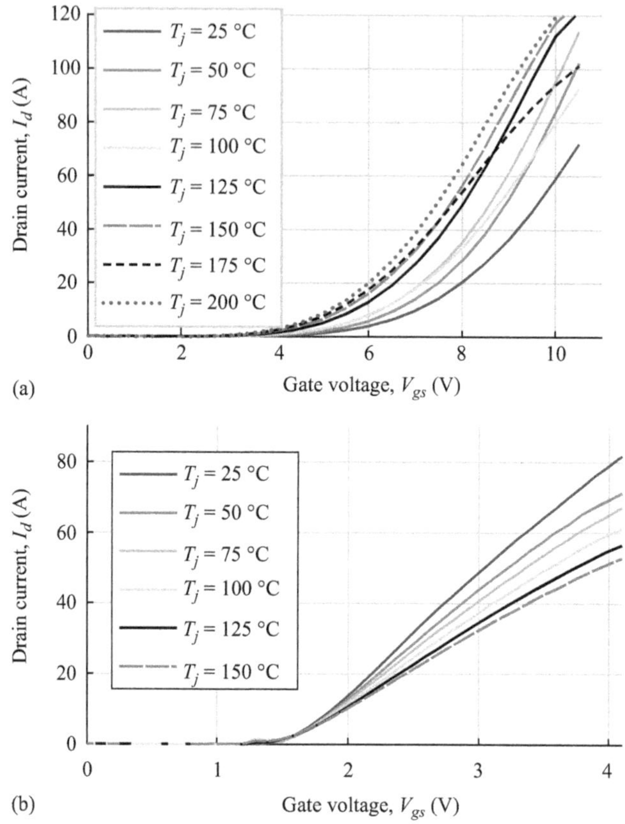

Figure 2.18 Example transfer characteristics for (a) SiC MOSFET and (b) GaN FET

This limit is based on the definition of the saturation region in a MOSFET, so it may not be strictly true for a GaN FET. However, this is easily verified for each unique DUT. An effective choice of drain voltage should be selected from the output characteristic. This can be any drain voltage in the saturation region (as opposed to ohmic) for the desired range of gate voltages, which is typically apparent from reduced slope of I_d–V_{ds}. Figure 2.18(a) and (b) shows example transfer functions for a SiC MOSFET and GaN FET, respectively.

In contrast, Figure 2.19 shows a transfer function for a GaN FET with an applied drain voltage that is too low to ensure saturation across the full v_{gs} range. This curve can be misleading. In an output characteristic, the linear rising portion of the curve indicates ohmic operation, whereas the flatter portion indicates saturation. The transfer characteristic is the opposite. The flatter portion here is undesirable, because the goal of the transfer characteristic is to capture the cutoff and saturation regions only. The ohmic portion of a transfer characteristic does not provide useful information and should not generally be included.

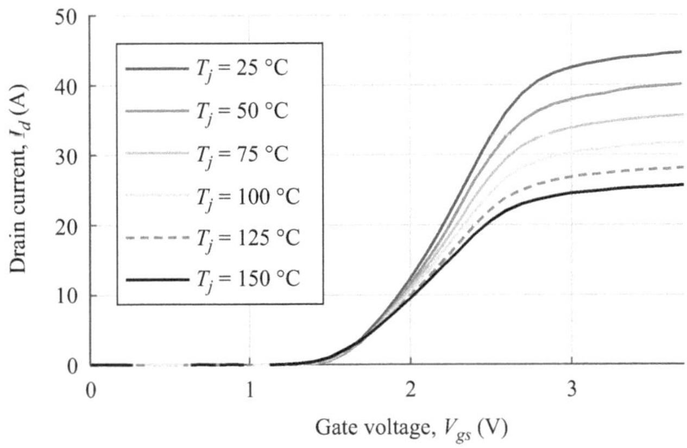

Figure 2.19 Example transfer characteristics for a GaN FET with insufficient drain voltage

On the other hand, the saturation and cutoff region portions of the curve can be very useful for identifying three key parameters. As shown in Figure 2.18, the slope of the curve represents the forward transconductance, known as either g_m (typically representing the small-signal slope) or g_{fs} (typically representing the average or large-signal slope). Second, the threshold voltage can be captured from this curve based on selection of a drain current threshold marking the transition from cutoff to saturation regions. This selection can be somewhat arbitrary and vary widely between devices with different voltage and current ratings, and for different intended applications. Values in the mA range are commonly used. Third, the Miller voltage can be identified for any given current level. V_{Miller} is the gate voltage at which a device transitions between ohmic and saturation regions. It can be found as the gate voltage corresponding to the desired current level in the transfer characteristic. In designing a gate drive circuit for a particular device, it is important to choose a driving voltage that is sufficiently higher than the Miller voltage at the maximum anticipated load current. These three parameters are useful in modeling the device behavior and will also be applied for data processing of the dynamic characterization results in Chapter 8.

2.9 Gate current ($I_{g,ss}$–V_{gs}) characteristic

The vertical power MOSFET has a layer of oxide or other insulation between the gate terminal and the semiconductor layer. As such, no current is expected to flow between the gate and source other than capacitive charging current during switching transients. Any steady-state current in the gate is described as a "leakage" current. Characterization of this leakage current is useful for MOSFETs, particularly at the rated boundaries of temperature and gate voltage conditions.

For GaN FETs, this current may or may not actually be leakage. Some have an insulating layer beneath the gate, but others (e.g., the GIT) have no gate insulator. In this case, the gate current is intentional and may actually be the primary control

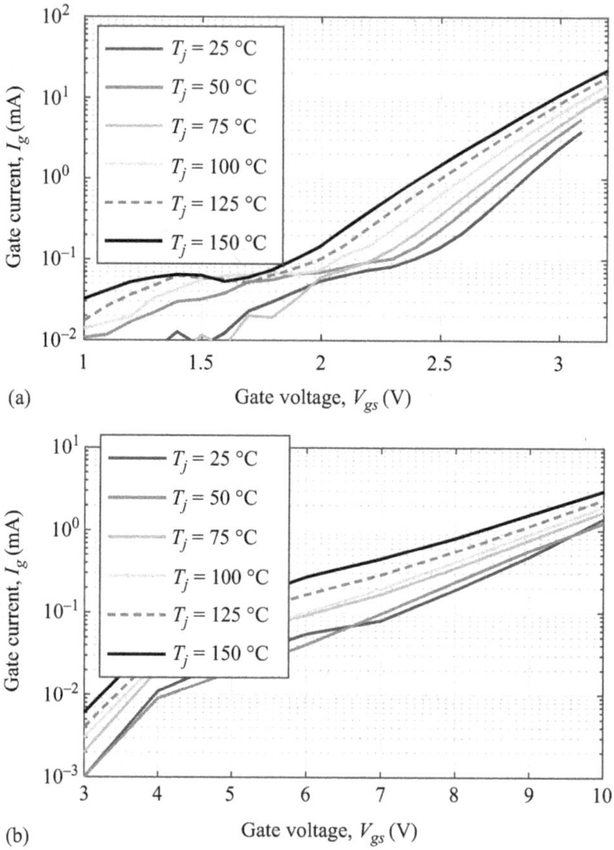

Figure 2.20 Example gate current characteristics for (a) GaN FET (with noninsulated gate) and (b) GaN FET (with insulated gate)

parameter for the on-state of the device. Whether leakage or intentional gate current, there is typically a strong relationship with both temperature and gate voltage, as demonstrated in the examples shown in Figure 2.20. Like the on-resistance characteristic, there may also be a relationship with drain current or voltage.

The gate leakage characteristic can be measured with a curve tracer during the *I–V* or transfer characteristic testing. If gate current measurement is not a feature of the curve tracer, a shunt resistor can be added in series with the gate to be measured with an oscilloscope (or with a multimeter, for some curve tracers that generate a nonpulsed gate voltage) as pictured in Figure 2.21(a). This requires manual recalculation of the actual gate voltage on the device terminals by subtracting the voltage drop across the shunt resistor. However, this test can also be performed with more conventional benchtop equipment using the test setup shown in Figure 2.21(b). This test can be performed on the device at any time, even after it is already soldered to a populated PCB with other components. Any connection on the board to the gate must be severed, which can often be accomplished by simply removing the gate resistor. In place of the gate drive circuit, a benchtop DC power supply and ammeter

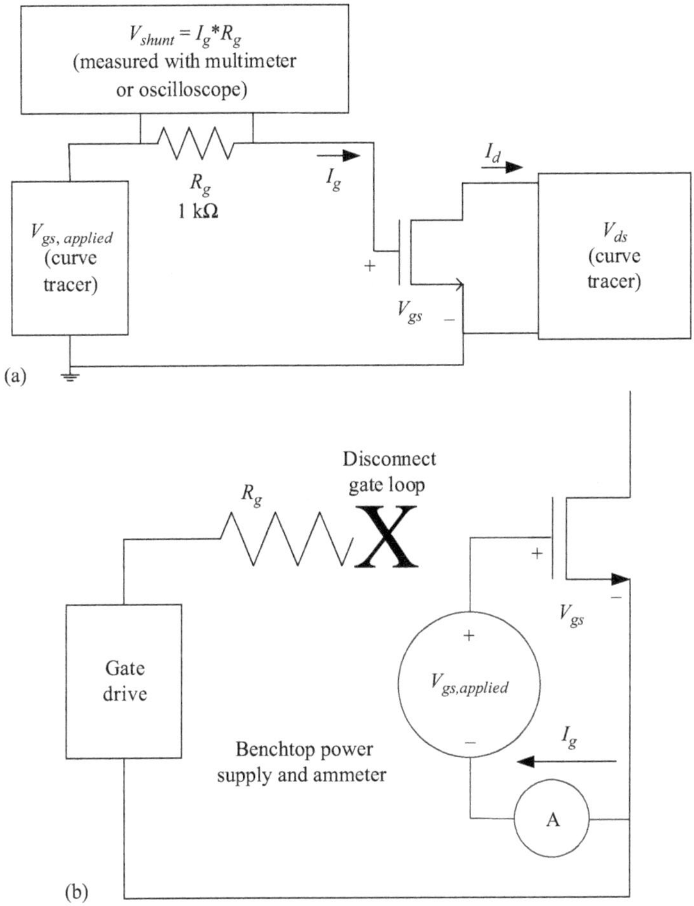

Figure 2.21 Test circuits for gate current: (a) as part of pulsed I–V test if gate current is not already measured; (b) in situ after added to converter/ DPT circuit

can be used to measure the gate voltage/current relationship. This is not pulsed, so particular attention must be paid to ensure that the device is not self-heating during the test. The steady-state gate current (or leakage) energy loss is typically in the μA or mA range, and is not likely to cause significant self-heating for most devices.

If a nonzero gate current is expected, particular caution must be exercised in the selection of gate resistance and test lead length/gauge. If these values are known, the gate voltage can be adjusted by subtracting the product of the gate current and the total loop resistance. Otherwise, it is important to ensure that the curve tracer is configured with zero gate resistance (sometimes a setting that can be selected in the software or hardware), and that a large gauge and reasonably short wire be used for the gate leads. The contact resistance of interconnections should also be considered. Soldered wires, screw terminals, and banana connectors are typically preferable over alligator clips.

2.10 Drain-source leakage ($I_{d,off}$–V_{ds}) characteristic

The drain-source leakage characteristic is also a measure of the I_d–V_{ds} relationship with a fixed V_{gs}. The two distinctive features of this test are (1) operation in the cutoff region and (2) elevation of V_{ds} up to or above the voltage rating of the device. Because of the potential for damage to the device or self-heating, it is recommended that this test also be pulsed. However, the gate voltage is typically not pulsed, and may in fact be directly shorted to the source in some cases. The $I_{d,off}$–V_{ds} relationship typically has a strong temperature dependency, although the leakage current may become higher or lower with elevated temperature. Similarly, applying a more negative gate voltage may cause higher or lower leakage current. The drain-source leakage current should theoretically always increase somewhat with higher drain voltage. The shape of this characteristic and its underlying physical mechanism is different for GaN FETs than it is for MOSFETs (whether Si or SiC).

Figures 2.22 and 2.23 show example drain leakage current characteristics for a SiC MOSFET and GaN FET, as functions of drain voltage, gate voltage, and

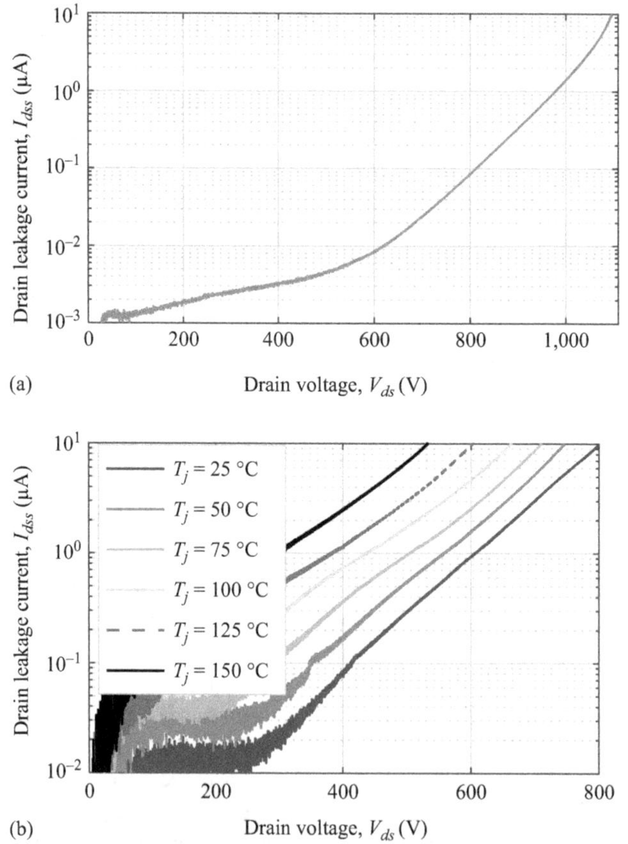

Figure 2.22 Example drain leakage current characteristics for (a) SiC MOSFET and (b) GaN FET

junction temperature. As shown, both curves indicate a positive trend between drain voltage and current. The breakdown voltage can typically be extracted from the leakage characteristic by setting a current threshold higher than the expected leakage below the avalanche voltage, which should be selected based on initial observations of the device's leakage characteristic. However, GaN FETs do not experience avalanche breakdown, so there may not be a clear "breakdown" voltage that can be extracted from this characteristic. The dielectric breakdown of lateral GaN FETs can occur gradually over an extended time of repeated stress, so device manufacturer guidelines should be followed when selecting a maximum allowable leakage current. Figure 2.22(b) indicates a logarithmically increasing leakage current over drain voltage, with no apparent "knee" where the current sharply increases, as would be expected when approaching the breakdown voltage.

It is important to consider the potential for irreversible device damage when selecting the pulse settings for the $I_{d,off}$–V_{ds} test. Similar to the issues discussed previously in Section 2.5, the pulse must be long enough to ensure an accurate and stable measurement point, but short enough to prevent self-heating or damage to the device.

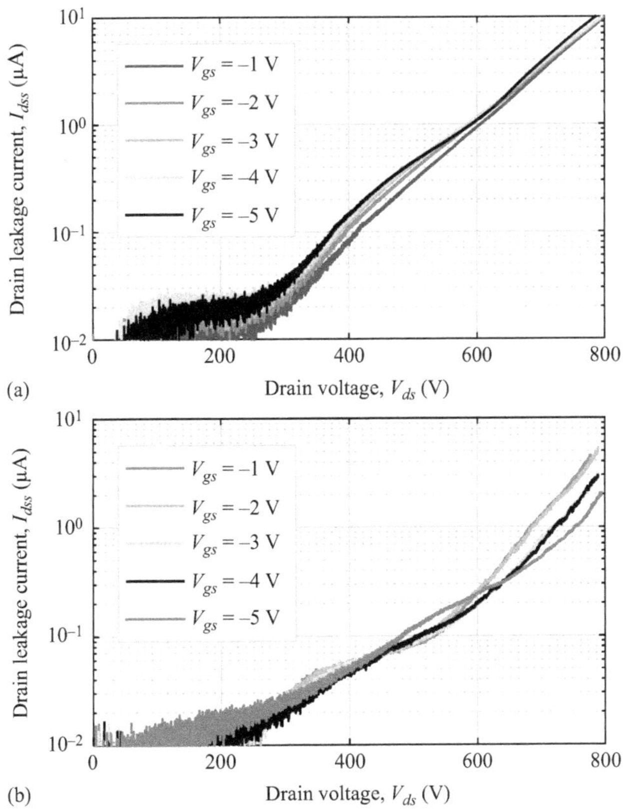

Figure 2.23 Example drain leakage current characteristics over gate voltage for two example GaN FETs

2.11 Summary

Static characterization is often the first step in modeling a new device or assessing a device for new application. The drain and gate leakage characteristics help identify the maximum voltage capabilities, and the output characteristics help identify a recommended operating current range. The on-resistance characteristic is particularly important for estimating the conduction loss in a converter, which can be used to meet an efficiency target or optimize a thermal design. This curve can also be used to choose a gate voltage or current for the gate drive design. The transfer characteristic is also an important consideration in gate drive design, as it indicates the threshold voltage and Miller voltage as well as the forward transconductance. With the addition of the junction capacitance characteristics presented in the next chapter, these static *I–V* curves also provide valuable input for the dynamic characterization, which will be discussed in Chapter 8.

References

[1] Keysight. (2004). B1505a power device analyzer/curve tracer [Online]. Available: http://www.keysight.com/en/pd-1480796-pn-B1505A/. Accessed in December 2017.

[2] E. A. Jones, F. Wang, and D. Costinett, "Review of commercial GaN power devices and GaN-based converter design challenges," *IEEE Journal of Emerging and Selected Topics in Power Electronics*, vol. 4, no. 3, pp. 707–719, September 2016.

Chapter 3

Junction capacitance characterization

This chapter will describe the procedures and methodology for characterizing the equivalent capacitances of a WBG power device, including the individual capacitances between each electrical terminal of the device and the conventionally reported lumped capacitance parameters. These capacitances, often in the form of curves as functions of operating conditions, have inherent value in selecting an appropriate device for a given application, as well as creating accurate simulation models and interpreting dynamic characterization data. They are also valuable for dynamic characterization design as will be explained in Chapter 4.

3.1 Fundamentals of *C–V* testing

Parasitic capacitances are formed within the power device between any two terminals at the die level and in the packaging. At the die level, these capacitances partly represent the charge separation that must occur to create the depletion region that blocks voltage, and also the recombination of these charges that restores the conductive channel. The capacitance representing this charge separation is highly nonlinear with respect to voltage. The nonlinearity depends on the geometry and material properties of the device. There are also additional sources of capacitance between contact metallizations, field plates, and the epitaxial layers of the die. Metal in the package may add capacitance as well. While it is not necessary to understand the physical origin of these capacitances to characterize them, it is helpful to remember that the physical structure of the device has a tremendous impact on these characteristics. For example, in a Super Junction MOSFET, the capacitance between drain and source (C_{ds}) may change by several orders of magnitude over the rated voltage range, while in others it may only change slightly. Similarly, the capacitance from gate to source (C_{gs}) typically changes more dramatically for GaN FETs over the rate gate voltage range than in typical MOSFETs. Figure 3.1 shows the physical origins of the junction capacitances in two example device structures, a vertical MOSFET (either Si or SiC) and a lateral GaN HFET.

Although it is more straightforward to understand the physical origins of capacitances between the three terminals of the transistor, they are more often described using the following definitions:

$$C_{oss} = C_{ds} + C_{gd} \tag{3.1}$$

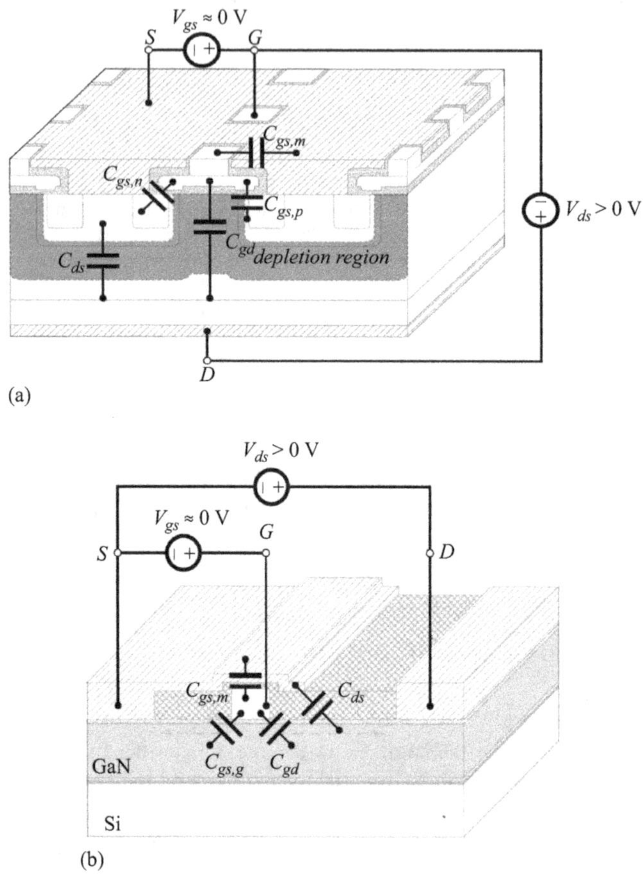

Figure 3.1 Examples of the physical origins of junction capacitances in (a) a MOSFET and (b) a GaN HFET [1]

$$C_{iss} = C_{gs} + C_{gd} \tag{3.2}$$

$$C_{rss} = C_{gd} \tag{3.3}$$

where C_{ds} is the capacitance between the drain and source terminals (often called the "drain capacitance"), C_{gs} is the capacitance between the gate and source terminals (often called the "gate capacitance"), and C_{gd} is the capacitance between the gate and drain terminals (often called the "Miller capacitance").

The lumped parameters are described as follows:

- Output capacitance, C_{oss}, represents the total capacitance seen at the drain terminal with the source and gate shorted together.
- Input capacitance, C_{iss}, represents the total capacitance seen at the gate terminal with the drain and source shorted together.
- Reverse transfer capacitance, C_{rss}, represents the capacitance seen between the gate and drain terminal with the source terminal floating [2].

The physically derived capacitances C_{ds}, C_{gs}, and C_{gd} can easily be calculated if all three of the lumped parameters are known. One of the reasons that capacitances are typically reported with these lumped values is ease of measurement. It is possible to directly measure the capacitance between any two terminals, but this may result in error due to the series connection of the remaining two capacitances. It may also be dangerous to apply a voltage to the devices with either the gate or source terminal floating as this may allow for accidental turn-on of the device.

3.2 Test equipment description

The primary equipment used for junction capacitance measurement is a precision impedance analyzer, although this may be a feature of some parametric curve tracers as well. This impedance analyzer must be precisely calibrated with the test fixture and any mechanical/electrical adapter sockets before measurement of the device. The complex impedance at the calibrated socket terminals is measured by the equipment at a particular DC voltage and AC frequency. This is then converted to capacitance using the complex impedance and phase measured at each point, assuming that any stray inductance in the measurement is in series with the capacitance of interest. The DC voltage range for this measurement is easily determined from the intended application range of the device. For example, all three lumped capacitances are typically measured with a DC voltage applied to V_{ds} between 0 V and the rated drain voltage of the device. The gate capacitance is measured with a DC voltage applied to V_{gs} from the negative rating to the positive rating of the gate voltage. The frequency may be a less obvious selection. The measurement frequency must be selected below the resonant frequency of the measurement circuit to accurately report the capacitance. In practice, 100 kHz or 1 MHz is conventionally selected as the measurement frequency. This selection is unrelated to equivalent frequency of the switching transient. The average frequency of a switching transient is often approximated by

$$f_{eq,on} = \frac{0.45}{t_{vf,90-10}} \tag{3.4}$$

for the turn-on transient, where $t_{vf,90-10}$ is the time required for v_{ds} to fall from 90% to 10% of the DC bus voltage. Similarly, the equivalent frequency of the turn-off transient is approximated as

$$f_{eq,off} = \frac{0.45}{t_{vr,10-90}} \tag{3.5}$$

where $t_{vr,10-90}$ is the time required for v_{ds} to rise from 10% to 90% of the DC bus voltage. The implications of this equivalent frequency on measuring and processing dynamic transient characteristics will be discussed further in Chapter 8. However, as previously mentioned, the most common practice is simply to use 1 MHz for this measurement [3, 4].

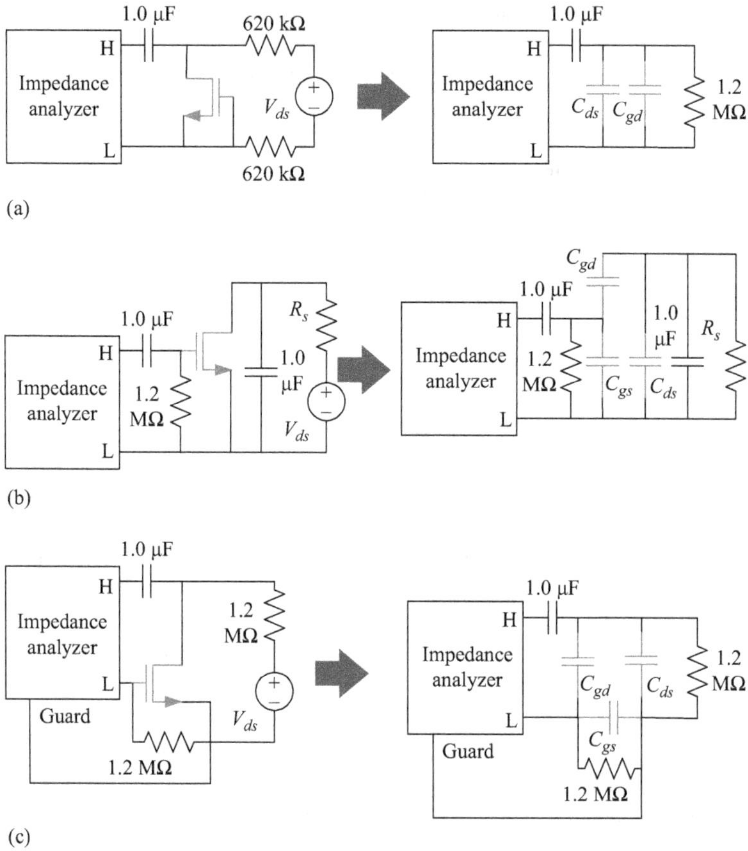

Figure 3.2 Test setups and AC equivalent circuits for characterization of junction capacitances, including (a) output capacitance C_{oss}, (b) input capacitance C_{iss}, and (c) reverse transfer capacitance C_{rss} [3]

Special steps may be necessary in order to safely apply a DC voltage to the device terminals without damaging the equipment. The data presented in this chapter was collected with two pieces of equipment: a Keysight B1505a curve tracer with up to 3 kV DC bias capability (and reconfigurable for up to 10 kV) or an Agilent 4294A precision impedance analyzer with a 40 V DC bias capability (and up to 200 V with an external adapter). If the available equipment has an insufficient DC bias capability for the intended measurement, external adapter circuits can be constructed as shown in Figure 3.2 [3]. In these circuits, relatively large capacitances (1 μF) act as AC shorts but DC open-circuits, thereby applying the high DC bias to the drain terminal while blocking that bias voltage from the equipment terminals. This capacitance causes an error in the measurement given by

$$C_j = \frac{C_{isolation} C_{measured}}{C_{isolation} - C_{measured}} \tag{3.6}$$

where C_j is the desired capacitance measurement, $C_{measured}$ is the value reported by the impedance analyzer, and $C_{isolation}$ is 1.0 μF. Because the junction capacitances are typically in the pF range, it is evident that $C_j \approx C_{measured}$ with an error of 0.1% or less, and compensation may not be necessary at all.

The resistances shown in Figure 3.2 are used to ensure that the DC bias voltage is fully applied to the device terminals rather than the isolation capacitor, and also to limit short-circuit current when the device is not connected. The values listed here for capacitors and resistors are only one example and can be substituted for similar components.

The Miller capacitance C_{gd} is typically much smaller than C_{gs} or C_{ds}, so it is critical that these capacitances not be included in parallel with C_{gd} during the reverse transfer capacitance measurement. This is accomplished by connecting the Guard terminal of the impedance analyzer to the source terminal of the device. The Guard is the circuit common of the impedance analyzer's internal bridge, and any AC current that flows into this terminal is neglected by the impedance measurement [5].

3.3 Test fixture selection/design and calibration

In order to measure these pF-scale capacitances, it is very important to precisely calibrate the equipment using the test fixture and any adapters necessary to connect the device. The impedance analyzer measures the complex impedance at the terminals used for calibration. If an additional fixture adapter is used between the calibration point and the device, the resulting measurement will include the impedance of the adapter as well. In some cases, the impedance of the adapter may be greater than the device impedance itself. Furthermore, if the adapter is modified or moved even slightly, the overall system impedance may change.

The equipment may come with some fixture adapters provided, such as the TO-package socket shown in Figure 3.3. Most surface-mount devices require a custom PCB fixture adapter for the particular package style and pin/pad positions. One such adapter is shown in Figure 3.4 for a GaN FET. The critical design requirement is that separate circuit paths be provided from each equipment terminal to the device (gate, drain, and source) without sharing any test leads or connectors. In this example, the PCB was designed to also function as the fixture adapter for the pulsed *I–V* tests in the previous chapter, with removable banana plugs to allow for transition between the two purposes. Figure 3.5 shows a third option for a capacitance test fixture adapter, customized for a large power module with screw terminals. This adapter is composed of a plastic board, rigidly connected banana plugs, and rigidly mounted test leads. These leads are solid core and heavy gauge, so that they cannot easily shift or reorient when the module is connected or disconnected. The wires are also taped firmly to the board. If the wires are allowed to shift in any way, the parasitic capacitance between them and their mutual inductance will change, and this may impact the measurement.

Two steps are involved in the equipment calibration procedure. The first is open-circuit calibration, which allows the impedance analyzer to measure the stray

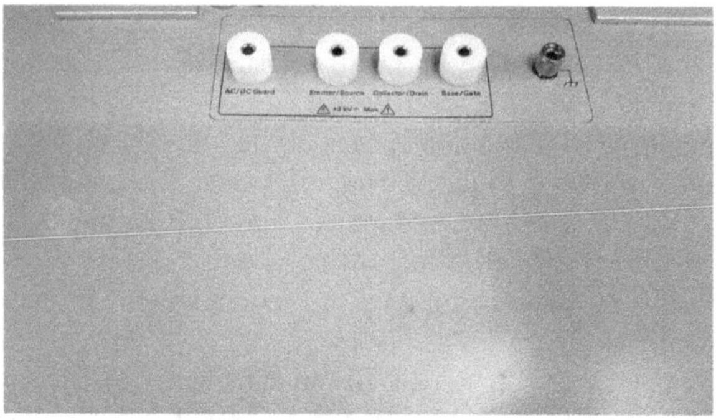

(a)

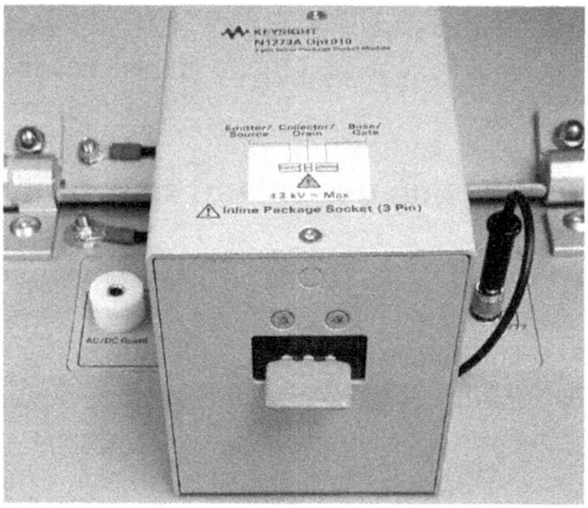

(b)

Figure 3.3 Test fixture for capacitance measurement in a Keysight B1505a, including (a) the bare test fixture and (b) the manufacturer-provided adapter for TO packaged devices

capacitance between the fixture terminals. The second is short-circuit calibration, which allows the impedance analyzer to measure the stray inductance between the terminals. After this calibration is performed once for the equipment, it does not need to be recalibrated again until the test fixture is modified.

3.4 Output capacitance (C_{oss}) characteristic

The output capacitance of a power device is highly nonlinear with respect to drain voltage, so it is important to characterize it across the full range of intended operation. However, this capacitance is most sensitive to voltage in the <100 V range, after which it drops much more gradually with voltage. Figure 3.6(a) and (b) shows example C_{oss} curves for a 1200 V SiC MOSFET and a 600 V GaN FET, respectively. It is clear in this example that the SiC MOSFET's output capacitance

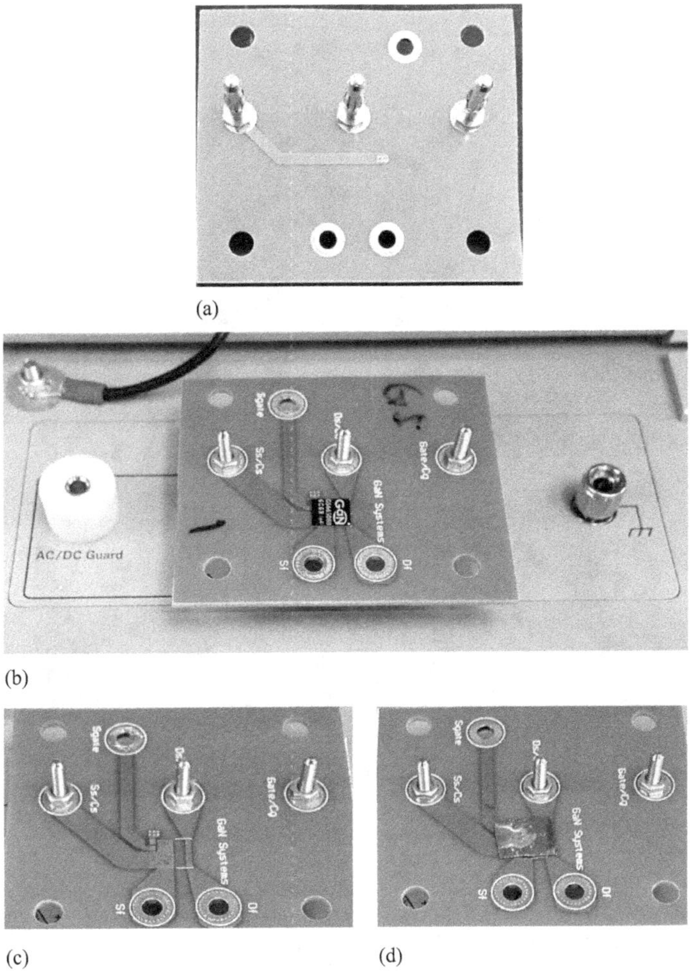

Figure 3.4 Test fixture adapters for capacitance measurement of a surface-mount GaN FET in a Keysight B1505a, including (a) bottom view with rigidly connected banana plugs, (b) top view with device installed, (c) top view for open-circuit calibration, and (d) top view for short-circuit calibration

is more strongly nonlinear in the low-voltage region than the GaN FET, and the C_{oss} curve is difficult to interpret in the higher voltage region as a result. For some devices like this one, it is more helpful to plot the curve with a semilog-*y* axis as shown in Figure 3.7. This semilog plot is not as useful for the GaN case, which is clearer in the linear plot of Figure 3.6.

As previously mentioned, the physical origin of this nonlinear capacitance is heavily dependent on the device structure or geometry, and material properties. The

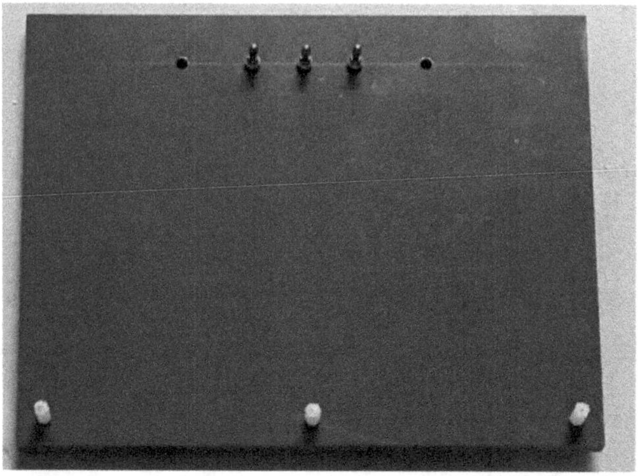

(a)

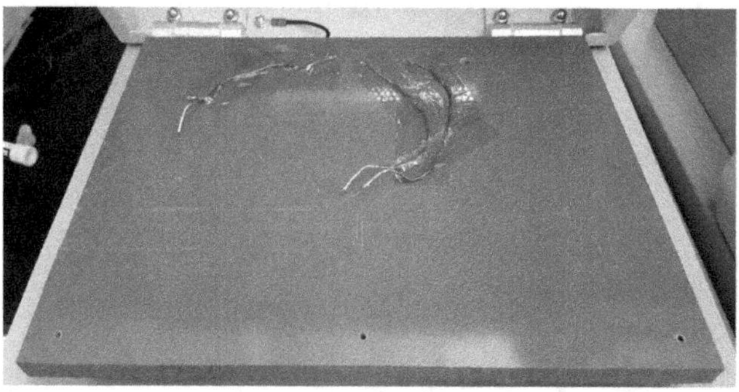

(b)

Figure 3.5 Test fixture adapters for capacitance measurement of a large power module in a Keysight B1505a, including (a) bottom view with rigidly connected banana plugs and (b) top view with rigidly mounted cables for device connection

lateral structure of the GaN FET leads to a special consideration for C_{oss} characterization. The substrate of a lateral GaN FET is not intended as an electrical connection since the current must flow through several buffer layers between the 2DEG channel and the Si substrate. However, the substrate is often electrically connected to the source terminal either within the package or on the PCB. This connection is recommended by some GaN manufacturers to reduce the impact of dynamic $R_{ds,on}$. However, the parasitic capacitance between the drain and substrate can be significant, and the source–substrate connection adds it in parallel with C_{oss}. Figure 3.8 compares an example C_{oss} curve of a GaN FET with the substrate floating and source-connected, with the difference between them indicating the additional capacitance between the drain and substrate.

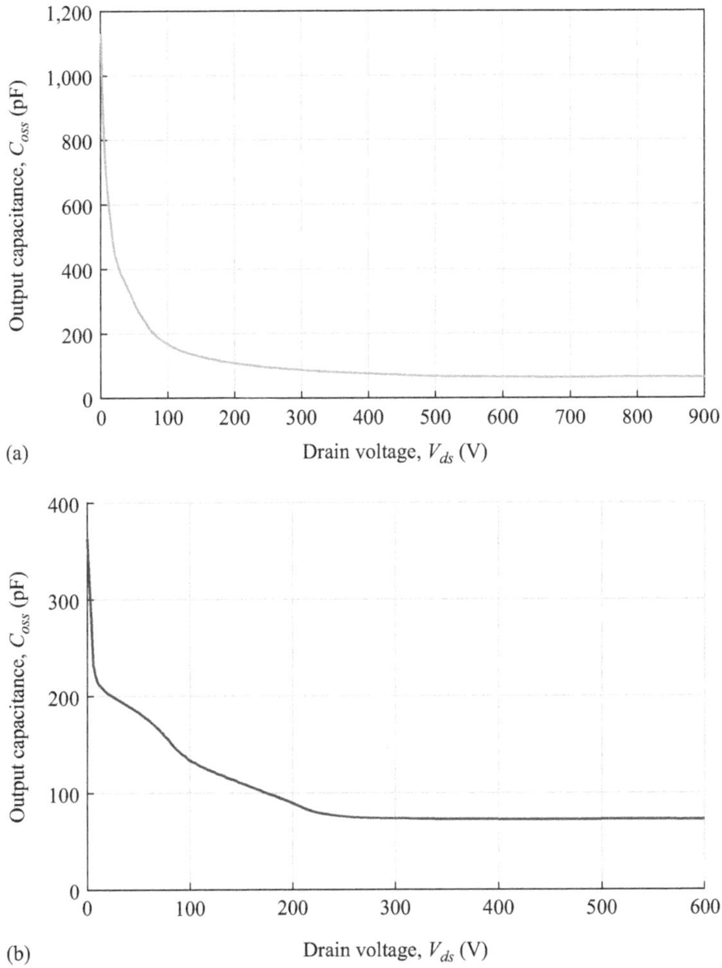

Figure 3.6 Output capacitance (C_{oss}) versus drain voltage (V_{ds}) for an example (a) SiC MOSFET and (b) GaN FET

3.5 Input capacitance (C_{iss}) characteristic

The input capacitance is typically given in device datasheets with respect to drain voltage. This input capacitance is the sum of the gate capacitance and Miller capacitance, as previously mentioned. The Miller capacitance is strongly dependent on drain voltage for both MOSFETs and GaN FETs, so C_{iss} is as well. However, the gate capacitance component is not significantly affected by variation in drain voltage. Therefore, the C_{iss}–V_{ds} curve is typically very flat with the exception of a small drop in the low-voltage range. Figure 3.9 shows an example of such a curve for a SiC MOSFET and for a GaN FET. This curve is typically measured with a DC

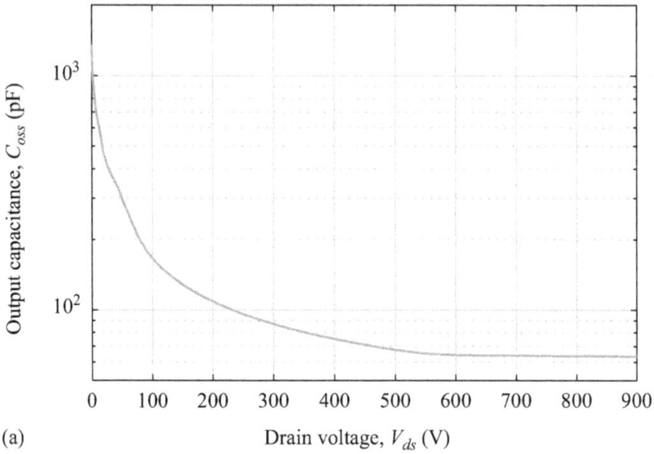

(a)

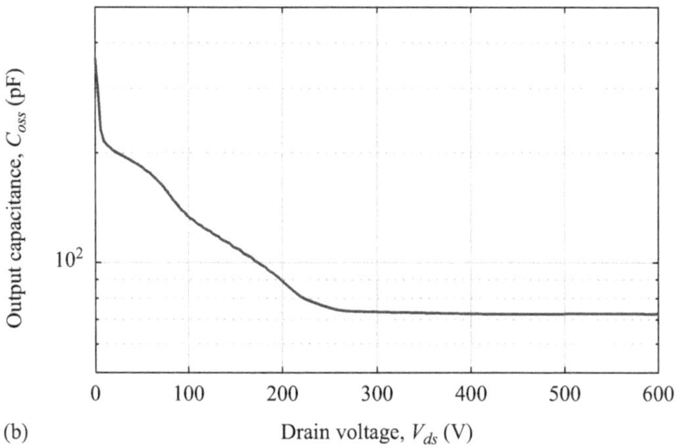

(b)

Figure 3.7 Output capacitance (C_{oss}) versus drain voltage (V_{ds}) on a semilog scale for an example (a) SiC MOSFET and (b) GaN FET

short (0 V) between the gate and source to maintain the off-state of the device. As with the output capacitance, it is useful to plot C_{iss} with a semilog-y axis as in Figure 3.10 for the SiC MOSFET, but not as much for the GaN FET.

While it may not vary with drain voltage, the gate capacitance C_{gs} is variable and nonlinear with respect to gate voltage V_{gs}. The physical origin of this V_{gs}–C_{gs} relationship has been well-reported for both MOSFETs and GaN FETs [6–8]. The mechanism is quite different for each, as is the shape of the characteristic. The examples in Figure 3.11 show a SiC MOSFET with a C_{gs} variation of approximately 50% across the gate voltage range, and a GaN FET with a C_{gs} variation closer to 200%. In both cases, the curve is steepest as V_{gs} approaches the threshold voltage and begins to enhance the channel, and relatively flat at much lower or much higher gate voltages. Notably, the MOSFET curve shows a negative trend in

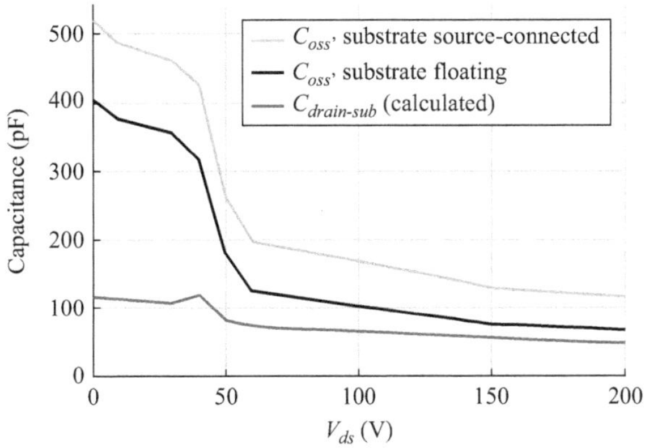

Figure 3.8 Comparison of example output characteristic versus drain voltage for a GaN HFET (GaN Systems GS66508P) with the substrate floating versus source-connected

the negative V_{gs} range, while the GaN FET curve shows a continuing positive trend with negative V_{gs}.

This C_{gs}–V_{gs} relationship is not unique to WBG devices, but the faster switching speed amplifies the impact of inaccurate characteristics in simulation and data analysis. The nonlinear C_{gs} characteristic plays an important role in predicting and interpreting dynamic V_{gs} switching waveforms (and indeed, all switching waveforms), as well as in calculating the total gate charge for gate drive selection and driving loss calculation.

3.6 Reverse transfer capacitance (C_{rss}) characteristic

The reverse transfer capacitance C_{rss}, which is equivalent to the Miller capacitance C_{gd}, is typically characterized with respect to drain voltage. For both SiC MOSFETs and GaN FETs, the reverse transfer capacitance is much smaller than the other two lumped capacitance parameters. At high voltage (>100 V), C_{rss} is typically in the range of 1–10 pF. C_{rss} increases at lower voltage, although typically still much lower than either C_{oss} or C_{iss}. Figure 3.12 shows example C_{rss}–V_{ds} characteristics for a SiC MOSFET and a GaN FET.

For this characteristic, the semilog-*y* axis is necessary in most cases in order to interpret the results. The reverse transfer capacitance typically drops by several orders of magnitude over the rated drain voltage range, and the minimum value may appear to be 0 pF in a linear plot. However, the semilog plots shown in Figure 3.13 indicate a minimum C_{rss} of 4.3 pF for the SiC example and 0.7 pF for the GaN example. The dynamic behavior of the device is particularly sensitive to this capacitance, even when it is this small, so this added detail can improve the accuracy of simulation models.

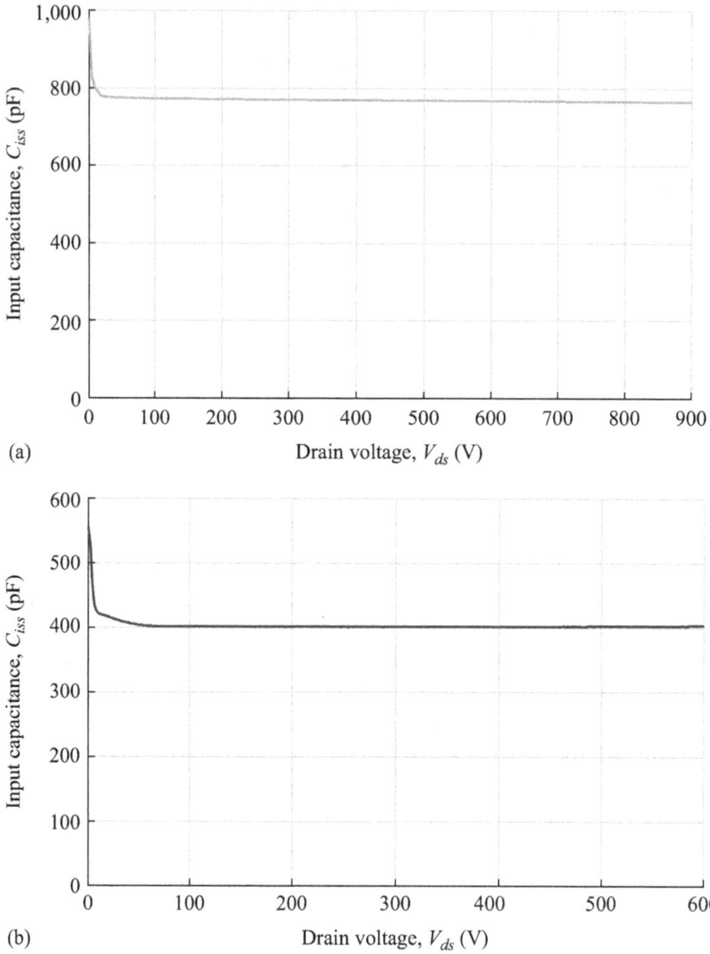

Figure 3.9 Input capacitance (C_{iss}) versus drain voltage (V_{ds}) for an example (a) SiC MOSFET and (b) GaN FET

3.7 Gate charge (Q_g) characteristic

The gate charge characteristic shows the relationship between Q_g and V_{gs} for a particular operating condition, including DC bus voltage, load current, and junction temperature. Notable features of the Q_g curve include the following:

- Total gate charge ($Q_{g,total}$) supplied by the gate drive
- Charge needed to bring V_{gs} to its threshold voltage ($Q_{gs,th}$)
- Miller voltage (V_{Miller})
- Gate charge needed to bring V_{gs} up to the Miller voltage ($Q_{gs,Miller}$)
- Miller charge (Q_{gd})

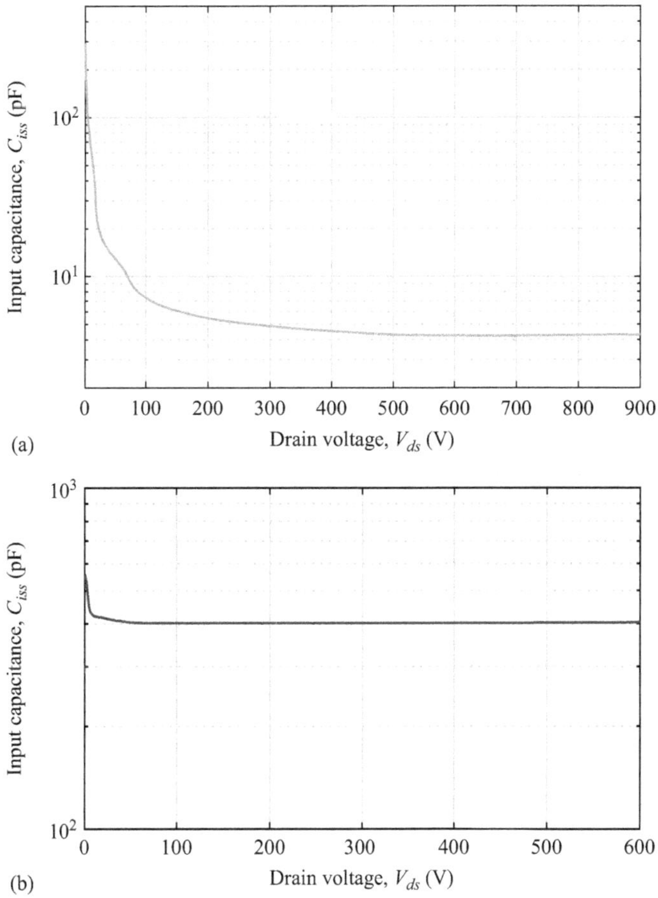

Figure 3.10 Input capacitance (C_{iss}) versus drain voltage (V_{ds}) on a semilog scale for an example (a) SiC MOSFET and (b) GaN FET

Each of these parameters can be calculated using the characterization results from Chapter 2 and the previous sections of this chapter. The Miller voltage can be extracted from the transfer characteristic, which is sensitive to the load current and the junction temperature. The threshold voltage can also be extracted from the transfer characteristic. Using these voltages, the remaining parameters can be calculated as

$$Q_{gs,th} = \int_{V_s}^{V_{gs,th}} C_{gs}(v)dv \tag{3.7}$$

$$Q_{gs,Miller} = \int_{V_s}^{V_{Miller}} C_{gs}(v)dv \tag{3.8}$$

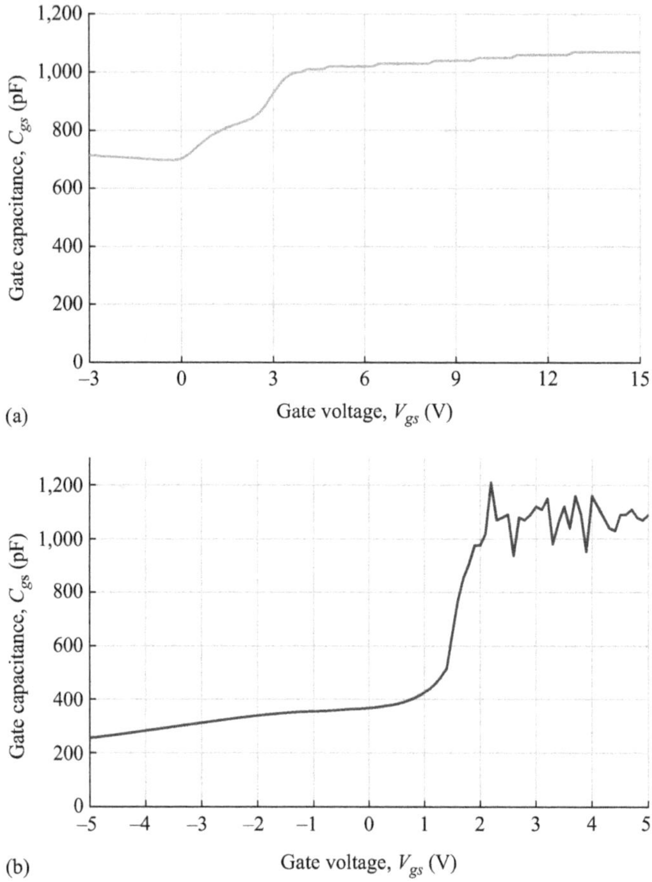

Figure 3.11 Gate capacitance (C_{gs}) versus gate voltage (V_{gs}) for an example (a) SiC MOSFET and (b) GaN FET

$$Q_{gd} = \int_0^{V_{dc}} C_{rss}(v)dv \tag{3.9}$$

$$Q_{g,total} = Q_{gs,Miller} + Q_{gd} + \int_{V_{Miller}}^{V_p} \left(C_{gs}(v) + C_{rss}(0\ V)\right)dv \tag{3.10}$$

where V_p is the positive gate driving voltage and V_s is the negative gate driving voltage (or 0 V if no negative gate driving voltage is supplied). In datasheets, 0 V is typically used for V_s, but many WBG devices benefit from a negative off-state driving voltage to prevent spurious switching events. Considering this in the Q_g calculation can have a significant impact on the results.

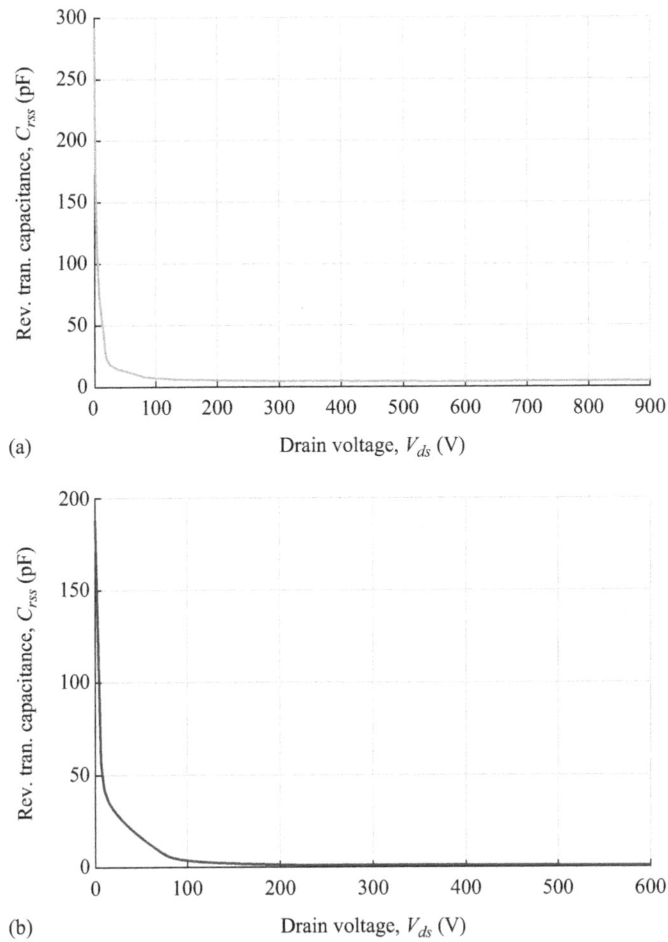

Figure 3.12 Reverse transfer capacitance (C_{rss}) versus drain voltage (V_{ds}) for an example (a) SiC MOSFET and (b) GaN FET

In fact, the entire Q_g curve can be extrapolated using the piecewise function

$$Q_g = \begin{cases} \int_{V_s}^{v_{gs}} C_{gs}(v_{gs})dv, v_{gs} < V_{Miller} \\ Q_{gs,Miller} + Q_{gd} + \int_{V_{Miller}}^{v_{gs}} (C_{gs}(v_{gs}) + C_{rss}(0\,V))dv, v_{gs} \geq V_{Miller} \end{cases} \tag{3.11}$$

This function is discontinuous at the point $v_{gs} = V_{Miller}$, indicating that v_{gs} is completely flat in the Q_g curve from $Q_{gs,Miller}$ to $Q_{gs,Miller} + Q_{gd}$. This region of the curve is often shown with a very slight slope rather than completely flat, based on

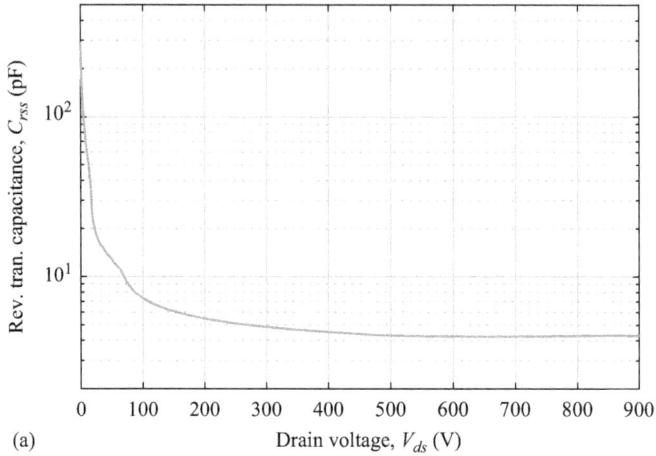

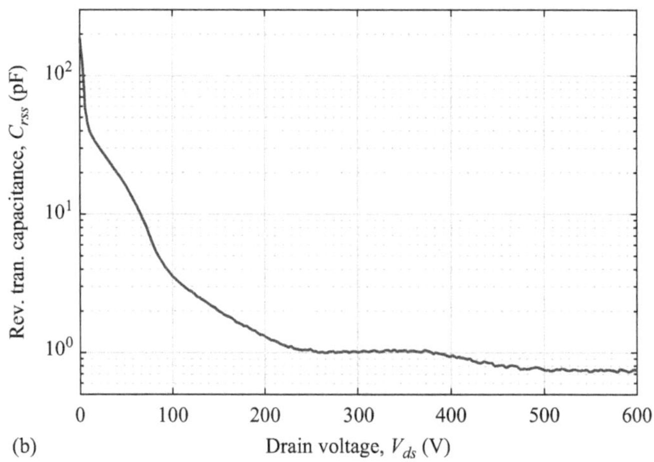

Figure 3.13 Reverse transfer capacitance (C_{rss}) versus drain voltage (V_{ds}) on a semilog scale for an example (a) SiC MOSFET and (b) GaN FET

the equipment used for the gate charge characteristic measurement. However, this behavior is more accurately captured in the dynamic characterization process, and it has no impact on the calculations in (3.7)–(3.10). Figure 3.14 demonstrates example Q_g characteristics generated using these equations for a SiC MOSFET and for a GaN FET. These curves are shown at three operating conditions, each a combination of DC bus voltage and load current, with a junction temperature of 25 °C. As shown, the impact of higher load current is a higher Miller voltage, and the final gate charge is the same. Although it is not shown in Figure 3.14, changing the junction temperature will also only affect the Miller voltage. The impact of higher DC bus voltage is a higher Miller charge, which increases the total gate charge by the same amount.

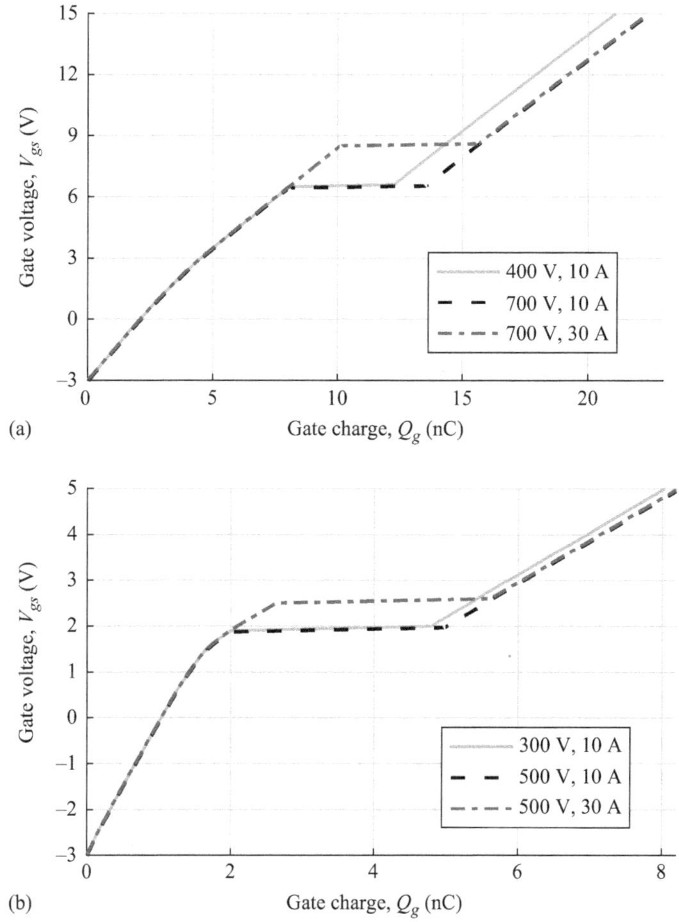

Figure 3.14 Example gate charge characteristics at several DC bus voltages and load currents and 25 °C for (a) a SiC MOSFET and (b) a GaN FET

Some curve tracers are capable of performing a gate charge characteristic directly. However, this may give somewhat arbitrary results near the $v_{gs} = V_{Miller}$ region of the curve, as it depends heavily on the method used to drive the gate.

3.8 Calculation of C_{oss}-related switching energies

Details on device switching and its characterization can be found in Chapter 4 and other chapters. Here, the discussion focuses only on C_{oss}-related switching energies. Datasheets often report three other parameters related to output capacitance, which are E_{oss}, $C_{oss,er}$, and $C_{oss,tr}$. These values are given to help users interpret the C_{oss} curves more easily. The parameter E_{oss} represents the energy required to bring the drain voltage up to a particular DC bus voltage. This energy can be calculated using

$$E_{oss} = \int_0^{V_{dc}} \left[C_{oss}(v) + C_{stray}\right] \times v\,dv \tag{3.12}$$

where C_{stray} is the stray capacitance in the circuit (other than inside the device package itself) that are in parallel with the switch. If the stray capacitance in the application circuit is not known, this term can be neglected, although practically it is often in the tens of pF range for discrete devices.

A second energy quantity can also be calculated from this curve, which is the additional energy that is lost when charging C_{oss} up to a given DC bus voltage, if the switch in question is operating as the synchronous switch at the time (see Chapter 4 for detailed description of device switching). This energy loss only occurs during a hard turn-on transient of the complementary switch, and the energy is lost in the channel of that same complementary switch. However, the value is solely dependent on the C_{oss} that is being charged, according to

$$E_{on,Coss,sync} = \int_0^{V_{dc}} \left[C_{oss}(v) + C_{stray}\right] \times (V_{dc} - v)dv \tag{3.13}$$

where C_{stray} is the stray capacitance in parallel with that synchronous switch.

It has previously been explained and demonstrated that the value of this energy is not related to the characteristics of the complementary switch turning on, which is actually incurring the loss [9]. Therefore, this energy can be considered another property of the device being characterized. Again, if the stray resistance in the intended application circuit is not known, it can be neglected or determined later during the dynamic characterization process.

These two calculated energies are shown in Figure 3.15 for the SiC and GaN examples, both as functions of the operating DC bus voltage with the stray capacitance assumed to be 0.

The two equivalent capacitances are linearized calculations using integrations similar to (3.12) and (3.13). One is the "energy-related" equivalent, $C_{oss,er}$, intended to indicate the energy storage at a particular drain voltage. The other is the "time-related" (sometimes called "charge-related) equivalent, $C_{oss,tr}$, meant to indicate the time needed to charge the device up to a particular drain voltage. These capacitances can be calculated as

$$C_{oss,er} = \frac{2}{V_{dc}^2} \int_0^{V_{dc}} C_{oss}(v) \times v\,dv \tag{3.14}$$

$$C_{oss,tr} = \frac{1}{V_{dc}} \int_0^{V_{dc}} C_{oss}(v)dv \tag{3.15}$$

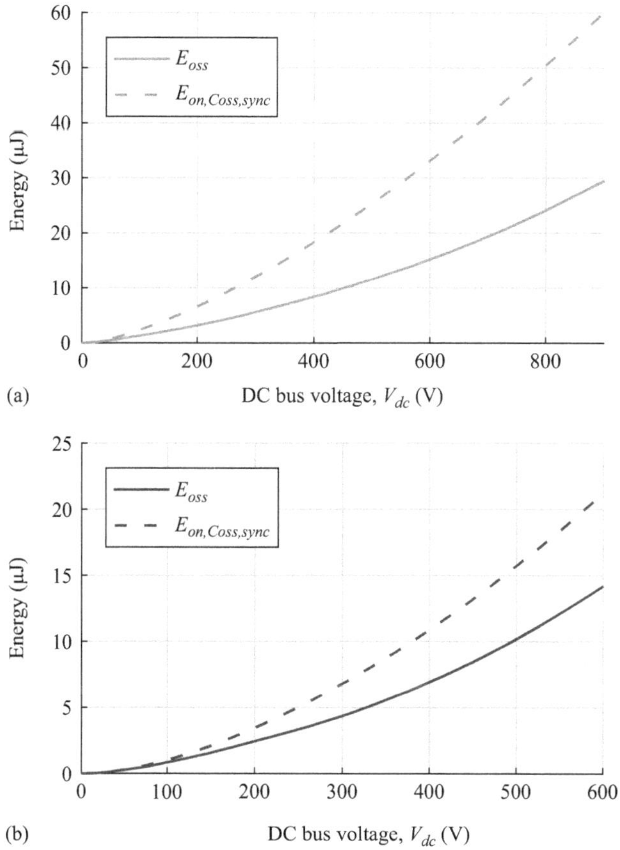

Figure 3.15 Calculated C_{oss}-related switching energies versus drain voltage for (a) SiC MOSFET and (b) GaN FET

These can be easily related back to the previously calculated energies as

$$E_{oss} = \frac{1}{2}\left(C_{oss,er} + C_{stray}\right)V_{dc}^2 \tag{3.16}$$

$$E_{on,Coss,sync} = \left(C_{oss,tr} - \frac{1}{2}C_{oss,er} + \frac{1}{2}C_{stray}\right)V_{dc}^2 \tag{3.17}$$

Much like the output capacitance characteristics, these equivalent capacitances are highly nonlinear with respect to voltage, so they are only valid approximations when calculated at the DC bus voltage intended for the application. However, as shown in Figure 3.16, they are most sensitive in the low-voltage range.

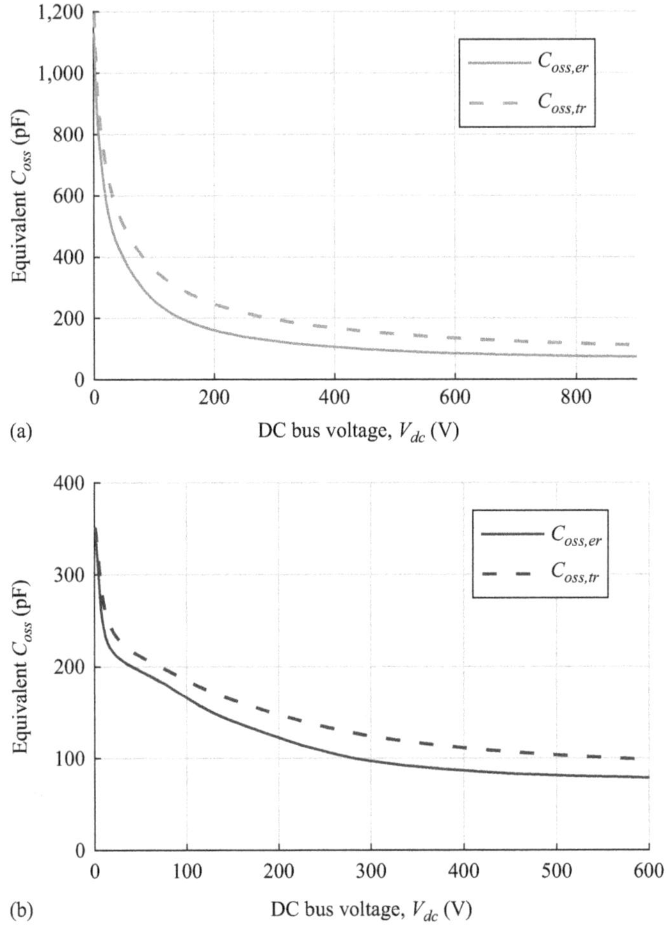

Figure 3.16 Calculated energy-related and time-related equivalent C_{oss} versus drain voltage for (a) SiC MOSFET and (b) GaN FET

3.9 Summary

This chapter described the methodology for static characterization of WBD device parasitic capacitances as functions of relevant voltages using a precision impedance analyzer or parametric curve tracer. The equipment may have a limited DC bias capability and can be supplemented by additional circuits to increase this range. Resulting capacitance curves give insight into the switching and driving losses, improve the accuracy of simulation models, and allow more in-depth analysis of switching waveforms collected in the dynamic characterization process.

References

[1] D. Costinett, "Analysis and design of high efficiency, high conversion ratio, dc-dc power converters," Ph.D. thesis, Dept. Elect., Comp., and Energy Eng., University of Colorado Boulder, Boulder, CO, 2013.

[2] V. Barkhordarian. (2005). Power MOSFET basics [Online]. Available: www.irf.com/technical-info/appnotes/mosfet.pdf. Accessed in December 2017.

[3] Z. Chen, "Characterization and modeling of high-switching-speed behavior of SiC active devices," M.S. Thesis, Department of Electrical and Computer Engineering, Virginia Tech, 2009.

[4] Z. Zhang, B. Guo, F. Wang, E. Jones, L. Tolbert, and B. Blalock, "Methodology for wide band-gap device dynamic characterization," *IEEE Transactions on Power Electronics*, vol. PP, no. 99, pp. 1–1, 2017.

[5] Agilent. (2009). Agilent impedance measurement handbook [Online]. Available: https://www.amt.pl/pl/faq/podstawy-pomiarow-impedancji/dnl/file. Accessed in December 2017.

[6] S. Khandelwal and T. A. Fjeldly, "A physics based compact model of gate capacitance in AlGaN/GaN HEMT devices," in *Proceedings of the International Caribbean Conference on Devices, Circuits and Systems* (ICCDCS), 2012, pp. 1–4.

[7] D. Neamen, *Semiconductor Physics and Devices*. McGraw-Hill, Inc., New York, NY, 2003.

[8] T. L. Wu, B. Bakeroot, H. Liang, *et al.*, "Analysis of the gate capacitance-voltage characteristics in p-GaN / AlGaN/GaN heterostructures," *IEEE Electron Device Letters*, vol. PP, no. 99, pp. 1–1, 2017.

[9] D. Costinett, D. Maksimovic, and R. Zane, "Circuit-oriented treatment of nonlinear capacitances in switched-mode power supplies," *IEEE Transactions on Power Electronics*, vol. 30, no. 2, pp. 985–995, 2015.

Chapter 4
Fundamentals of dynamic characterization

Dynamic characteristics illustrate the performance of power semiconductors during the switching transient. They are also termed as switching characteristics. The most important dynamic characteristics of a power semiconductor device include switching energy loss, switching transition parameters (e.g., switching time, *dv/dt*, *di/dt*), dynamic spikes (e.g., current/voltage overshoot/undershoot), and parasitic ringing. The DPT is a widely accepted method to characterize the dynamic performance of power semiconductors. This chapter introduces the fundamentals of dynamic characterization. Starting from the switching commutation analysis in an actual power electronics converter, the simplified test circuit for dynamic characterization is derived such that not the whole converter will be needed for the test. Then, fundamentals of DPT based on the simplified test circuit are introduced, including operation principle, configuration, and setup. Detailed DPT design and control are discussed. In the end, according to the design theory, a case study is given. Two-level voltage source converter is the focused topology in the following description, since it is the one of the most widely used converter types. Special consideration of dynamic characterization for other commonly used topologies will be discussed in Chapter 11.

4.1 Switching commutation analysis

Dynamic characterization aims at assessing the switching behavior of power semiconductors to support the design and optimization of power electronics converters. Thus, the methodology for dynamic characterization should be simple and generic, and conducted prior to the converter prototyping. With this, a good understanding of the switching commutation of power semiconductors in actual converters is essential.

An example is the voltage source inverter-fed motor drive, which is a typical representative of voltage source converter in industrial applications. As shown in Figure 4.1, the inverter consists of three phase-legs. For each phase-leg, it consists of two series-connected power semiconductor switches with a DC voltage source across the lower/upper terminals and an AC current source through its midpoint. It is worthwhile to notice that the freewheeling diode symbol illustrated in Figure 4.1 can physically be the body diode of the MOSFET, external PiN, or Schottky diode anti-paralleled with the active device, or "diode-like behavior" mechanism of a GaN FET.

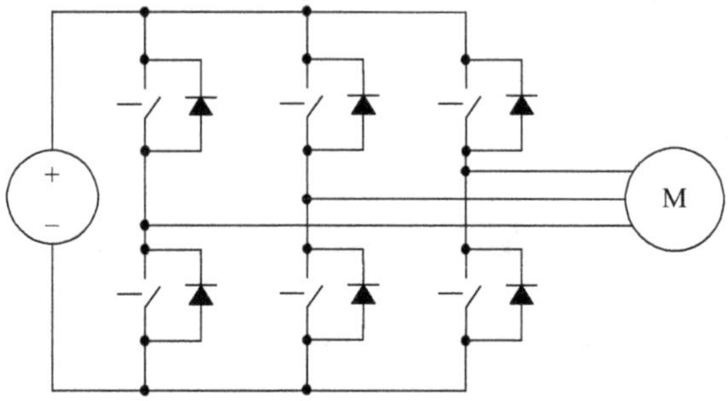

Figure 4.1 PWM voltage source inverter-fed motor drive

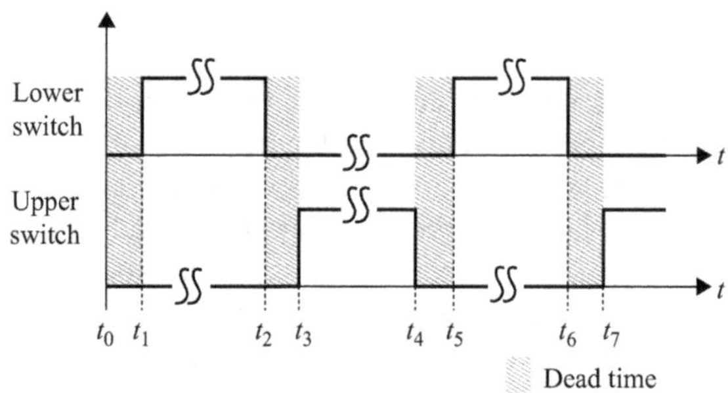

Figure 4.2 PWM control signal of upper and lower switches

Figure 4.2 illustrates the typical pulse width modulation (PWM) control signals of lower and upper switches in a phase-leg. These two signals are complementary with the dead time embedded to avoid short circuit of the DC voltage source.

Considering the symmetry of three phase-legs and their similar switching commutation, only one phase-leg is focused on in the following description. Regarding the motor load as AC current source, there are two scenarios: one is the inductive load current flowing into the midpoint of the phase-leg; the other is the inductive load current flowing out of it. Detailed analysis of switching commutation for these two cases are presented as follows.

First consider the case of the load current flowing into the midpoint of the phase-leg, as shown in Figure 4.3.

[$t_0 - t_1$]: Both the lower and upper switches are off. During this dead time, load current flows through the freewheeling diode of the upper switch (see Figure 4.3(a)).

[$t_1 - t_2$]: Lower switch turns on. Load current commutates from the upper diode to the lower switch while the DC voltage shifts to the upper diode from the lower switch. Thus, a hard turn-on switching commutation occurs between the upper diode and lower switch (see Figure 4.3(b)).

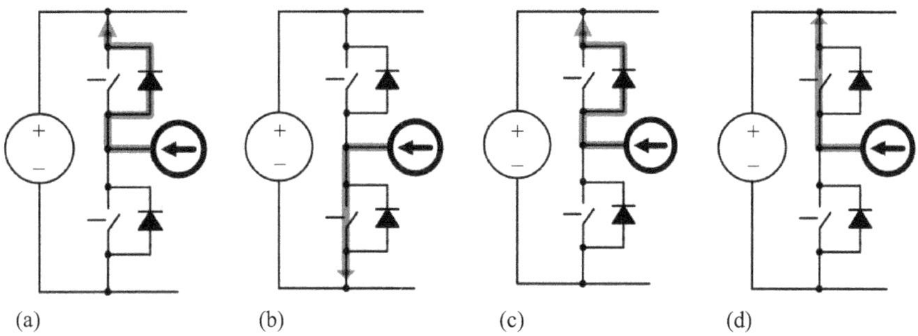

Figure 4.3 Switching commutation analysis when load current flows into the phase-leg: (a) $[t_0 - t_1]$; (b) $[t_1 - t_2]$; (c) $[t_2 - t_3]$; (d) $[t_3 - t_4]$

$[t_2 - t_3]$: Lower switch turns off. Load current commutates from the lower switch to the upper diode while the DC voltage shifts to the lower switch from the upper diode. Again, a hard turn-off switching commutation occurs between the upper diode and lower switch (see Figure 4.3(c)).

$[t_3 - t_4]$: Upper switch turns on. Before the turn-on of the upper switch, the load current already flows through its freewheeling diode. Turn-on of the upper switch only offers another current path in paralleled with the upper diode if the switch is capable of conducting reverse current. The impedance of this current path offered by the switch channel may be lower than the impedance of its freewheeling diode, as in the case of WBG devices. Thus, this turn-on transition only transfers the load current (part or all depending on the impedance ratio) from the upper diode to the upper switch when both devices are under zero voltage; there is no hard switching commutation between the upper and lower devices (see Figure 4.3(d)). This is termed as the synchronous turn-on switching commutation and no hard switching commutation loss exists.

$[t_4 - t_5]$: Upper switch turns off. Similar to the aforementioned analysis for $[t_3 - t_4]$, during this turn-off transient, the load current transfers from the upper switch back to the upper diode. As a result, there is no hard turn-off switching commutation along with its associated turn-off switching energy loss (see Figure 4.3(a)). This is called synchronous turn-off switching commutation.

In summary, when the load current flows into the midpoint of the phase-leg, the hard switching commutation only occurs between the upper diode and lower switch. Upper switch in this case behaves as the synchronous switch.

Consider the case of the load current flowing out of the midpoint of the phase-leg, as shown in Figure 4.4.

$[t_0 - t_1]$: Both the lower and upper switches are off. During this dead time, load current flows through the freewheeling diode of the lower switch (see Figure 4.4(a)).

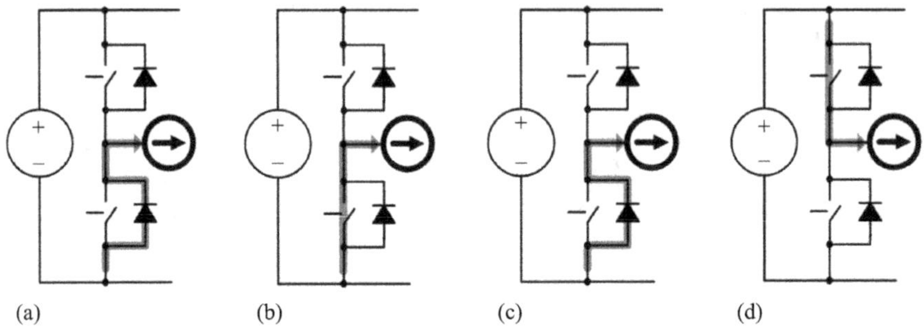

Figure 4.4 Switching commutation analysis when load current flows out of the phase-leg: (a) $[t_0 - t_1]$; (b) $[t_1 - t_2]$; (c) $[t_2 - t_3]$; (d) $[t_3 - t_4]$

$[t_1 - t_2]$: Lower switch turns on. Load current transfers from the lower diode to the lower switch (see Figure 4.4(b)). This is a synchronous turn-on switching commutation.

$[t_2 - t_3]$: Lower switch turns off. Load current transfers from the lower switch back to the lower diode (see Figure 4.4(c)). This is a synchronous turn-off switching commutation.

$[t_3 - t_4]$: Upper switch turns on. Load current commutates from the lower diode to the upper switch. This is a hard turn-on switching commutation between the lower diode and upper switch (see Figure 4.4(d)).

$[t_4 - t_5]$: Upper switch turns off. Load current commutates from the upper switch to the lower diode. This is also a hard turn-off switching commutation between the lower diode and upper switch (see Figure 4.4(a)).

In summary, when the load current flows out of the midpoint of the phase-leg, the hard switching commutation only occurs between the lower diode and upper switch. The lower switch in this case behaves as the synchronous switch.

According to the aforementioned discussion, to analyze the switching commutation of the three-phase voltage source inverter-fed motor drive, there is no need for employing the three-phase voltage source inverter illustrated in Figure 4.1, and the simplified circuits in Figure 4.5(a) and (b) are practically adequate. Considering the symmetry of these two simplified circuits, dynamic characterization usually employs the configuration in Figure 4.5(a) consisting of a lower switch and upper diode because of the stable measurement ground for the lower switch. It should be noted that the circuit in Figure 4.5(a) is employed to test and obtain dynamic characteristics by device manufacturers for device datasheets. Although the aforementioned analysis uses the voltage source inverter-fed motor drive as an example, it is valid for other voltage source-based power electronics converters as well.

In the rest of this book, without loss of generality, unless specified otherwise, the lower switch in the phase-leg is designated as the active switch while the upper one is the synchronous switch.

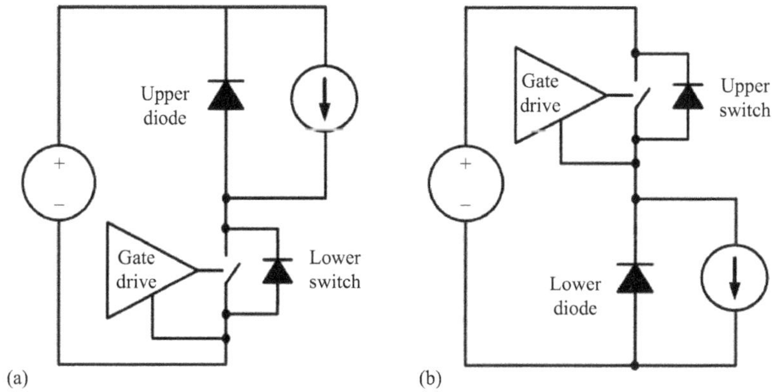

Figure 4.5 The simplified circuit for dynamic characterization for voltage source converter: (a) load current flows into the phase-leg; (b) load current flows out of the phase-leg

4.2 Fundamentals of DPT

Based on the simplified circuit in Figure 4.5(a), DPT is used to assess the dynamic performance of power devices [1–7]. Two pulses are sent to the lower switch (i.e., DUT) in a clamped inductive load circuit, as shown in Figure 4.6(b). During the first pulse, DUT is on, DC bus voltage applies across the load inductor, and load current linearly increases assuming the load inductor works in its unsaturated region with a constant inductance. When the current reaches the desired value I_L, the first pulse ends, the DUT is turned off, and its switching performance is obtained. After the first pulse and the DUT is off, the inductor current flows through the upper freewheeling diode and remains almost constant during the short interval between the first and second pulses. At the beginning of the second pulse, the DUT is turned on and the turn-on switching behavior is assessed under essentially identical operating condition (e.g., voltage and current) under which the turn-off switching behavior is evaluated. After the second pulse, DUT turns off. The inductor current flows through the upper diode and naturally decays to zero due to the loss dissipation of the load inductor and diode. DPT completes.

By adjusting the DC bus voltage and the first pulse duration, the DUT's switching transients can be captured under different desired voltage and current conditions. Since the DUT switches only twice for each test and also the pulses are short, the device junction temperature rise due to the switching and conduction losses should be negligibly small even without deliberate cooling.

Note that the DPT circuit in Figure 4.6(a) features a diode as the sole upper device. In a real voltage source converter, the device can be a switch with freewheeling diode or simply a switch. Thus, as shown in Figure 4.7, there are two typical DPT configurations: switch/diode pair and switch/switch pair. Device manufacturers generally rely on the switch/diode pair for the dynamic characterization. However, it should be noted that the selection of the complementary switch (i.e. a switch/diode or switch/switch pair) has a significant impact on the results of the DPT, owing to the different

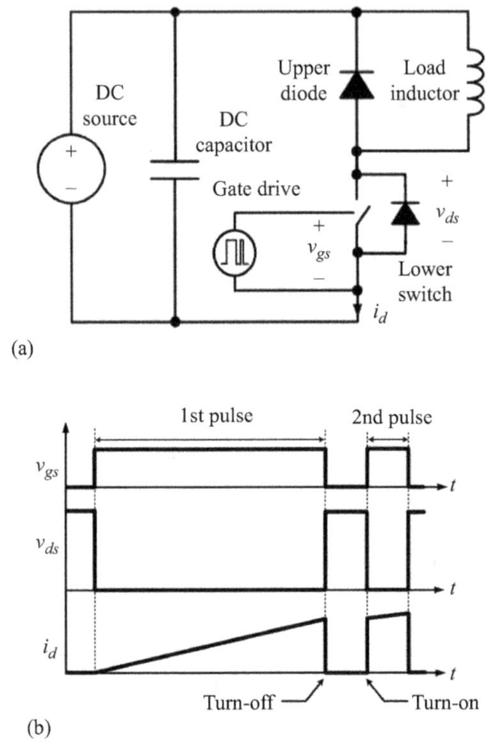

Figure 4.6 DPT: (a) schematics of DPT; (b) typical waveforms of DPT

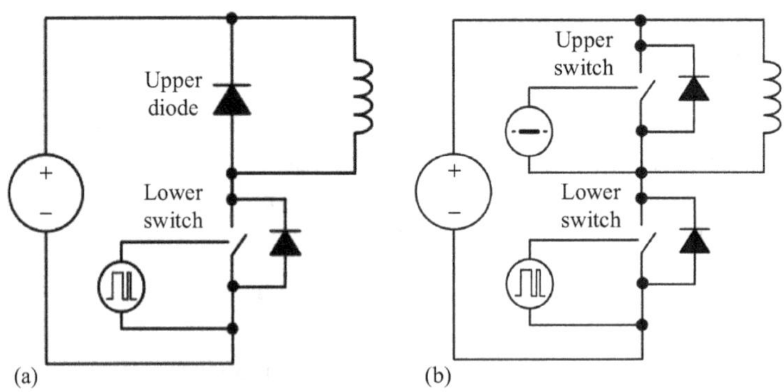

Figure 4.7 DPT configuration: (a) switch/diode pair; (b) switch/switch pair

output capacitance and dynamic characteristics of the complementary switch during the switching transition.

In Si-based power electronics converters, to mitigate the reverse recovery introduced by anti-paralleled Si PiN diodes for IGBT and/or MOSFET body diode, and improve the switching performance, a SiC Schottky diode can be employed.

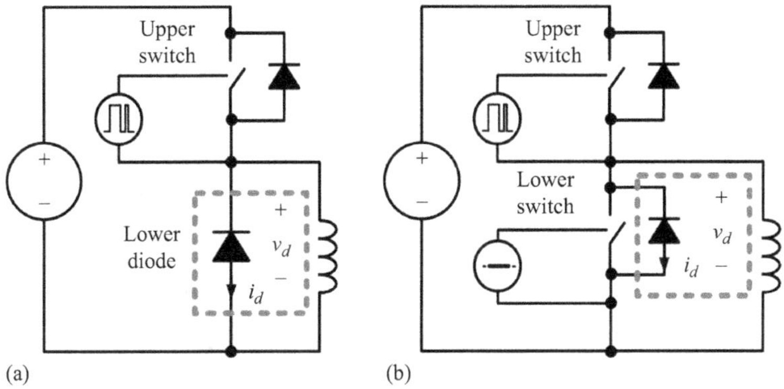

Figure 4.8 Dynamic characterization of WBG diode based on DPT: (a) WBG diode characterization; (b) WBG device body diode or "diode-like behavior" characterization

Thus, in addition to WBG-based active devices, it is also worthwhile to evaluate the dynamic characteristics of the diode. Furthermore, in WBG-based power electronics converters, instead of external anti-paralleled diode, a body diode (e.g., SiC MOSFETs), or "diode-like behavior" (e.g., GaN HEMTs) is typically utilized as the freewheeling diode. Therefore, the dynamic performance of body diode or "diode-like behavior" of WBG devices should be characterized.

Based on the similar DPT circuit discussed earlier, the dynamic characterization configuration for a WBG diode and WBG device body diode or "diode-like behavior" is illustrated in Figure 4.8. As can be observed, diode under test is located on the lower side of the phase-leg due to the stable measurement ground. In this case, the upper switch becomes the active switch and is responsible for generating the switching action by double-pulse control. The load inductor is placed in parallel with the lower diode under test. Since the methodology for dynamic characterization of a diode is the same as that of the active switch, in the following description, a WBG active switch is the main focus.

A typical implementation of DPT is illustrated in Figure 4.9. The core is the DPT board, which includes DC capacitor bank, gate drive, protection circuit, power devices, and peripheral interfaces to connect with double-pulse generator, measurement kits, DC voltage supply, auxiliary power supply, load inductor, and temperature controller. The double-pulse signals with adjustable pulse widths are sent from the microcontroller or function generator, which is controlled by a personal computer (PC). The switching waveforms are measured by oscilloscope and probes, and then the data are transmitted back to the PC. The PC is responsible for programming the double-pulse signals, collecting switching waveforms, and post-processing the raw data. Operating conditions of the DUT, such as DC voltage, load current, and junction temperature, are regulated with the DC voltage supply, pulse widths, load inductance, and temperature controller. An auxiliary power supply is used to power the gate drive and other auxiliary circuits on the DPT board.

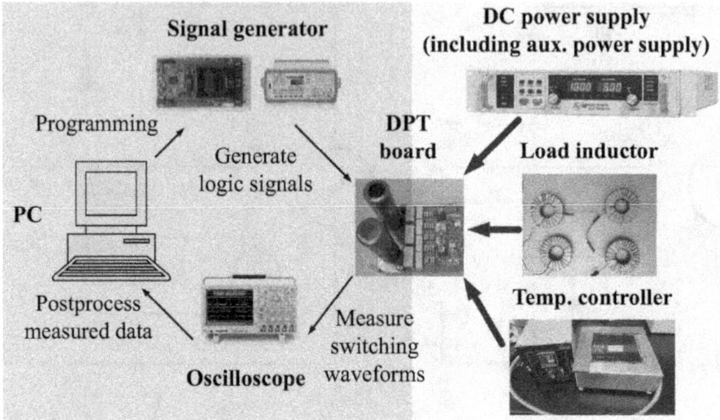

Figure 4.9 Physical implementation of DPT

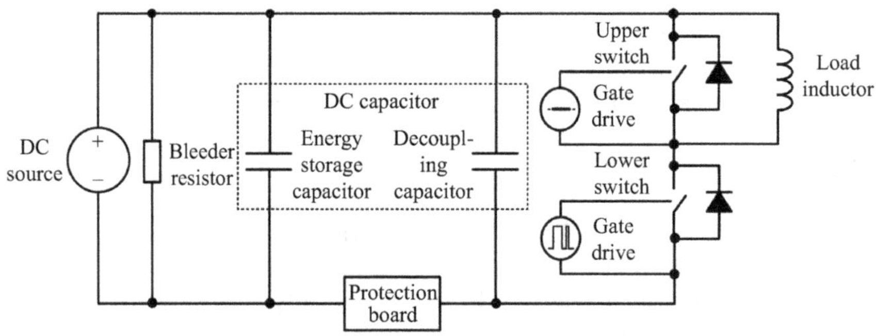

Figure 4.10 Key components of DPT circuit

The key elements to an accurate assessment of dynamic characteristics of DUT include DPT components design and selection, DPT board physical layout, control, protection, measurement, and data processing.

4.3 DPT design

As shown in Figure 4.10, in addition to power devices, key components of DPT circuit include a load inductor, DC source, DC capacitor (including energy storage capacitor and decoupling capacitor), bleeder resistor, gate drive, and short circuit protection circuit. The design criteria of load inductor, DC source, DC capacitor, and bleeder resistor are presented in this section. Gate drive and protection circuit will be introduced in details in Chapters 5 and 7, respectively. Also, a temperature controller is critical to stabilizing the device junction temperature for temperature-dependent device dynamic assessment. Basic design consideration and physical implementation of temperature controller discussed in Chapter 2 for static characterization can be leveraged and will not be repeated here. Some special considerations for high-voltage WBG dynamic characterization are summarized in Chapter 7.

4.3.1 Load inductor

The load inductor is used to establish the desired current during the first pulse and keep this current nearly constant during the subsequent turn-off and turn-on transients. Therefore, the selected load inductance L should be large enough to limit the inductive current variation ΔI_L during the switching transient, yielding:

$$L \geq \frac{V_{DC}}{\Delta I_L} t_{sw} = \frac{V_{DC}}{k_{\Delta i} \times I_L} t_{sw} \tag{4.1}$$

where t_{sw} is the time to complete a switching event and is assumed to be known and accurate on the order of magnitude from the device manufacturer before the DPT; $k_{\Delta i}$ represents the current variation percentage during the switching transition, which is practically selected to be 1%–5%; and V_{DC} and I_L refer to the DC bus voltage and load current, respectively. Usually, the switching performance of the DUT is characterized under several different V_{DC} and I_L points. From (4.1), the worst condition for load inductance selection is under the maximum operating voltage and minimum load current. Also note although (4.1) only gives the low limit of the required inductance, inductance cannot be too large; otherwise, longer duration of the first pulse has to be used to achieve the same operating current, which may cause temperature rise and thermal concern during the DPT, especially for very small WBG devices. It is suggested to select the inductance close to or slightly larger than the value calculated based on (4.1).

According to this required inductance along with the maximum operating current, the inductor is then physically designed. It is often preferred to construct the load inductor by connecting several small inductors in series. Each inductor has only one layer of windings to minimize its equivalent parallel capacitance (EPC). WBG devices with fast switching speed are highly sensitive to parasitics, so load inductor EPC minimization is critical. Also, the winding insulation must be capable of withstanding maximum DC bus voltage.

4.3.2 DC source

The DC source is used to establish the DC bus voltage. Thus, its rated voltage should be higher than the maximum DC bus voltage needed for device characterization. On the other hand, the current rating of the DC source can be much lower than the maximum characterization current for the DUT, since most of the energy for DPT comes from the DC capacitor bank. It is possible to use the DC source to charge the DC capacitor prior to the DPT and then disconnect the DC source from the DPT circuit before the test. The charged DC capacitor can be directly used or reconnected for higher voltages to power the DPT circuit. The practice is often preferred for high-voltage DPT for safety concerns or to avoid the need for a high-voltage DC source.

4.3.3 DC capacitor

The DC capacitors are employed to maintain a stable DC bus voltage across the phase-leg. One bank of bulk capacitors with large capacitance supplies energy to establish the inductive current during the first pulse, while the decoupling capacitor with low parasitic equivalent series inductance (ESL) is responsible for supplying transient current during the switching transition.

The capacitance of the bulk DC energy storage capacitor C_{bulk} is designed to limit the bus voltage variation ΔV_{DC} when the energy transfers from C_{bulk} to load inductor L during the first pulse, that is,

$$C_{bulk} \geq \frac{LI_L^2}{(2V_{DC} - \Delta V_{DC})\Delta V_{DC}} = \frac{LI_L^2}{(2 - k_{\Delta v})k_{\Delta v} V_{DC}^2} \approx \frac{LI_L^2}{2k_{\Delta v} V_{DC}^2} \tag{4.2}$$

where $k_{\Delta v}$ is the voltage variation percentage, which is practically selected to be 1%–5%. Unlike the load inductance, the worst condition for DC energy storage capacitance selection is under the minimum operating voltage and maximum load current. According to this required capacitance together with the maximum operating voltage to be tested, the capacitor is then physically selected. Generally speaking, aluminum electrolytic capacitors or film capacitors are employed for the DC capacitor bank C_{bulk} due to their high available capacitances and high energy density. In high-current high-voltage-based DPT, to achieve large capacitance with high current capability while in the meantime have sufficient voltage capability, it may be necessary to parallel several series-connected capacitors. Although the ESL due to series configuration becomes larger, it can be compensated by the decoupling capacitors.

The capacitance of the decoupling capacitors C_{dec} is selected to supply transient current and mitigate over-voltage due to the loop parasitics during the switching transient. Two examples are given in Figure 4.11, illustrating the tested over-voltage during the switching transient as a function of the decoupling capacitance. For both Si IGBT and SiC MOSFET, a similar trend can be observed,

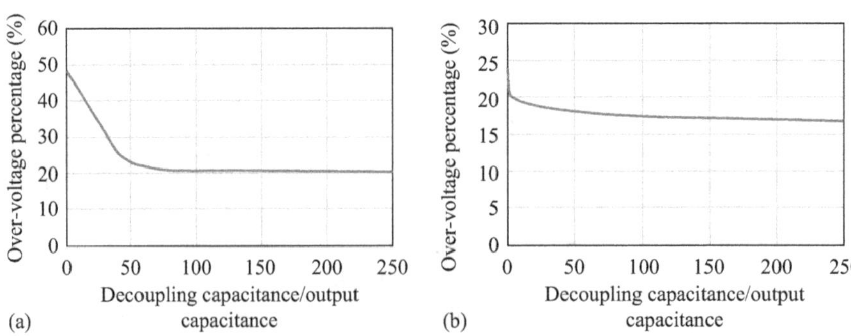

Figure 4.11 Over-voltage dependence on decoupling capacitance: (a) Si IGBT [8]; (b) SiC MOSFET [9]

that is, as decoupling capacitance increases, over-voltage becomes lower, but when decoupling capacitance is more than 100× larger than the output capacitance of DUT, the benefit of further mitigating over-voltage is negligible.

Accordingly, C_{dec} is practically chosen as

$$C_{dec} \geq 100 \times C_{oss} \tag{4.3}$$

where C_{oss} is the output capacitance of the DUT. Although (4.3) gives a low limit, larger decoupling capacitance is not necessary for the sake of over-voltage reduction. Moreover, under the short circuit fault condition, the protection circuit will be tripped to isolate the DC source and energy storage capacitor from the DPT circuit, but not the decoupling capacitor, since the decoupling capacitor is placed after the protection circuit to minimize insertion parasitics introduced by the protection circuit. Thus, the energy stored in the decoupling capacitance has to be dissipated by DUT under fault condition. Therefore, larger decoupling capacitance than necessary turns out to be worse. Detailed analysis with respect to protection circuit can be found in Chapter 7.

In addition to sufficient capacitance and voltage rating, minimized ESL for decoupling capacitors is critical. Ceramic capacitors with low ESL are often the capacitors of choice. Their ESL is determined by chip size [10]. Practically, maximum testing voltage determines the rated voltage of decoupling capacitors, and then the smallest footprint is selected, considering the clearance/creepage limitation of PCB. According to the required capacitance from (4.3), rated voltage, and footprint, decoupling capacitors are physically selected. Several approaches can be utilized to further reduce the ESL: (1) adopt conformal coating and/or PCB slot to improve the voltage insulation capability, and as a result, the capacitor footprint and its resultant ESL become smaller; and (2) use multiple capacitors in parallel to reduce the ESL. Series connection of capacitors as the decoupling capacitor should be avoided if possible.

4.3.4 Bleeder resistor

A bleeder resistor is introduced across the DC bus to dissipate the remaining energy stored at DC capacitor and output capacitor of DC source after the DPT. Generally speaking, there are two design criteria for the bleeder resistor to follow: (1) power dissipation of bleeder resistor during DPT should be reasonably small, and practically, a few watts of loss are acceptable; and (2) the remaining DC capacitor energy release after the DPT should be adequately quick, and practically, a discharge time constant of tens of seconds is reasonable. Based on the tradeoff between less power loss and fast discharge time, the bleeder resistor can be selected.

4.4 DPT control

Based on the DPT operation principle described in Section 4.2, the DPT control is straightforward. As illustrated in Figure 4.12, three intervals (i.e., t_1, t_2, and t_3) need

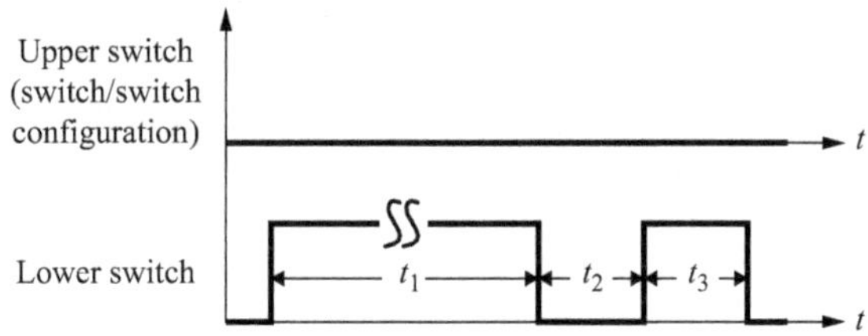

Figure 4.12 DPT control signals

to be determined. The interval t_1 is set to enable load current to reach the desired value with the given load inductance L under DC bus voltage V_{DC}, as in (4.4). Thus, to characterize DUT under different operating voltages and currents, t_1 should be set accordingly. The interval t_2 is set to allow the completion of the turn-off transition before the second pulse. Thus, it needs to be longer than the maximum turn-off time $t_{off(max)}$, as shown in (4.5). Similarly, in (4.6), the interval t_3 should be longer than the maximum turn-on time $t_{on(max)}$. On the other hand, t_2 should be short enough to avoid significant change of I_L, and t_3 should be sufficiently short to avoid too much conduction loss to affect junction temperature. For WBG devices with switching times in the range of nanoseconds to hundreds of nanoseconds, practically t_2 and t_3 are set to be 1–2 μs:

$$t_1 = L\frac{I_L}{V_{DC}} \tag{4.4}$$

$$t_2 > t_{off(max)} \tag{4.5}$$

$$t_3 > t_{on(max)} \tag{4.6}$$

In switch/switch configuration, usually the upper switch is controlled to be constantly off, as shown in Figure 4.12. Sometimes the upper switch is controlled with a signal complementary to the lower switch to characterize the reverse conduction behavior of the synchronous switch. One example is the characterization of "diode-like behavior" of GaN HEMTs. Control signals in this case are illustrated in Figure 4.13. Compared with the previous case, t_2 becomes longer and is determined based on (4.7):

$$t_2 = 2\,t_{dt} + t_{syn} \tag{4.7}$$

where t_{dt} is the dead time, which should be longer than $t_{off(max)}$, and t_{syn} is the on-state time of synchronous switch (e.g., upper switch in this case).

Programmable function generator and micro-controller are two practical approaches to generate the double-pulse signals. Also, it is helpful to use a human–machine interface (HMI), for example, National Instruments LabVIEW, to conveniently tune the pulse durations for different testing conditions.

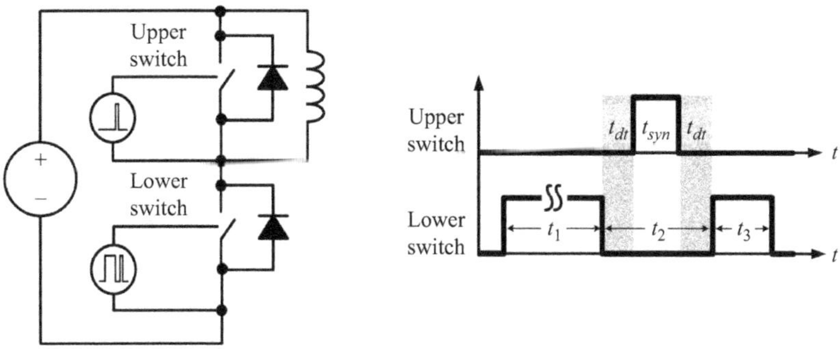

Figure 4.13 Complementary control signal for switch/switch pair

Table 4.1 Summary of DUT

Vendor/model	Breakdown voltage	Rated current	Max. switching time	Output capacitance	Testing voltage range	Testing current range
Wolfspeed C2M0080120D	1,200 V	24 A @ 100 °C	50 ns	80 pF	200–800 V	5–24 A

4.5 Case study

A SiC MOSFET from Wolfspeed is used to illustrate the design and implementation of DPT key components and control discussed earlier. Table 4.1 summarizes the basic information and desired testing conditions of the DUT. This design method can be applied to other WBG devices, including GaN HEMTs and high-voltage (>3.3 kV) SiC devices.

4.5.1 Load inductor

Step no. 1: Determine the maximum testing voltage $V_{DC(max)}$ and minimum current $I_{L(min)}$. According to Table 4.1

$$V_{DC(max)} = 800 \text{ V}$$

$$I_{L(min)} = 5 \text{ A}$$

Step no. 2: Determine the maximum switching time $t_{sw(max)}$ based on known information of DUT (e.g., preliminary datasheet from the vendor or a conservative estimate) and select the current variation percentage $k_{\Delta i}$.

$$t_{sw(max)} = 50 \text{ ns}$$

$$k_{\Delta i} = 2.5\%$$

Table 4.2 Design summary of load inductor

Core	Winding	Number of turns per inductor	Configuration	Total load inductor EPC
Fluxsan™ powder toroid FS-300060-2	10–12 AWG magnet wire	42	Four inductors in series	2 pF

Step no. 3: Calculate the low limit of load inductance according to (4.1)

$$L \geq \frac{V_{DC(max)}}{k_{\Delta i} \times I_{L(min)}} t_{sw(max)} = \frac{800}{0.025 \times 5} 50 \times 10^{-9} = 320\ \mu\text{H}$$

Step no. 4: Design the physical load inductor based on inductance L, maximum current $I_{L(max)}$, and maximum testing voltage $V_{DC(max)}$.

$$\begin{aligned} L &= 400\ \mu\text{H} \\ I_{L(max)} &= 24\ \text{A} \\ V_{DC(max)} &= 800\ \text{V} \end{aligned}$$

The design results are summarized in Table 4.2. AWG of winding is selected based on the maximum testing current. For 24 A current, 10–12 AWG magnet wire is used. Note that in DPT, the load inductor does not operate continuously, so the thermal design should not be a concern. Thus, winding selection in this design is conservative. Regarding the magnetic core selection, it is imperative to consider the flux density under the maximum current, to avoid saturation of the inductor. Fluxsan™ powder toroid FS-300060-2 is used. Also note in this design that the load inductor consists of four 100 μH inductors in series to minimize the EPC below 2 pF. Compared with 80 pF output capacitance of the DUT, the impact of nonideal high-frequency characteristics of the load inductor is negligible.

4.5.2 DC source

Step no. 1: Determine the low limit of the output voltage V_{DC_source} based on the maximum testing voltage $V_{DC(max)}$ of the DUT.

$$V_{DC\_source} \geq V_{DC(max)} = 800\ \text{V}$$

Step no. 2: Select DC power supply. As discussed earlier, current capability of the DC power supply is not required as large as the maximum testing current. A SL series programmable DC power supply from Magna-Power is selected with detailed specification listed in Table 4.3.

Table 4.3 Summary of DC source

Vendor/model	Output voltage range	Output current range
Magna-Power/SL1000-1.5/UI	0 1,000 V	0–1.5 A

Table 4.4 Summary of DC capacitor

DC capacitor type	Vendor/model	Specification	Configuration
DC bulk capacitor	Vishay/ MKP1848720454Y5	200 μF/450 V metallized polypropylene film capacitor	Two capacitors in series
DC decoupling capacitor	Vishay/ VJ2220A332JXGAT	3.3 nF/1,000 V C0G/NP0 ceramic capacitor	Three capacitors in parallel

4.5.3 DC capacitor bank

Step no. 1: Determine the minimum testing voltage $V_{DC(min)}$ and maximum testing current $I_{L(max)}$. According to Table 4.1

$$V_{DC(min)} = 200\ \text{V}$$

$$I_{L(max)} = 24\ \text{A}$$

Step no. 2: Assume the voltage variation percentage $k_{\Delta v}$

$$k_{\Delta v} = 3\%$$

Step no. 3: Calculate the low limit of DC bulk capacitance C_{bulk} according to (4.2)

$$C_{bulk} \geq \frac{L I_{L(max)}{}^2}{2k_v V_{DC(min)}{}^2} = \frac{400 \times 10^{-6} \times 24^2}{2 \times 0.03 \times 200^2} = 96\ \mu\text{F}$$

Step no. 4: Select a commercially available DC bulk capacitor based on low limit of capacitance C_{bulk} and maximum testing voltage $V_{DC(max)}$. The design results are summarized in Table 4.4. In this design, DC bulk capacitor consists of two 200 μF/450 V film capacitors in series.

Step no. 5: Calculate the low limit of DC decoupling capacitance C_{dec} according to (4.3)

$$C_{dec} > 100 \times C_{oss} = 100 \times 80 \times 10^{-12} = 8\ \text{nF}$$

Step no. 6: Determine the smallest footprint of decoupling capacitor based on the maximum testing voltage $V_{DC(max)}$.

$$V_{DC(max)} = 800\ \text{V}$$

Required distance of insulation for FR4-based PCB $D_{insulation}$ at sea level

$$D_{insulation} > 4 \text{ mm}$$

Accordingly, 2220 footprint with a distance of 4.5 mm between two soldering pads is selected.

Step no. 7: Select DC decoupling capacitor based on lower limit of capacitance C_{dec}, maximum testing voltage $V_{DC(max)}$, and footprint. The design results are summarized in Table 4.4. In this design, the DC decoupling capacitor consists of three 3.3 nF/1000 V ceramic capacitors in parallel.

4.5.4 Bleeder resistor

Step no. 1: Determine maximum testing voltage $V_{DC(max)}$ of the DUT.

$$V_{DC(max)} = 800 \text{ V}$$

Step no. 2: Determine bleeder resistance $R_{bleeder}$ based on the tradeoff between power loss and discharge time

$$R_{bleeder} = 180 \text{ k}\Omega$$

Power loss dissipated by bleeder resistor under the maximum testing voltage is

$$P_{bleeder} = \frac{{V_{DC(max)}}^2}{R_{bleeder}} = \frac{800^2}{180 \times 10^3} = 3.56 \text{ W}$$

Discharge time constant $T_{discharge}$ determined by bleeder resistor and DC energy storage capacitor is

$$T_{discharge} = R_{bleeder} \times C_{bulk} = 180 \times 10^3 \times 100 \times 10^{-6} = 18 \text{ s}$$

The physical design results are summarized in Table 4.5. In this design, the bleeder resistor consists of two 360 kΩ, 2 W metal oxide film resistors in series. Also, these two resistors are paralleled with two series-connected energy storage capacitors individually to ensure the voltage balance of the two energy storage capacitors.

Table 4.5 Summary of bleeder resistor

Vendor/model	Specification	Configuration
Yageo RSF200JB-73-360K	360 kΩ/2 W metal oxide film resistor	Two bleeder resistors in series. Each resistor paralleled with one of the energy storage capacitors

4.5.5 DPT control

Step no. 1: Determine operating conditions of DUT to be tested. In this example, assume

$$V_{DC} = 800 \text{ V}$$
$$I_L = 24 \text{ A}$$

Step no. 2: Calculate/determine durations of t_1, t_2, and t_3 according to (4.4)–(4.6)

$$t_1 = L\frac{I_L}{V_{DC}} = 400 \times 10^{-6}\frac{24}{800} = 12 \text{ µs}$$
$$t_2 = 1 \text{ µs}$$
$$t_3 = 1 \text{ µs}$$

Step no. 3: Generate double-pulse control signal. Function generator and micro-controller are two practical implementations. Detailed setting and programming are not included here, since they are instrument-dependent, and the readers are advised to refer to the user manual. Several examples of double-pulse generators, which have been used by authors for WBG device characterization, are given next as reference.

Function generator: Agilent 33522A, Tektronix AFG1062
Micro-controller: Texas Instruments TMS320F28335

Based on TMS320F28335 with 150 MHz system clock, an example of DSP code is presented next to convey the basic idea for double-pulse generation via micro-controller based on the double-pulse signal illustrated in Figure 4.14.

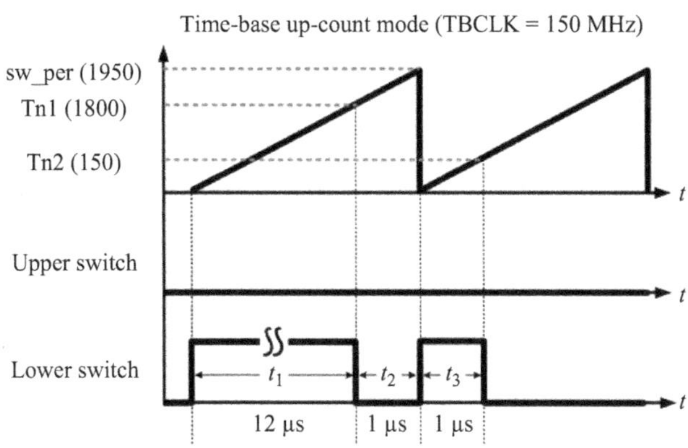

Figure 4.14 Double-pulse control signal example based on micro-controller

```
void InitEPwm1Example ()                                  // Initialize ePWM1 register
{
  EPwm1Regs.TBCTR = 0x0000;                               // Clear TBCTR

 // Setup counter mode

  EPwm1Regs.TBCTL.bit.CTRMODE = TB_COUNT_UP;              // Count up
  EPwm1Regs.TBCTL.bit.PHSEN = TB_DISABLE;                 // Disable phase loading
  EPwm1Regs.TBCTL.bit.PRDLD = TB_SHADOW;                  // Enable shadow register for TBPRD
  EPwm1Regs.TBCTL.bit.HSPCLKDIV = TB_DIV1;                // TBCLK = SYSCLKOUT
  EPwm1Regs.TBCTL.bit.CLKDIV = TB_DIV1;

 // Setup shadow mode

  EPwm1Regs.CMPCTL.bit.SHDWAMODE = CC_SHADOW;             // Enable shadow register for ePWM1A
  EPwm1Regs.CMPCTL.bit.SHDWBMODE = CC_SHADOW;             // Enable shadow register for ePWM1B
  EPwm1Regs.CMPCTL.bit.LOADAMODE = CC_CTR_ZERO;           // Load when TBCTR = 0 for ePWM1A
  EPwm1Regs.CMPCTL.bit.LOADBMODE = CC_CTR_ZERO;           // Load when TBCTR = 0 for ePWM1A

 // Setup interrupt

  EPwm1Regs.ETSEL.bit.INTSEL = ET_CTR_ZERO;               // Enable event when TBCTR = zero
  EPwm1Regs.ETSEL.bit.INTEN = 1;                          // Enable INT generation
  EPwm1Regs.ETPS.bit.INTPRD = ET_1ST;                     // Generate INT on 1st event
  epwm1_info.EPwmTimerIntCount = 0;                       // Zero the interrupt counter
  epwm1_info.EPwmRegHandle = &EPwm1Regs;                  // Set the pointer to ePWM module
}
```

```
interrupt void epwm1_isr(void)                            // interrupt for double – pulse generation
{

 // Set double-pulse parameter

  sw_per=1950;                                            // Period = sw_per/TBCLK (150 MHz)=13 μs
  Tn1=1800;                                               // 1st pulse comparator = Tn1/ TBCLK=12 μs
  Tn2=150;                                                // 2nd pulse comparator = Tn2/ TBCLK=1 μs
}
```

(Continues)

(*Continued*)

```
// Generate the first pulse

 if (DPT_count == 1)                    // DPT_count is the pulse number account, initial value is 1
 {
   EPwm1Regs.TBPRD = sw_per;            // set TBPRD for EPWM1

  // set EPWM1A reg. for lower switch (double-pulse)

   EPwm1Regs.CMPA.half.CMPA = Tn1;         // set CMPA = Tn1
   EPwm1Regs.AQCTLA.bit.CAU = AQ_CLEAR;    // force low when TBCTR = CMPA
   EPwm1Regs.AQCTLA.bit.ZRO = AQ_SET;      // force high when TBCTR = 0

  // set EPWM1B reg. for upper switch (always low)

   EPwm1Regs.CMPB = 0;                     // set CMPB = 0
   EPwm1Regs.AQCTLB.bit.CBU = AQ_CLEAR;    // force low when TBCTR = CMPB
   EPwm1Regs.AQCTLB.bit.ZRO = AQ_CLEAR;    // force low when TBCTR = 0
   DPT_count=2;                            // For 2nd pulse
 }
else                                       // Generate the second pulse
 {
   if (DPT_count == 2)
     {
      EPwm1Regs.CMPA.half.CMPA = Tn2;      // set CMPA = Tn2
      DPT_count = 3;                       // Complete double-pulse generation
     }
   else                                    // After 2nd pulse, disable PWM output
     {
       EPwm1Regs.CMPA.half.CMPA = 0;          // set CMPA = 0
       EPwm1Regs.AQCTLA.bit.ZRO = AQ_CLEAR;   // force low when TBCTR = 0
     }
 }
```

4.6 Summary

1. In phase-leg-based voltage source power electronics converters, depending on the load current direction, hard switching commutation only occurs between the upper diode and lower switch (when current flows into the phase-leg) or lower diode and upper switch (when current flows out of the phase-leg) while the remaining switching action is synchronous switching with negligible loss during the switching transient. Note that the diode could be the body-diode or "body-diode like behavior" of the switch.
2. Simplified circuit consisting of switch/diode pair or switch/switch pair is practically adequate for dynamic characterization. DPT is a widely accepted method to assess the dynamic characteristics.
3. In DPT, dynamic characteristics of lower and upper switches should be the same due to the circuit symmetry. The lower switch is commonly selected as the DUT because of the stable measurement ground of the lower switch.
4. Key components of the DPT circuit consist of load inductor, DC source, DC capacitor bank, gate drive, and short circuit protection circuit. In this chapter, design criteria and method of load inductor, DC source, and DC capacitor bank are introduced.
    (a) Load inductance is determined based on the maximum operating voltage and minimum load current. Then its physical design depends on this required inductance and maximum operating current. Series-connected configuration is preferred due to minimized EPC.
    (b) DC source is selected according to the DUT maximum operating voltage. Current rating of the DC source can be much lower than the maximum characterization current since most of the energy for DPT comes from the DC capacitor bank.
    (c) DC capacitor bank consists of bulk capacitor and decoupling capacitor. Bulk capacitance is determined based on the minimum operating voltage and maximum load current. Electrolytic capacitor or film capacitor is commonly employed. Decoupling capacitance is designed based on DUT output capacitance. Surface mount ceramic capacitors are preferred. Parallel-connected configuration is suggested due to its reduced ESL.

References

[1] Z. Zhang, B. Guo, F. Wang, E. Jones, L. M. Tolbert, and B. J. Blalock, "Methodology for wide band-gap device dynamic characterization," *IEEE Transactions on Power Electronics*, vol. 32, no. 12, pp. 9307–9318, December 2017.

[2] C. Zhen, "Characterization and modeling of high-switching-speed behavior of SiC active devices," M.S thesis, Dept. Electrical and Computer Engineering, Virginia Polytechnic Institute and State Univ., Blacksburg, USA, 2009.

[3] J. B. Witcher, "Methodology for switching characterization of power devices and modules," M.S thesis, Dept. Electrical and Computer Engineering Virginia Polytechnic Institute and State Univ., Blacksburg, USA, 2003.

[4] E. Jones, F. Wang, D. Costinett, Z. Zhang, and B. Guo, "Characterization of an enhancement-mode 650-V GaN HFET," in *Proceedings of the IEEE Energy Conversion Congress and Exposition*, September 2015, pp. 400–407.

[5] S. Ji, S. Zheng, Z. Zhang, F. Wang, and L. Tolbert, "Protection and temperature-dependent switching characterization of latest generation 10 kV SiC MOSFETs," in *Proceedings of the IEEE Applied Power Electronics Conference and Exposition*, March 2017, pp. 783–788.

[6] M. Danilovic, Z. Chen, R. Wang, F. Luo, D. Boroyevich, and P. Mattavelli, "Evaluation of the switching characteristics of a gallium-nitride transistor," in *Proceedings of the IEEE Energy Conversion Congress and Exposition 2011*, 17–22 September 2011, pp. 2681–2688.

[7] B. Callanan, SiC MOSFET double pulse fixture [Online]. Available: http://www.cree.com. Accessed in February 2011.

[8] L. Qian, S. Wang, A. C. Baisden, F. Wang, and D. Boroyevich, "EMI suppression in voltage source converters by utilizing dc-link decoupling capacitors," *IEEE Transactions on Power Electronics*, vol. 22, no. 4, pp. 1417–1428, 2007.

[9] Z. Chen, D. Boroyevich, P. Mattavelli, and K. Ngo, "A frequency-domain study on the effect of DC-link decoupling capacitors," in *IEEE Energy Conversion Congress and Exposition (ECCE)*, 2013, pp. 1886–1893.

[10] J. Cain. Parasitic inductance of multilayer ceramic capacitors [Online]. Available: https://www.avx.com/docs/techinfo/CeramicCapacitors/parasitc.pdf.

Chapter 5
Gate drive for dynamic characterization

Gate drive is the main link between the world of control and the world of power in a power conversion system. The world of control includes 1.8, 3.3, or 5 V with limited current generated by a micro-controller, while the world of power includes up to thousands of volts and thousands of amperes delivered by power devices. A good gate drive is an absolute necessity for a power electronics converter. As clearly illustrated in Figure 5.1, which uses motor drives as an example, the gate drive serves as the interface between logic-level control signals and power devices.

For device dynamic characterization, the gate drive is critical, because it greatly affects the switching behavior of the DUT. In this chapter, we focus on the gate drive design in switch/switch-based DPT circuit for dynamic characterization of WBG devices, as shown in Figure 5.2. First, gate drive fundamentals and unique design considerations for WBG devices are presented. Second, key power device characteristic parameters used for gate drive design are highlighted, which are some of the characteristics already obtained in static characterization described in previous chapters. Third, detailed gate drive design is introduced based on the basic functional blocks. Finally, leveraging the gate drive design criteria, a case study based on a 1,200 V SiC MOSFET is given. It is noted that although gate drive illustrated in this chapter is for WBG device dynamic characterization, its basic theory and design consideration can be utilized for the gate drive design in actual WBG-based power electronics converters.

5.1 Gate drive fundamentals

A good gate drive is able to efficiently and reliably control power devices under static, dynamic, and fault conditions. Specifically, in the static state, a good gate drive can keep the device in on-state with minimized conduction loss and keep the device safely in off-state to prevent spurious change of the switch state due to external or internal disturbances. During the dynamic transient, a good gate drive is capable of driving the device from on to off and off to on states with fast speed and low switching losses, acceptable electromagnetic interference, and low-voltage spike and parasitic ringing. In case of any hazardous situations, the gate drive can quickly and reliably protect the device, including typically shoot-through, over current, over-voltage, and over temperature faults. This chapter focuses on the gate

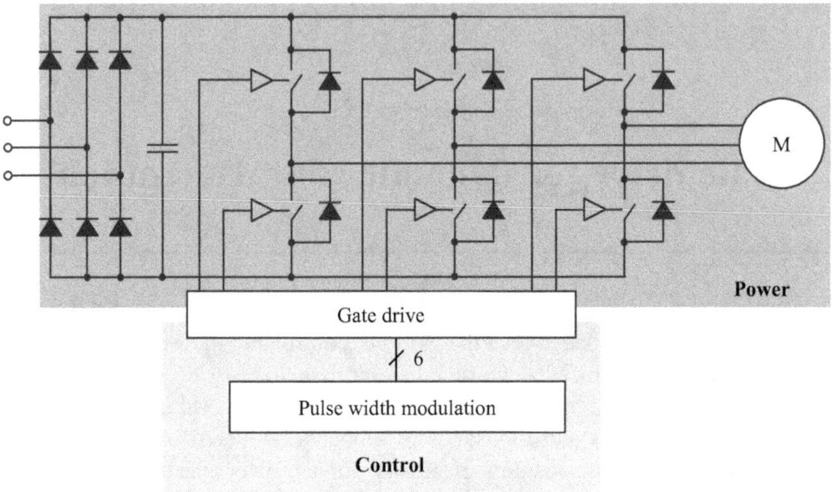

Figure 5.1 Gate drive as interface between the world of control and the world of power in motor drives

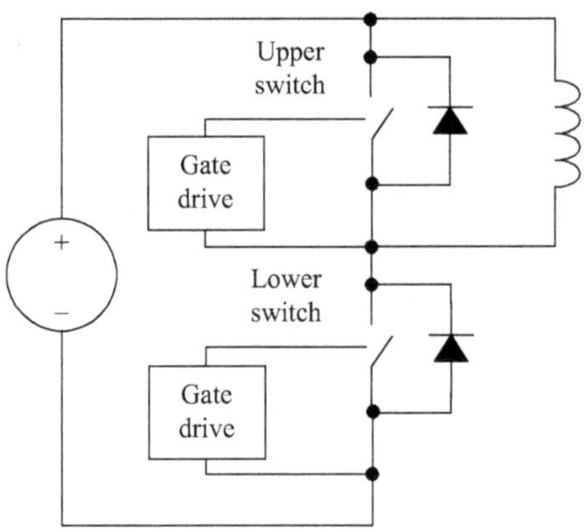

Figure 5.2 Gate drive in DPT circuit with switch/switch configuration

drive design for the dynamic characterization of WBG devices, and protection function of the gate drive will not be covered here. The protection scheme of DPT will be introduced in Chapter 7.

To achieve the basic function of a gate drive, the following functional blocks are necessary. As illustrated in Figure 5.3, it includes gate drive IC/buffer circuit, signal isolator, and isolated power supply. Gate drive IC is used to amplify the driving capability of logic-level control signal from micro-controller to drive power device. In case the driving capability provided by gate drive IC is not sufficient, an extra buffer circuit is added. This can be commonly observed in gate drive design

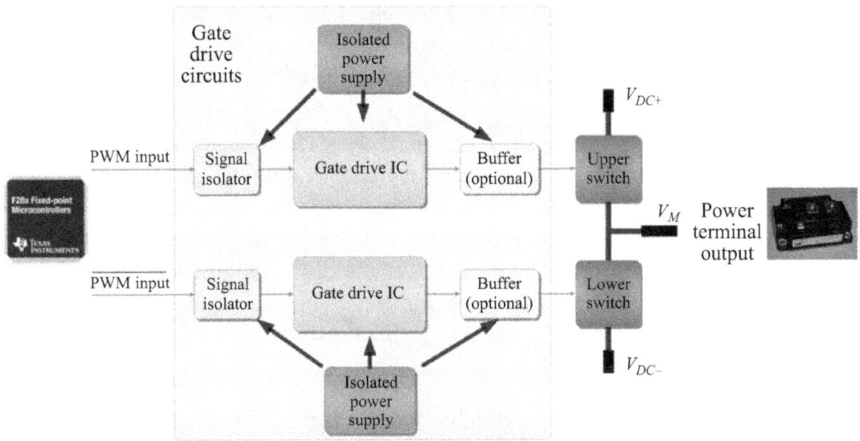

Figure 5.3 Basic functional blocks of a gate drive

of power modules for high-power applications. The signal isolator provides galvanic isolation between the world of control and the world of power due to their potential differences and noise interference. The isolated power supply powers the secondary side of the signal isolator, gate drive IC, and buffer circuit if applicable. Detailed introduction and design criteria will be presented in Section 5.3. Note that the gate drive isolation strategy and architecture presented here represents the most popular scheme used in power converters today. The scheme is adequate and convenient for WBG device characterization gate drive needs. Other isolation strategies like using transformers for both the signal and power isolation are also used for converters.

As already discussed in previous chapters, WBG materials have superior properties as compared to those of Si, including high breakdown electric filed, high doping density, high drift velocity, and high thermal conductivity. Consequently, WBG devices are capable of blocking high operating voltage, achieving low on-state resistance, accelerating switching speed, and allowing low thermal resistance. As a result, high-voltage application, high *dv/dt*, *di/dt*, severe parasitic ringing, and high-temperature operation become unique for WBG devices. These features pose new and special requirements for WBG device gate drive as it is the neighboring component of power devices. The special requirements include high galvanic isolation capability, high common mode transient and ringing immunity capability, and high operating temperature capability. Dedicated design criteria have to be considered for the WBG gate drive design.

5.2 Gate drive-related key device characteristics

An ideal gate drive should be designed for the specific device it is to drive. That is, the gate drive should be carefully designed for the specific device characteristics. In practice, people may prefer a more generic gate drive design for various devices to save time and cost. However, it is definitely desirable for the gate drive to be matched with devices especially for the new emerging WBG devices that are fast and potentially more sensitive to driving control. Thus, it is critical to understand the impact of the device on the gate drive design and identify the gate drive-related key device characteristics before the design of the gate drive. As discussed earlier, in addition to the protection that will be covered in a subsequent chapter, the gate drive should meet static and dynamic requirements imposed by the device. Thus, it is necessary to review the static and dynamic characteristics of power devices and identify the key device parameters needed for gate drive design. Note again that the dynamic characteristics here refer to parameters obtained from static characterization that will impact device switching characteristics.

5.2.1 Gate drive design considering device static characteristics

Starting from the static state, to minimize the on-state resistance or voltage and conduction loss when the device is on, and in the meantime, not overstress the device gate, device static characteristics significantly affect the on-state gate voltage selection. Specifically, output characteristics of the device are critical to identify the proper on-state gate voltage. For example, Figure 5.4 illustrates the output characteristics of two example devices; one is Si MOSFET (APT34M120J from Microsemi), and the other is SiC MOSFET (C2M0080120D from Wolfspeed). It shows clearly that for Si MOSFET the change of *V–I* curves as the gate voltage increase from 7.5 to 15 V becomes smaller, indicating it is not necessary to further increase the gate voltage of the Si MOSFET beyond 15 V to reduce the on-state resistance/voltage, even if the maximum allowable positive gate voltage of this Si MOSFET is up to 30 V (see Table 5.1). On the other hand, for SiC MOSFET, according to the *V–I* curve, it is obvious that as the gate voltage increases up to 20 V, the on-state resistance/voltage can still reduce significantly. However, further increase of the gate voltage is limited by its maximum positive gate voltage rating, which is 25 V for this specific case (see Table 5.1).

Additionally, when device is in off-state, to prevent spurious change of the switch state due to external or internal disturbances, it is preferred to introduce negative off-state gate voltage, especially for fast switching WBG devices. However, concerning the reverse conduction characteristics along with the negative gate voltage rating of WBG devices, off-state gate voltage selection becomes a tradeoff between power loss, reliability, and noise immunity capability. For example, as shown in Figure 5.5 for a GaN HEMT, due to the unique "diode-like behavior," as the negative off-state gate voltage amplitude increases, the voltage drop for reverse conduction becomes higher, leading to more conduction loss during the dead time. On the other hand, the higher negative gate voltage is better with regard to the noise immunity capability.

For SiC MOSFET, its intrinsic body diode dominates the reverse conduction characteristics, which is regardless of the off-state gate voltage. However, due to the relatively high-voltage forward drop of SiC MOSFET's body diode, it is interesting to observe that as the off-state gate voltage decreases from 0 V, the

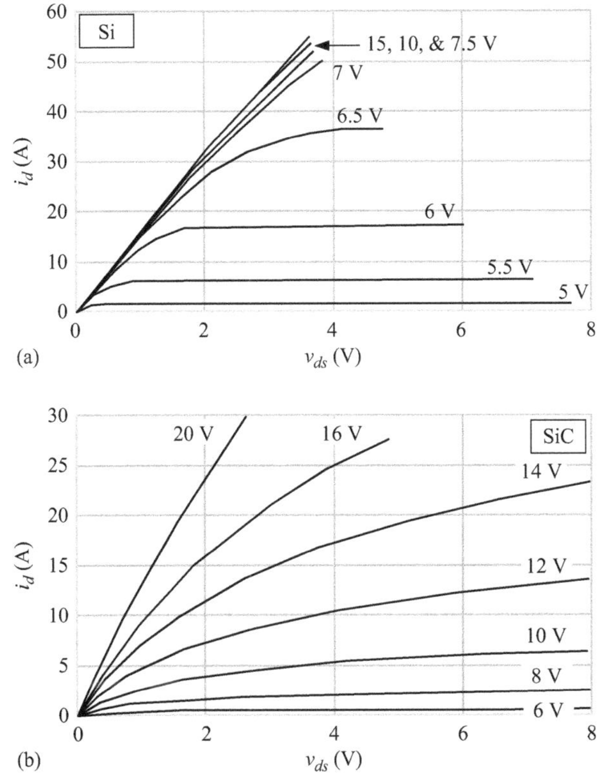

Figure 5.4 Output characteristics of Si and SiC MOSFETs [1, 2]: (a) Si MOSFET APT34M120J; (b) SiC MOSFET C2M0080120D

Table 5.1 Gate voltage ratings of Si and SiC MOSFETs [1, 2]

Type	**Manufacturer**	**Model**	**Gate voltage maximum ratings**
Si MOSFET	Microsemi	APT34M120J	+30 V/–30 V
SiC MOSFET	Wolfspeed	C2M0080120D	+25 V/–10 V

reverse conduction voltage drop slightly increases [2], leading to additional reverse conduction loss. Thereby, similar to the GaN transistor, this phenomenon affects the selection of off-state gate voltage for SiC MOSFET.

Furthermore, for both GaN HEMTs and SiC MOSFETs today, the maximum allowable negative gate voltage is limited as compared to their Si counterparts. For example, in Table 5.1, as compared to –30 V negative gate voltage rating of Si MOSFET, SiC MOSFET is only –10 V. A similar situation can be found for GaN HEMTs. Thus, the negative gate voltage rating is also an important device characteristics for the off-state gate voltage selection of the gate drive design.

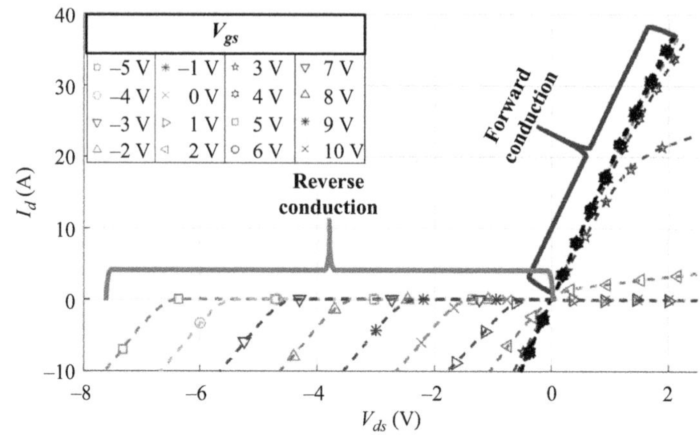

Figure 5.5 Output characteristics of GaN HEMT (GS66508P from GaN Systems) [3]

In summary, with respect to device static characteristics, output characteristics, including forward and reverse conduction performance, as well as the gate voltage rating of the power device play a significant role for the gate drive design of WBG devices.

5.2.2 Gate drive design considering device dynamic characteristics

Device dynamic characteristics are primarily determined by the gate drive, the device itself, and operating condition. Therefore, to achieve optimal dynamic performance of DUT, a good gate drive design should be based on consideration of those device parameters that affect dynamic characteristics. Taking switch/diode-based DPT configuration as an example illustrated in Figure 5.6, the key device parameters that affect the switching performance and are critical for gate drive design include Miller capacitance C_{rss} (i.e., C_{gd} in Figure 5.6(b)), input capacitance C_{iss} (i.e., sum of C_{gd} and C_{gs} in Figure 5.6(b)), output capacitance C_{oss} (i.e., sum of C_{gd} and C_{ds} in Figure 5.6(b)), internal gate resistance of the device $R_{g(in)}$, threshold voltage V_{th}, and transconductance g_{fs}. Specific impacts of these device parameters on the gate drive design will be discussed in Section 5.3.

5.3 Gate drive design

Based on gate drive basic functional blocks in Figure 5.3, the typical gate drive circuit is illustrated in Figure 5.7. In addition to the signal isolator, isolated power supply, and gate drive IC, the gate resistor and decoupling capacitor are necessary components for a gate drive. Detailed introduction of each subcomponent is presented next.

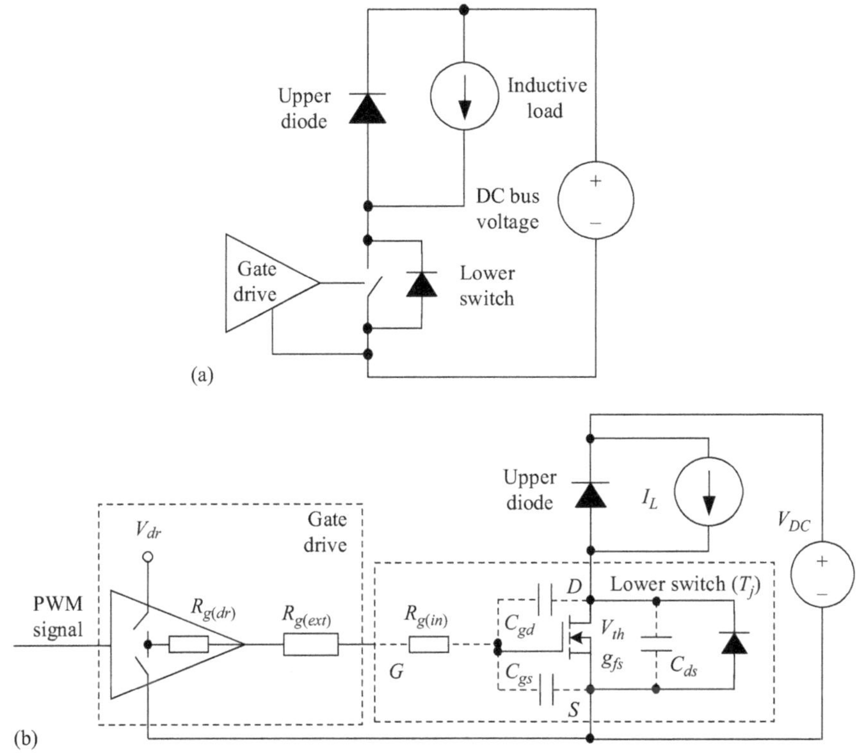

Figure 5.6 Device dynamic performance dependence on device, gate drive, and operating condition: (a) typical DPT configuration diagram; (b) DPT circuit with detailed device and gate drive models

5.3.1 Signal isolator

Using a gasoline powered car as an analogy for a gate drive, the signal isolator is the "steering wheel" of a gate drive. As illustrated in Figure 5.8, the signal isolator provides galvanic isolation between the ground of micro-controller and the ground of gate drive and safely transfers the control and fault signals from the input to the output of the gate drive. Galvanic isolation capability and common mode transient immunity (CMTI) are the key criteria for signal isolator selection.

Galvanic isolation capability indicates the maximum potential difference between the primary and secondary sides of the signal isolator to be handled. For the gate drive design in DPT circuit (i.e., phase-leg configuration), the worst case is when the lower switch in the phase-leg turns off, applying the potential difference across the gate drive signal isolator of the upper switch. In steady state, this potential difference is approximately equal to the DC bus voltage; considering the turn-off transition, and its resultant parasitic ringing and over-voltage across the

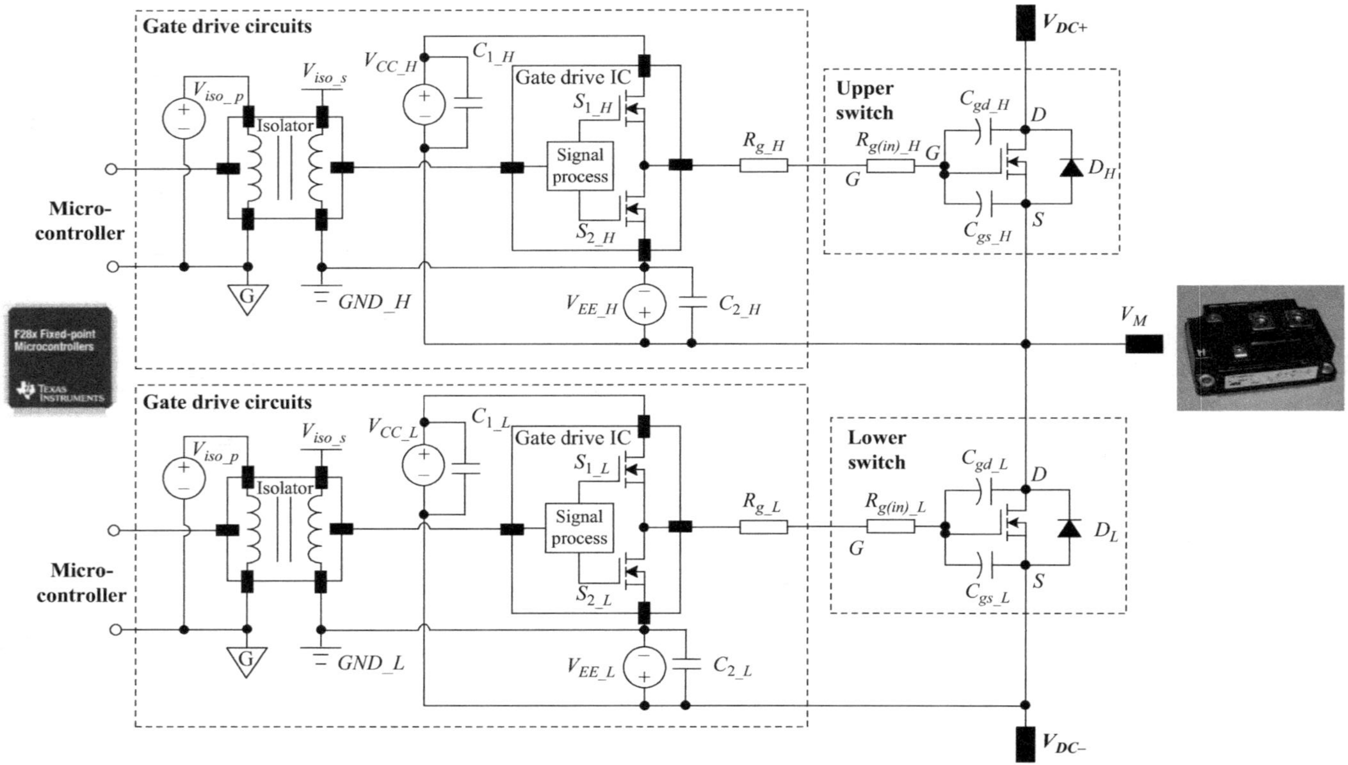

Figure 5.7 Typical gate drive circuit with detailed components

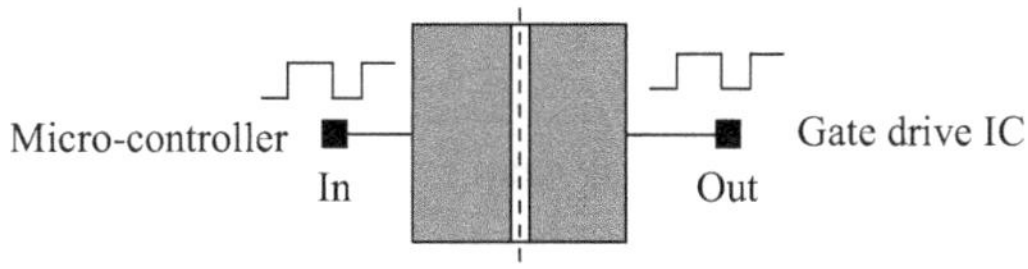

Figure 5.8 Signal isolator in gate drive circuit

lower switch, the dynamic potential difference across the upper signal isolator can be even higher. With the conservative consideration, the galvanic isolation capability of the signal isolator should be greater than the voltage rating of the DUT.

In the datasheet of the signal isolator, multiple terminologies can be found to define the galvanic isolation capability in different manners. Here, a clarification is given and the correct value for signal isolator selection in DPT test is recommended according to [4].

First, basic isolation versus reinforced isolation. Basic isolation refers to isolation that provides sufficient protection against electrical shock as long as the insulation barrier is intact. However, practically, some safety regulations require the basic isolation to be supplemented with a secondary isolation barrier for redundancy, so that the additional barrier provides shock protection, even if the first barrier fails. This is called double isolation. To make systems compact and save cost, it is desirable to have only one level of isolation that has the required electrical strength, reliability, and shock protection of two levels of basic isolation. This is called reinforced isolation. Therefore, with a given signal isolator, the basic isolation voltage is always higher than the reinforced isolation corresponding voltage. For the signal isolator selected for DPT, which is usually a lab test with no safety regulation requirement, basic isolation-related voltage should be acceptable. However, it is better to ensure that the reinforced isolation voltage also meets the required galvanic isolation requirement.

Second, multiple terminologies are used to describe isolation voltage capability, including maximum repetitive peak voltage, working voltage, maximum transient isolation voltage, isolation withstand voltage, and maximum surge isolation voltage. Different terminologies indicate the isolation voltage capability of the signal isolator according to different voltage profiles, as summarized in Table 5.2. For the signal isolator in DPT circuit, before the first pulse and after the second pulse, the lower switch turns off, and the whole DC bus voltage is applied across the upper signal isolator continuously. Thus, the maximum repetitive peak voltage should be selected as the required galvanic isolation capability for the signal isolator selection.

In addition to the galvanic isolation, CMTI is another important selection criterion for the signal isolator. As illustrated in Figure 5.9, for the upper switch signal isolator, the ground of the secondary side is connected with the source terminal of the upper switch, which is the middle point of the phase-leg. During the switching transient of the lower switch, the potential of the phase-leg middle point varies between the positive and negative DC bus with a high *dv/dt*. Considering the inevitable coupling capacitance (input-to-output capacitance) across the signal isolator *dv/dt* coupled through the parasitic paths may lead the isolator to lose control by false triggering a function or

Table 5.2 Summary of terminology for signal isolator voltage isolation

Terminology	Symbol	Definition
Maximum repetitive peak voltage	V_{IORM}	Continuous, peak voltage
Working voltage	V_{IOWM}	Continuous, RMS voltage
Maximum transient isolation voltage	V_{IOTM}	Transient (in a range of seconds), peak voltage
Isolation withstand voltage	V_{ISO}	Transient (in a range of seconds), RMS voltage
Maximum surge isolation voltage	V_{IOSM}	Transient (in a range of microseconds), peak voltage

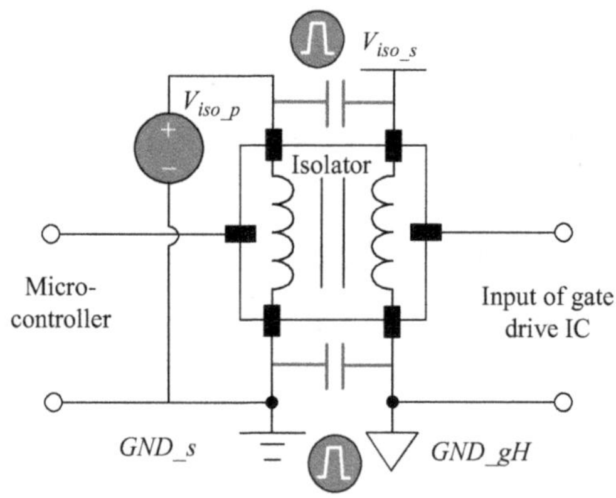

Figure 5.9 CMTI of the signal isolator

causing erroneous feedback. CMTI illustrates the maximum *dv/dt* that the single isolator can endure without potential malfunction. Since this *dv/dt* is determined by the switching speed, fast switching WBG devices makes the selection of signal isolator significantly more challenging.

Table 5.3 lists the tested peak *dv/dt* of several WBG devices. Typically, more than 100 kV//μs peak *dv/dt* can be easily observed for fast switching WBG devices. Note that instead of average *dv/dt*, peak *dv/dt* should be used here as the worst scenario. Accordingly, the signal isolator should be selected with CMTI capability greater than the peak *dv/dt* of the DUT. Table 5.4 lists the minimum CMTI of the state-of-the-art commercially available signal isolators with different isolation technologies, including optical (i.e., opto-coupler), capacitive, magnetic (i.e., transformer), and radio-frequency (RF)-based approaches. It is the minimum CMTI rather than the typical CMTI in the datasheet that should be used to identify the capability of the signal isolator. According to Tables 5.3 and 5.4, it can be seen that the CMTI of state-of-the-art signal isolators barely meets the WBG requirement. With the advancement of WBG devices, CMTI capability of signal isolator should be further improved.

Table 5.3 Tested peak dv/dt for WBG devices

Device under test	Model number	Typical *dv/dt*
SiC discrete switch	CMF20120D from Wolfspeed	~30 kV/μs
SiC power module	CPM212000025B from Wolfspeed	~80 kV/μs
GaN e-mode HEMT	GS66508P from GaN Systems	~100 kV/μs

Table 5.4 Minimum CMTI of state-of-the-art commercially available signal isolators [5–8]

Isolation techniques	Opto-coupler	Capacitive	Transformer	RF-based
Minimum CMTI	30 kV/μs	100 kV/μs	75 kV/μs	200 kV/μs

Additionally, signal transmission frequency range, propagation delay, and distortion of the signal isolator should also be taken into account, especially considering that the designed gate drive for dynamic characterization will eventually be used for power electronics converters. Specifically, maximum and minimum signal transmission frequencies should cover required the switching frequency range of the future power converter to be developed. Propagation delay and distortion introduced by the signal isolator should be sufficiently short to not introduce unacceptable control error. In general, these two selection criteria are relatively easy to meet according to the requirement of the state-of-the-art power electronics converters.

5.3.2 Isolated power supply

The power supply is the “gas tank” of a gate drive. It provides the required voltage for the operation of the gate drive circuit. Also, it offers necessary voltage and power for driving the switch into on-state and off-state.

Similar to the signal isolator, the isolated power supply also should be capable of providing sufficient galvanic isolation and CMTI. Specifically, with conservative consideration, galvanic isolation capability should be greater than the voltage rating of the DUT. In the meantime, CMTI should be greater than peak *dv/dt* during the switching transient. In most of the isolated power supply datasheet, input-to-output capacitance instead of CMTI is provided. Equivalently, small input-to-output capacitance indicates high CMTI capability. Practically, for fast switching WBG devices, this capacitance should be limited in a range of a few pF.

Additionally, output voltage and power rating are another two significant design criteria for the isolated power supply. Output voltage of the isolated power supply is determined by the turn-on and -off gate voltage for the DUT, while the turn-on and -off gate voltage is designed based on the device static and dynamic requirements. As discussed in Sections 5.1 and 5.2, the static requirement includes the minimization of on-state resistance or on-state voltage and conduction loss, and the suppression of spurious switching due to the external or internal disturbance. During the switching transient, gate voltage should be selected to turn-on and -off power device fast, and, in the meantime, maintain sufficient margin to not exceed gate voltage maximum ratings of the DUT. Gate voltages for both turn-on and -off should be selected based on the tradeoff between power loss, noise immunity, and gate terminal reliability (i.e., voltage margin). Device output characteristics and gate voltage ratings significantly affect the selection of the gate voltage. Practically, the preliminary datasheet of the DUT may provide recommended turn-on and -off gate voltage, which should be a good start for the output voltage determination of the isolated power supply.

Also, in practice, it is not easy to find an isolated power supply with an output voltage that directly meets the design requirement. In this case, another non-isolated voltage regulator (e.g., low-dropout or LDO regulator) can be introduced to tune the output voltage of the isolated power supply. Additionally, a dedicated 5 V non-isolated voltage regulator can be employed to power the secondary of the signal isolator, if necessary.

The power rating requirement of the isolated power supply for DPT is determined by the needed power for the secondary side of the signal isolator and gate drive IC, which is usually very small. DPT involves only two pulses and consumes negligible driving power. Considering that the selected isolated power supply can be leveraged for the future power converter, sufficient power for the gate drive output stage to drive the DUT continuously should be taken into account. In total, the power rating requirement of the isolated power supply can be expressed as

$$P_{out} > P_{iso} + P_{gd} + P_{sw} \tag{5.1}$$

where P_{iso} refers to the power dissipated by secondary side of isolator that can be obtained from the datasheet of signal isolator, P_{gd} indicates the power dissipated by gate driver IC that can be found based on the datasheet of the gate drive IC, and P_{sw} is the output power to drive the power device for an intended converter application. Detailed derivation for P_{sw} calculation is illustrated as follows.

Assume the gate charge of the DUT is Q_g. During the turn-on as shown in Figure 5.10(a), input capacitance of the device C_{iss} is charged from V_{EE} to V_{CC},

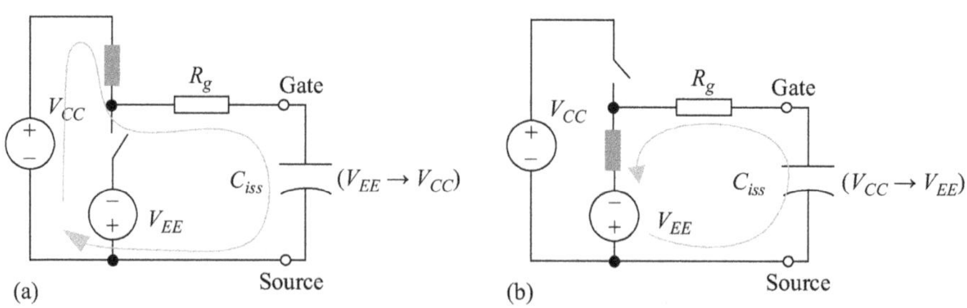

Figure 5.10 Output power to drive the device under test: (a) turn-on; (b) turn-off

where V_{EE} is the off-state gate voltage, which is zero or negative value, and V_{CC} is the on-state gate voltage. Then turn-on energy supplied by the power supply E_{PS_on} is

$$E_{PS\_on} = V_{CC} \times Q_g \tag{5.2}$$

Similarly, during the turn-off as shown in Figure 5.10(b), C_{iss} is discharged from V_{CC} to V_{EE}. Then turn-off energy supplied by the power supply E_{PS_off} is

$$E_{PS\_off} = V_{EE} \times (-Q_g) \tag{5.3}$$

In total, during one switching cycle, the energy provided by power supply, E_{PS_tot}, is

$$E_{PS\_tot} = (V_{CC} - V_{EE}) \times Q_g \tag{5.4}$$

Therefore, the needed power to drive the device P_{sw} with switching frequency f_s is

$$P_{sw} = (V_{CC} - V_{EE}) \times Q_g f_s \tag{5.5}$$

Note that P_{SW} is a function of V_{CC}, V_{EE}, Q_g, and fs, and is regardless of the gate resistance R_g. Besides, during one switching cycle, there is no net energy stored in C_{iss}. Therefore, according to the energy conservation, power provided by the isolated power supply to drive the device is dissipated by R_g. Thus, the power dissipation by gate resistor P_{rg} is the same as P_{SW}, and also independent of the gate resistance.

For SiC or some GaN transistors, a negative turn-off gate voltage is used to enhance switching speed, while also providing sufficient voltage margin to prevent the external or internal disturbance. In practice, it is not necessary to design two isolated power supplies for both positive gate voltage V_{CC} and negative gate voltage V_{EE}. One isolated power supply along with a non-isolated power supply is adequate. Starting from a two-isolated power supply solution in Figure 5.11(a), two options are illustrated in Figure 5.11(b) and (c) for V_{CC} and V_{EE} generation with one isolated power supply: one is to use the non-isolated voltage regulator, for example, LDO regulator; the other is to employ Zener diode.

5.3.3 Gate drive IC

The gate driver IC is the "engine" of a gate drive. It provides the voltage to drive the switch into on-state and off-state, with sufficient current to charge and discharge the switch input capacitance, and low loop impedance with quick response for fast switching. Accordingly, the primary selection criteria of the gate drive IC include adequate operating voltage range, sufficient peak source/sink driving current, and small pull-up and pull-down resistances of the gate drive IC buffer output.

Specifically, the operating voltage range of gate drive IC V_{GDIC} should be higher than the potential difference between the turn-on gate voltage V_{CC} and turn-off gate voltage V_{EE}, yielding

$$V_{GDIC} > V_{CC} - V_{EE} \tag{5.6}$$

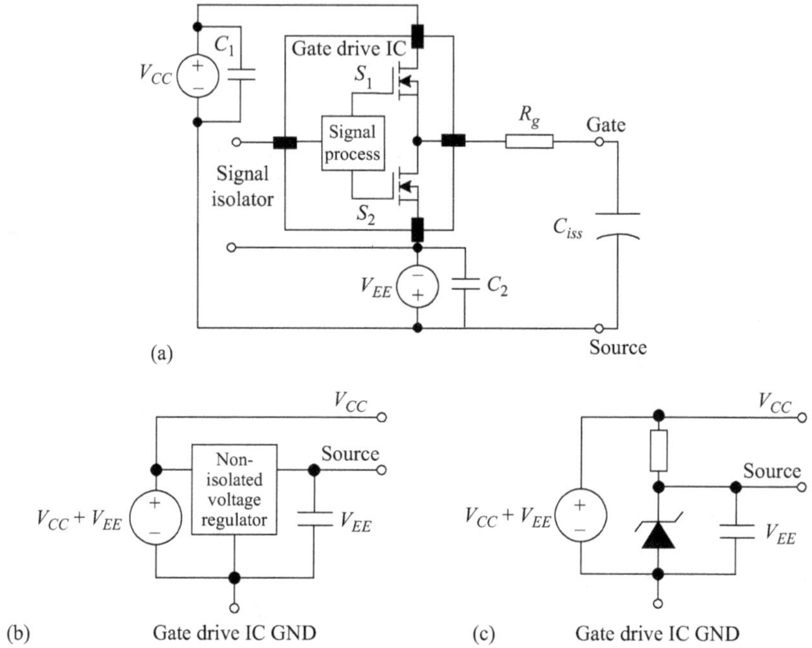

Figure 5.11 Isolated power supply configuration with negative turn-off gate voltage: (a) two isolated power supplies; (b) one isolated power supply with nonisolated voltage regulator; (c) one isolated power supply with Zener diode

Second, the peak source and sink driving current of the gate drive IC $I_{source/sink}$ should be larger than the peak gate current I_{g_peak} during the turn-on and turn-off transients. In the voltage source-based gate drive circuit to charge the capacitive load, as shown in Figure 5.12, the initial gate current has the maximum amplitude, and its value is determined by

$$I_{g\_peak} = \frac{V_{CC} - V_{EE}}{R_g} \tag{5.7}$$

where R_g is the total gate loop resistance, including the pull-up or pull-down resistance of the gate drive IC $R_{g(dr)}$, external gate resistance $R_{g(ext)}$, and internal gate resistance of the device $R_{g(in)}$, as shown in Figure 5.13. In addition, different turn-on and turn-off gate resistances are usually used to optimize the switching performance during both the turn-on and turn-off transients. Thus, the required peak source and sink currents of the gate drive IC can be different, and therefore should be calculated separately. Note that (5.7) is only valid without considering the gate loop parasitic inductance. Practically, due to this inductance, the peak gate current would be smaller. Therefore, the peak source and sink driving current

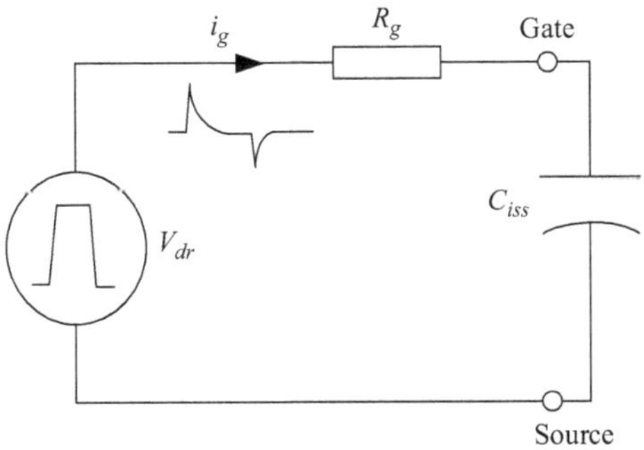

Figure 5.12 Transient gate current in voltage source-based gate drive

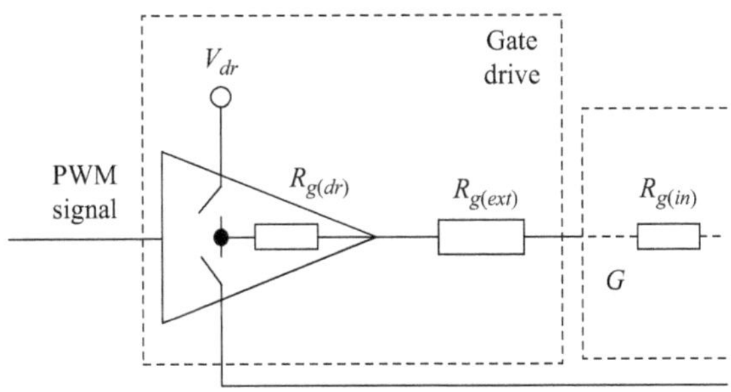

Figure 5.13 Total gate loop resistance

selection based on (5.7) should be for the worst scenario with sufficient margin to satisfy the practical need.

Third, the pull-up and pull-down resistances of the gate drive buffer stage should be adequately smaller than the desired gate resistance; otherwise, the switching speed would be limited. Practically, it is recommended to select the gate drive IC with the pull-up and pull-down resistances much smaller than the internal gate resistance of the device. Then the contribution of the pull-up and pull-down resistances to the total gate resistance is small, and its impact on the device dynamic characterization is negligible.

Similar to the signal isolator, several additional selection criteria of the gate drive IC should be taken into account: signal transmission frequency range, propagation delay, and distortion. Specifically, maximum and minimum signal transmission frequencies should cover the required switching frequency range of the future power converter to be developed. Propagation delay and distortion introduced by the gate drive IC should be sufficiently small to avoid introducing unacceptable control error.

5.3.4 Gate resistor

The gate resistor is the "gas pedal" of a gate drive. It controls the speed of switching transients. Although a smaller gate resistance is preferred, the gate resistor must be designed properly to accommodate tradeoffs among various switching behaviors, such as switching speed, switching loss, cross-talk suppression, and switch stresses. For dynamic characterization, gate resistance is a critical variable for users to adjust to evaluate the DUT performance.

Furthermore, it is difficult to achieve the optimal turn-on and turn-off behaviors with a single gate resistance. Figure 5.14 shows measured switching waveforms of a Wolfspeed 1,200 V/20 A SiC MOSFET operating at 600 V/10 A with gate resistance of 0 Ω [9]. Obviously, the over-voltage and ringing during the turn-on transient is more severe as compared to that during the turn-off transient. Thus, it is highly recommended to design the gate drive circuit capable of tuning the turn-on and turn-off gate resistances separately. Figure 5.15 illustrates a common approach to set different turn-on and -off gate resistances by using diodes. Note that low forward voltage drop and fast reverse recovery are the two key selection criteria for

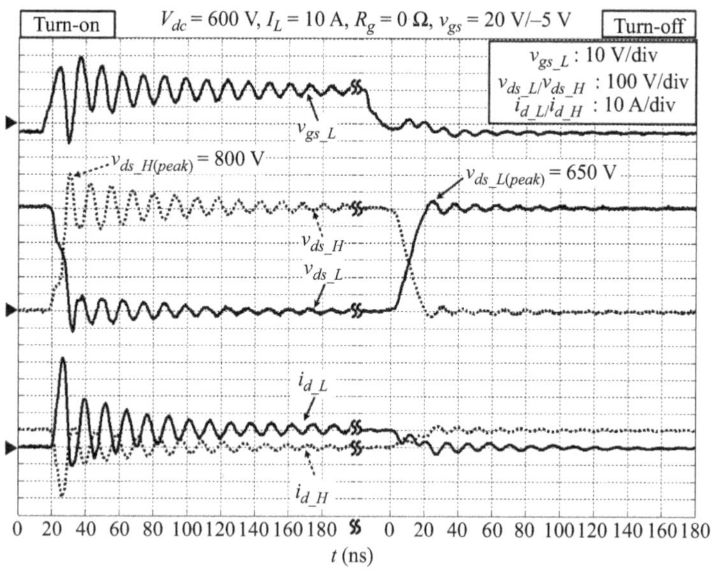

Figure 5.14 Measured switching waveforms of SiC devices

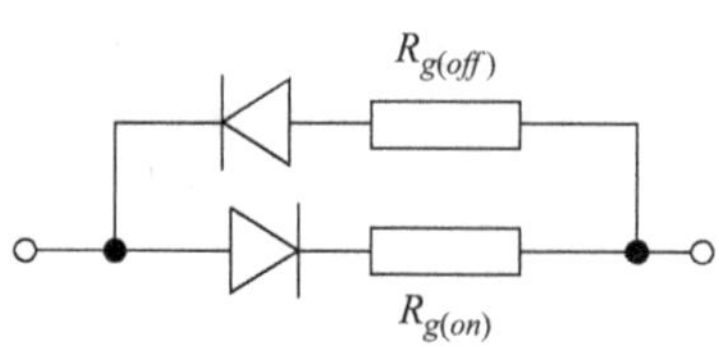

Figure 5.15 Using diode to set different turn-on and turn-off gate resistances

the diode. Additionally, some commercial gate drive ICs have the independent turn-on and -off output pins; as a result, different turn-on and -off gate resistances can be easily set.

In addition to the gate resistance, power rating of the gate resistor is another selection criterion to consider. As discussed earlier, power dissipation by gate resistor P_{rg} is the same as P_{SW} and is proportional to gate voltage ($V_{CC} - V_{EE}$), gate charge of the device Q_g, and switching frequency f_s as in (5.5). For the dynamic characterization, the power rating is not crucial since only two pulses are applied. However, it is worthwhile to mention, especially if the designed gate drive for dynamic characterization is intended to be eventually used for power electronics converters. In addition, it is important to notice that the power dissipation of the gate resistor is independent of the gate resistance.

5.3.5 Decoupling capacitor

The decoupling capacitor is the "fuel injector" of the gate drive. It provides the energy for the gate current pulse during the switching transient. The gate driving current originates from the power supplies V_{CC} and V_{EE}. However, the power supply cannot be placed in close proximity to the gate drive and DUT due to the relatively large volume. As a result, large parasitics are inevitably introduced, as shown in Figure 5.16. Therefore, it is necessary to place the decoupling capacitor locally with small footprint

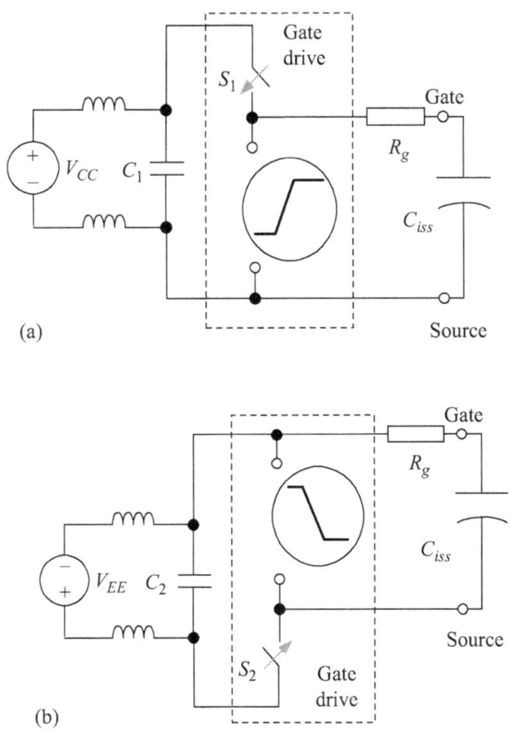

Figure 5.16 Decoupling capacitance for dynamic gate current: (a) turn-on transient; (b) turn-off transient

and low ESL to bypass the power supply resultant parasitics and improve the gate drive dynamic performance.

Three criteria should be taken into account for the decoupling capacitor selection, including sufficient voltage rating, adequate capacitance, and minimized ESL. Specifically, the voltage rating should meet

$$V_{C1} > V_{CC}, V_{C2} > |V_{EE}| \tag{5.8}$$

where V_{C1} is the voltage rating of turn-on-related decoupling capacitor C_1, and V_{C2} is the voltage rating of turn-off-related decoupling capacitor C_2.

Second, the capacitance should be sufficient to ensure a stable gate voltage during the switching transient. In other words, with the conservative assumption, that is, all the gate charge during the switching transient is provided by the decoupling capacitor, the voltage across the capacitor should remain almost constant. Therefore, the required capacitance should satisfy

$$C_1 > \frac{Q_g}{\Delta V_{CC}} = \frac{Q_g}{k_{VCC} \times V_{CC}}, C_2 > \frac{Q_g}{|\Delta V_{EE}|} = \frac{Q_g}{k_{VEE} \times |V_{EE}|} \tag{5.9}$$

where ΔV_{CC} and ΔV_{EE} are the turn-on and turn-off gate drive output voltage variations, respectively; k_{VCC} and k_{VEE} are coefficients to indicate the two voltage variation percentages, respectively. Practically, 1%–5% is selected for k_{VCC} and k_{VEE}.

Third, to minimize the ESL introduced by the decoupling capacitor, surface-mount ceramic capacitor is usually utilized. Also, multiple capacitors with small capacitance in parallel are preferred since ESL can be further reduced due to this paralleling configuration.

5.4 Case study

A SiC MOSFET from Wolfspeed is used to demonstrate the design and implementation of gate drive based on the configuration illustrated in Figure 5.17. Table 5.5 summarizes the basic information and desired testing conditions of the DUT. This design method can be applied to other WBG devices, including GaN HEMTs and high-voltage SiC devices.

5.4.1 Signal isolator

Step no. 1: Determine the required galvanic isolation capability V_{iso_si} based on the DUT's breakdown voltage V_{BD_DUT}.

$$V_{iso\_si} > V_{BD\_DUT} = 1200 \text{ V}$$

Step no. 2: Determine the required CMTI based on the estimated *dv/dt* according to the datasheet of DUT [2]. As summarized in Table 5.6, CMTI of the signal isolator should satisfy

$$CMTI > max(dv/dt\ (on), dv/dt\ (off)) = 47 \text{ V/ns}$$

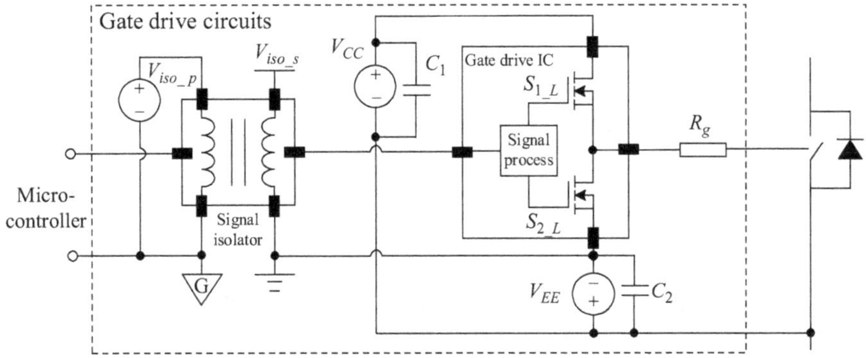

1. Signal isolator
2. Isolated power supply (V_{CC} & V_{EE})
3. Gate drive IC
4. Gate resistor (R_g)
5. Decoupling capacitor (C_1 & C_2)

Figure 5.17 Gate drive circuits in the case study

Table 5.5 Summary of DUT [2]

Vendor/model	Breakdown voltage	Rated current	Gate charge	Turn-on gate voltage	Turn-off gate voltage	Internal gate resistance
Wolfspeed C2M0080120D	1,200 V	24 A @ 100 °C	49.2 nC	20 V	–5 V	4.6 Ω

Table 5.6 Estimation of dv/dt based on datasheet

Fall time	Rise time	Test conditions
18.4 ns *Estimated dv/dt* (on) 35 V/ns	13.6 ns *Estimated dv/dt* (off) 47 V/ns	$V_{DD} = 800$ V $I_D = 20$ A

Step no. 3: Assume the switching frequency f_s based on the potential application.

$$f_s = 100 \text{ kHz}$$

Step no. 4: Determine the allowable propagation delay t_{pd_si} based on the switching frequency with the assumption that propagation delay has to be less than 1% of the switching period

$$t_{pd\_si} < 1\% \times \frac{1}{100 \text{ kHz}} = 100 \text{ ns}$$

Table 5.7 Design summary of signal isolator

Vendor/model		Texas Instruments ISO7841x	
Max. isolation working voltage	2121 V_{DC}	Max. frequency	100 MHz
CM transient immunity	>100 V/ns	Propag. delay	<16 ns
Max. secondary side power dissipation			30 mW

Step no. 5: Select a signal isolator to meet the requirements discussed earlier. In this case study, a capacitive-based signal isolator ISO7841x from Texas Instruments is selected with the key parameters summarized in Table 5.7. Note that the maximum secondary side power dissipation of the selected signal isolator is listed here for the power rating calculation of the isolated power supply in the next subsection.

5.4.2 Isolated power supply

Step no. 1: Determine the required galvanic isolation capability V_{iso_ips} based on the DUT's breakdown voltage V_{BD_DUT}.

$$V_{iso\_ips} > V_{BD\_DUT} = 1200 \text{ V}$$

Step no. 2: Determine the required CMTI based on the estimated *dv/dt* according to the datasheet of the DUT. Similar to the requirement estimated in signal isolator, CMTI of the isolated power supply should meet

$$CMTI > 47 \text{ V/ns}$$

As mentioned earlier, in most datasheets for isolated power supplies, input-to-output capacitance, instead of CMTI, is provided. Equivalently, small input-to-output capacitance indicates high CMTI capability. Practically, for fast switching WBG devices, this capacitance should be limited in a range of a few pF.

Step no. 3: Determine the output voltage based on the required driving voltage of the DUT. Also note the output voltage of the isolated power supply usually is not the same as the required driving voltage of the DUT. Then additional nonisolated voltage regulator has to be introduced.

$$V_{CC} = 20 \text{ V}$$
$$V_{EE} = -5 \text{ V}$$

Step no. 4: Determine the output power requirement based on (5.1)

$$P_{out} > P_{iso} + P_{gd} + P_{sw} = 30 + 23 + 123 = 176 \text{ mW}$$

where P_{iso}, power dissipation of the secondary side signal isolator, is obtained from the datasheet of the signal isolator; P_{gd}, power dissipated by gate driver IC, is

Table 5.8 Design summary of isolated power supply

Vendor/model	Murata Power Solutions MEJ1 series DC/DC converter		
Isolation voltage	5.2 kV (DC)	Input–output capacitance	3 pF
Power rating	1 W	Output voltage	24 V/5 V

obtained from the datasheet of the gate driver IC; and P_{sw}, power dissipated during switching transition, is calculated by (5.5)

$$P_{sw} = (V_{CC} - V_{EE}) \times Q_g f_s = (20 + 5) \times 49.2 \times 10^{-9} \times 100 \times 10^3 = 123 \text{ mW}$$

Step no. 5: Select an isolated power supply to meet the requirements discussed earlier. In this case study, a Murata Power Solutions MEJ1 series DC/DC converter is selected with the key parameters summarized in Table 5.8.

5.4.3 Gate drive IC

Step no. 1: Determine the operating voltage range based on (5.6)

$$V_{GDIC} > V_{CC} - V_{EE} = 20 - (-5) = 25 \text{ V}$$

Step no. 2: Determine the peak source/sink driving current based on (5.7)

$$I_{g\_peak} = \frac{V_{CC} - V_{EE}}{R_g} < \frac{V_{CC} - V_{EE}}{R_{g(in)}} = \frac{25}{4.6} = 5.4 \text{ A}$$

Step no. 3: Assume the switching frequency f_s based on the potential application

$$f_s = 100 \text{ kHz}$$

Step no. 4: Determine the allowable propagation delay t_{pd_gd} based on the switching frequency with the assumption that propagation delay has to be less than 1% of the switching period

$$t_{pd\_gd} < 1\% \times \frac{1}{100 \text{ kHz}} = 100 \text{ ns}$$

Step no. 5: Determine the acceptable pull-up and pull-down resistance $R_{g(dr)}$ based on the desired gate resistance or internal gate resistance of the device

$$R_{g(dr)} < R_{g(desire)} \ll R_{g(in)} = 4.6\ \Omega$$

Step no. 6: Select gate drive IC to meet the requirements discussed earlier. In this case study, a IXYS IXDN_609 gate drive IC is selected with the key parameters summarized in Table 5.9.

Table 5.9 Design summary of gate drive IC

Vendor/model	IXYS IXDN_609	Operating voltage	4.5–35 V
Peak source/sink current	9 A	Max. frequency	2 MHz
Propagation delay	40 ns @ 25 V	Pull-up/pull-down resistance	0.5/0.33 Ω @ 25 V
Quiescent power dissipation	23 W @ 20 °C		

Table 5.10 Design summary of decoupling capacitor

	Vendor/model	Description
C_1	TDK Corporation C2012X7R1H474K125AB	0.47 μF/50 V X7R ceramic capacitor
C_2	TDK Corporation CGJ4J2X7R1C155K125AA	1.5 μF/16 V X7R ceramic capacitor

5.4.4 Gate resistor

Step no. 1: Calculate the power rating of the gate resistor P_{rg} based on (5.5)

$$P_{rg} = P_{sw} = 123 \text{ mW}$$

Step no. 2: Prepare a set of gate resistors with different resistances and sufficient power rating for dynamic characterization of the DUT. As discussed earlier, it is preferred to allow gate drive circuit capable of tuning turn-on and turn-off gate resistances separately. Use of diode or gate drive IC with independent source and sink output terminals is recommended.

5.4.5 Decoupling capacitor

Step no. 1: Determine voltage rating of the decoupling capacitor according to (5.8)

$$V_{C1} > V_{CC} = 20 \text{ V}, \ V_{C2} > |V_{EE}| = 5 \text{ V}$$

Step no. 2: Determine required capacitance according to (5.9) where k_{VCC} and k_{VEE} are selected to be 1%.

$$C_1 > \frac{Q_g}{\Delta V_{CC}} = \frac{Q_g}{k_{VCC} \times V_{CC}} = \frac{49.2 \times 10^{-9}}{0.01 \times 20} = 0.25 \ \mu\text{F}$$

$$C_2 > \frac{Q_g}{|\Delta V_{EE}|} = \frac{Q_g}{k_{VEE} \times |V_{EE}|} = \frac{49.2 \times 10^{-9}}{0.01 \times 5} = 0.98 \ \mu\text{F}$$

Step no. 3: Select decoupling capacitor to meet the requirements discussed earlier. In this case study, TDK ceramic capacitors are selected with the key parameters summarized in Table 5.10. Note that the rated capacitance is larger than the required value to compensate for the capacitance reduction due to the increase of DC voltage for X7R ceramic capacitors. Another option is to use multiple capacitors in parallel to compensate for the capacitance reduction, and in the meantime, further reduce the ESL.

5.5 Summary

1. A good gate drive can best serve power devices under static state, dynamic transition, and fault condition. It is noted that although gate drive illustrated in this chapter is for device dynamic characterization, its basic theory and design consideration can be utilized for the gate drive design in actual WBG-based power electronics converters.
2. Usually, a gate drive consists of five components with the key selection consideration(s) summarized next:
   (a) Signal isolator is the "steering wheel" of a gate drive. It provides galvanic isolation between the ground of micro-controller and the ground of gate drive, and safely transfer the control and error signals from the input to the output of the gate drive. Its CM transient immunity capability is critical.
   (b) Isolated power supply is the "gas tank" of a gate drive. It provides necessary voltage for the operation of gate drive circuit. Also, it offers necessary voltage and power for driving the switch into on-state and off-state. Low coupling capacitance between the primary and secondary sides is the key.
   (c) Gate driver IC is the "engine" of a gate drive. It provides necessary voltage to drive the switch into on- and off-state, sufficient current to charge and discharge the switch input capacitance, and low loop impedance with quick response for fast switching. Accordingly, adequate operating voltage range, sufficient peak source and sink driving currents, and small pull-up and -down resistances of the gate drive IC buffer output are crucial.
   (d) Gate resistor is the "gas pedal" of a gate drive. It controls the speed of switching transients. Resistance is determined by the tradeoff between switching speed or loss with ringing or overshoot, and is a critical variable for users to adjust for the dynamic characterization. Power rating requirement of the gate resistor cannot be ignored and is independent of the resistance.
   (e) Decoupling capacitor is the "fuel injector" of the gate drive. It provides the gate current pulse during the switching transient. Surface-mount ceramic capacitor with small ESL is preferred.

References

[1] Microsemi APT34M120J Si MOSFET datasheet, Rev. C. (2011) [Online]. Available: https://www.microsemi.com/existing-parts/parts/60368#resources. Accessed in July 2011.

[2] Wolfspeed C2M0080120D SiC MOSFET datasheet, Rev. C. (2015) [Online]. Available: https://www.wolfspeed.com/media/downloads/167/C2M0080120D.pdf. Accessed in October 2015.

[3] F. Wang, Z. Zhang, and E. Jones, "Wide bandgap device characterization," *IEEE Applied Power Electronics Conference and Exposition*, Long beach, CA, March 2016.
[4] High-voltage reinforced isolation: definitions and test methodologies [Online]. Available: http://www.ti.com/general/docs/lit/getliterature.tsp?baseLiteratureNumber=slyy063. Accessed in November 2014.
[5] Avago ACPL-4800 high CMR intelligent power module and gate drive interface optocoupler datasheet. (2007) [Online]. Available: https://docs.broadcom.com/docs/AV01-0193EN. Accessed in June 2007.
[6] Texas Instruments ISO7841x high-performance, 8000-VPK reinforced quad-channel digital isolator datasheet. (2017) [Online]. Available: http://www.ti.com/lit/ds/symulink/iso7841.pdf. Accessed in March 2017.
[7] Analog Devices, ADuM120N/ADuM121N datasheet, Rev. C. (2017) [Online]. Available: http://www.analog.com/media/en/technical-documentation/data-sheets/ADuM120N_121N.pdf. Accessed in September 2017.
[8] Silicon Labs, Isolated gate drives [Online]. Available: https://www.silabs.com/products/isolation/isolated-gate-drivers.
[9] Z. Zhang, J. Dix, F. Wang, B. J. Blalock, L. M. Tolbert, and D. Costinett, "Intelligent gate drive for fast switching and cross-talk suppression of SiC devices," *IEEE Transactions on Power Electronics*, vol. 32, no. 12, pp. 9319–9332, 2017.

Chapter 6
Layout design and parasitic management

The high-speed switching performance of WBG devices poses unique challenges for circuit layout design and parasitics management. This chapter aims at understanding the impact of parasitics on the switching behavior, investigating the layout design principle for a DPT board, and presenting several case study examples to illustrate the way to apply this design theory in actual implementations.

6.1 Impact of parasitics on the switching performance

Circuit parasitics have a significant impact on the fast switching speed of WBG devices [1–6]. Figure 6.1 shows the primary parasitics involved in switching transitions, including three parasitic inductances and three parasitic capacitances, and the internal gate resistance of power devices. These parasitics can be divided into three groups according to their positions in the switching commutation loop, namely parasitics in the gate loop (i.e., gate loop inductance L_{gs}, gate-source capacitance C_{gs}, and internal gate resistance $R_{g(in)}$), parasitics in the power loop (i.e., power loop inductance L_{ds} and drain-source capacitance C_{ds}), and common parasitics shared in both gate and power loops (i.e., common source inductance L_{cm} and Miller capacitance C_{gd}).

6.1.1 Gate loop parasitics

As can be observed in Figure 6.2, parasitic inductance, junction capacitance, and gate resistance associated in the gate loop form an *LRC* series resonant network and cause oscillation in the gate-source voltage [7]. For GaN devices with low gate voltage rating, such as EPC GaN transistors, this gate loop ringing easily causes the voltage overshoot at gate-source terminals to exceed the maximum allowable gate voltage imposed by devices [8]. In this case, a large external gate resistance $R_{g(ext)}$ has to be selected to suppress the gate loop resonance with the penalty of switching speed reduction. Also, the phenomenon of the self-sustained oscillation during the turn-off transient due to relatively large gate loop inductance L_{gs} was investigated for fast switching SiC JFETs in [9]. Fortunately, parasitics in the gate loop have limited interaction with the parasitics of the power loop. Also, due to the small chip size of SiC devices as compared to their Si counterparts, the internal gate resistance $R_{g(in)}$ of SiC devices becomes large, which to a certain extent suppresses the ringing induced by the *LRC* resonance network in the gate loop [10]. In the end, gate loop

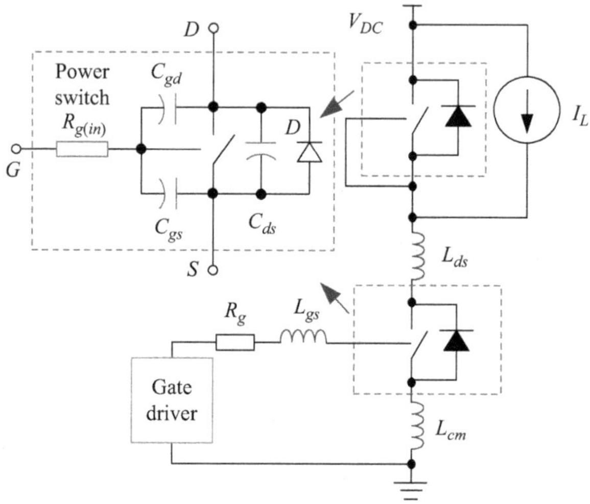

Figure 6.1 Parasitics involving in the switching commutation loop

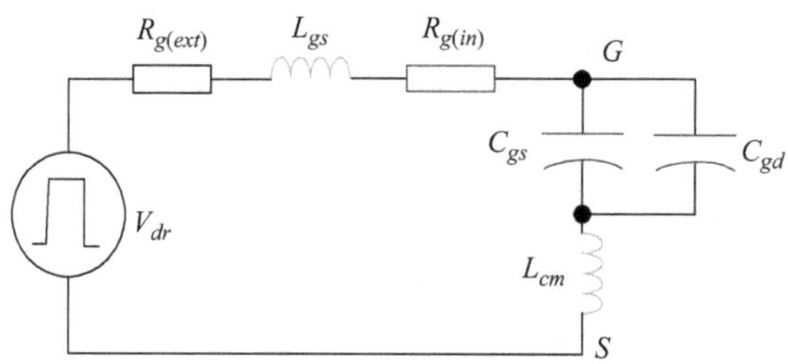

Figure 6.2 Equivalent circuit of gate loop with parasitics

parasitics may influence the switching performance, but the impact on resonances in the power loop is limited.

6.1.2 Power loop parasitics

Compared with parasitics in the gate loop, power loop parasitics play a greater role in the switching performance of most power devices [5, 7]. It is the power loop inductance L_{ds} resonance with drain-source capacitance C_{ds} that causes parasitic ringing in the drain current and drain-source voltage, as shown in Figure 6.3.

Specifically, during the turn-on transient of the lower switch, a voltage spike occurs during the *dv/dt* subinterval (see in Figure 6.3). As shown in Figure 6.4(a), the lower switch can be represented as a voltage source with the voltage falling rate of *dv/dt*, and the upper switch is modeled as an output capacitance, while the body diode has started to block the DC bus voltage. Thus, the over-voltage across

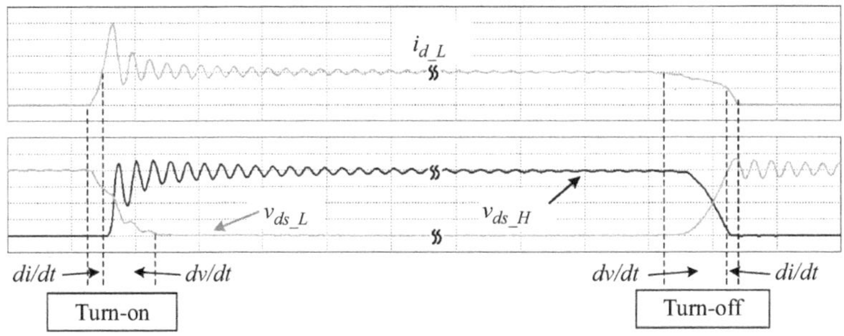

Figure 6.3 Typical power loop waveform during the switching transient

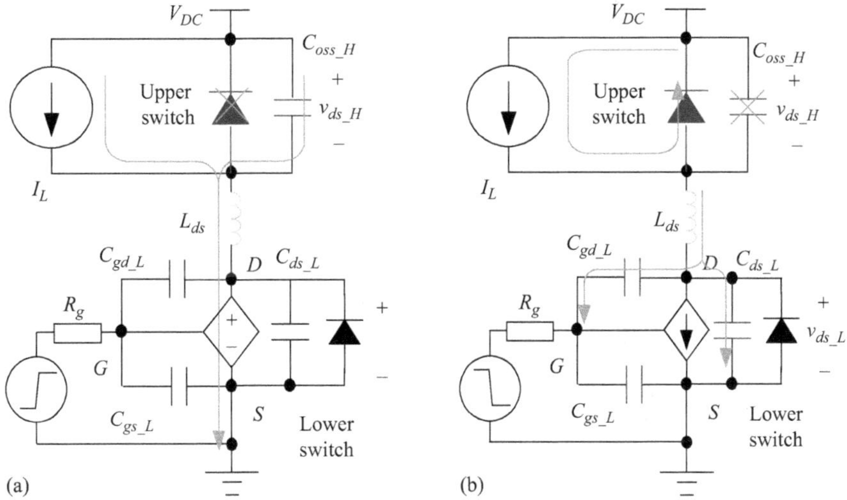

Figure 6.4 Mechanisms causing voltage overshoot: (a) during turn-on transient; (b) during turn-off transient

the drain-source terminals of the upper switch is originated by a *dv/dt* voltage source across the resonant network formed by the power loop parasitic inductance L_{ds} in series with the output capacitance of the upper switch C_{oss_H}. Figure 6.5(a) displays the simplified model for the turn-on over-voltage, where R_{stray} is the stray resistance in the power loop. Fast switching speed leads to high *dv/dt*, resulting in serious resonance and high over-voltage. In addition to the excitation, L_{ds} is a critical parameter that significantly affects the resonant period and damping factor.

During the turn-off transient of the lower switch, the over-voltage will be observed during the *di/dt* subinterval (see in Figure 6.3). As illustrated in Figure 6.4(b), the lower switch is modeled as a current source with the current falling rate of *di/dt*, and the upper

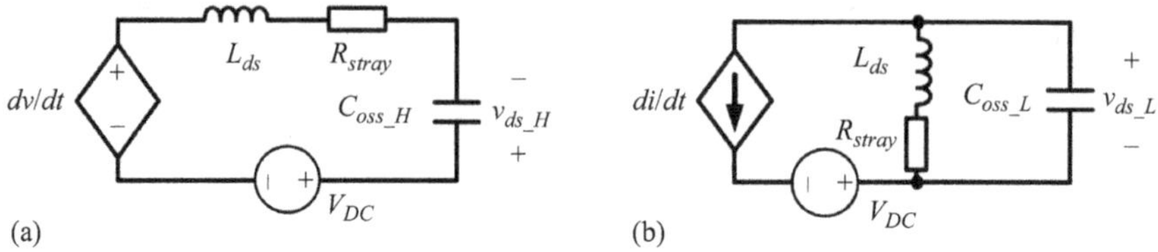

Figure 6.5 Model of drain-source over-voltage: (a) Turn-on over-voltage; (b) turn-off over-voltage

switch is modeled as a body diode operating in conduction mode, and its output capacitance C_{oss_H} is bypassed. Therefore, a *di/dt* current source is the excitation added across the parallel resonant network consisting of L_{ds} and the output capacitance of the lower switch C_{oss_L} that induces the turn-off over-voltage, as shown in Figure 6.5(b). Similar to the turn-on over-voltage, the turn-off switching speed determines *di/dt*, and then influences the turn-off over-voltage. L_{ds} is the main contributor to this resonance.

In the end, although the mechanisms causing turn-on and turn-off over-voltages are different, they both become worse as switching speed increases and damping factors in resonant networks decrease. WBG devices with high switching-speed capability and small on-state resistance make this issue more serious. To guarantee drain-source voltage of power devices remains below their breakdown voltage, the slew rate of drain current has to be controlled within certain boundary. In other words, the switching speed will have to be limited as a result of power loop parasitics.

6.1.3 Common parasitics

The last group of parasitics that also have the most impact on the switching speed is the common parasitics shared by both the gate and power loops, that is, common source inductance L_{cm} and Miller capacitance C_{gd}. During the *di/dt* transient (i.e., current rise subinterval during turn-on and current fall subinterval during turn-off), the voltage across the common source inductance L_{cm} reduces the effective gate drive output voltage, and then decreases the gate current that is used to charge or discharge gate-source capacitance C_{gs}. In the end, the switching speed is reduced [4, 7]. Similarly, during the *dv/dt* transient (i.e., voltage fall subinterval during turn-on and voltage rise subinterval during turn-off), the displacement current induced by Miller capacitance C_{gd} diverts part of the gate current that originally flows to gate-source capacitance C_{gs} [11–13]. Therefore, the speed of the switch state transition is slowed down. The aforementioned qualitative explanations can be supplemented and perhaps better understood with the following quantitative analysis.

During the *di/dt* transient, the voltage across common source inductance v_{Lcm} is expressed as

$$v_{Lcm} = L_{cm}\frac{d}{dt}\left(i_d - i_g\right) = L_{cm}\left(\frac{di_d}{di_g} - 1\right) \times \frac{di_g}{dt} = L_{cm\_eff}\frac{di_g}{dt} \tag{6.1}$$

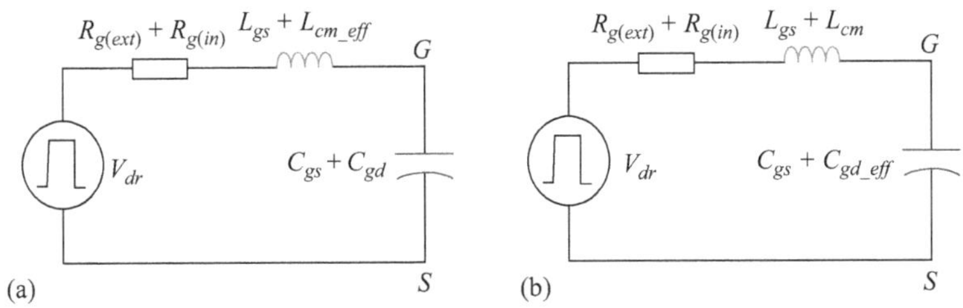

Figure 6.6 Equivalent circuit of gate loop considering the effect of common source inductance and Miller capacitance: (a) during di/dt transient (b) during dv/dt transient

where L_{cm_eff} is the effective common source inductance in the gate loop during *di/dt* transient. As $di_d/di_g \gg 1$, L_{cm_eff} is much larger than L_{cm}, which means the equivalent gate loop impedance increases, especially at high frequency. Therefore, the switching speed during *di/dt* decreases.

During the *dv/dt* transient, the current flowing into the Miller capacitance i_{Cgd} is given by

$$i_{Cgd} = C_{gd}\frac{d}{dt}\left(v_{ds} - v_{gs}\right) = C_{gd}\left(\frac{dv_{ds}}{dv_{gs}} - 1\right) \times \frac{dv_{gs}}{dt} = C_{gd\_eff}\frac{dv_{gs}}{dt} \tag{6.2}$$

where C_{gd_eff} is the effective Miller capacitance in the gate loop during the *dv/dt* transient. As $dv_{ds}/dv_{gs} \gg 1$, C_{gd_eff} is much larger than C_{gd}, which is the mechanism causing the Miller effect [13]. The speed of the *dv/dt* transient is inherently limited by this feedback mechanism. Figure 6.6 shows the equivalent circuit of the gate loop considering the effect of common source inductance and Miller capacitance.

6.2 DPT layout design

The purpose of a DPT is to experimentally gather the current and voltage waveforms of the DUTs to evaluate their switching behaviors in an actual power converter. Therefore, the primary guideline for the DPT layout is the similarity between the DPT and the intended converter. Second, the parasitics minimization should be considered.

In general, the principle of DPT layout design includes: (1) for power device-related parasitics, including C_{gs}, C_{ds}, and C_{gd}, we have to accept the existing parasitics and avoid adding additional equivalent parasitics externally, especially for Miller capacitance; (2) interconnection-related parasitics L_{gs}, L_{ds}, and L_{ss} must be minimized to be as small as possible; and (3) similarly, parasitics coupled between power stage and signal stage, for example, coupling capacitance of signal isolator and isolated power supply, need to be minimized as much as possible.

In the DPT configuration, as shown in Figure 6.7, the switching loop-associated parasitic inductance is physically contributed by three main parts: (1) interconnection

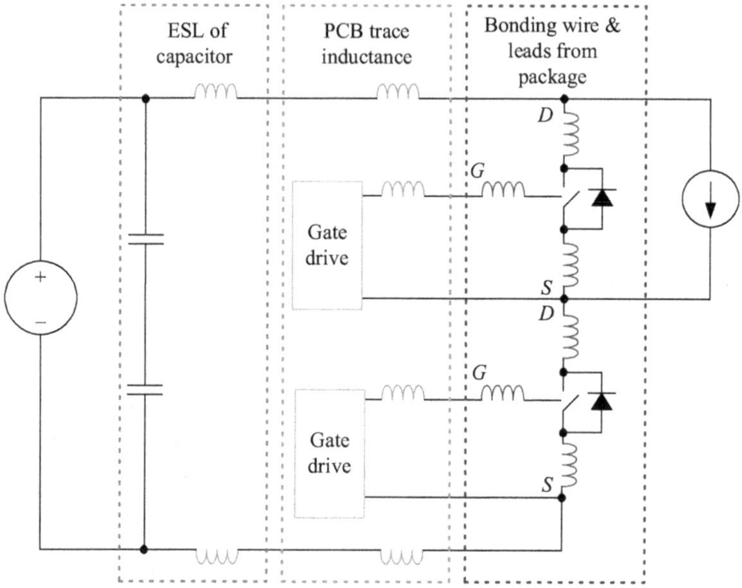

Figure 6.7 Distribution of interconnection parasitics

among power devices' bare dies, specifically including bonding wires and leads-related parasitics inductance from package; (2) PCB trace-induced inductance; and (3) ESL of decoupling capacitors. Note that PCB trace-introduced parasitics and ESL of the decoupling capacitors can be associated with both power loop and gate drive.

Among different interconnections, bonding wires or other device bonds are embedded in the devices' package. As device users, we have to accept their induced parasitic inductances. The leads of power devices act as connectors for assembling the power device, gate drive, and power stage board. To minimize the effective length of the lead in the switching loop, the placement of power device, gate drive, and power stage board needs to be investigated. The PCB trace-induced inductance greatly depends on the PCB layout design. Hence, the optimal PCB layout design of power stage and gate drive should be studied, and special attention should be paid to the power stage board since the impact of the power loop parasitic inductance on the switching behavior is more severe than that due to gate loop parasitics. For decoupling capacitors, the surface-mount ceramic capacitors are suggested. Their ESLs are determined by chip size, given by [14]:

$$ESL = 394.727 \times 1.052^{L} \times 1.317^{L/W} \tag{6.3}$$

where L and W refer to the length and width of the ceramic capacitor's package. As listed in Table 6.1, ESL of commonly used chip styles of 100 nF/1,000 V ceramic capacitors is around 1–1.5 nH, which is much smaller than 30 nH, the package-induced parasitic inductance of the commercially available SiC power module [15]. Moreover, the decoupling capacitor bank per phase-leg usually

Table 6.1 ESL of 100 nF surface-mount ceramic capacitors with 1,000 V voltage rating

Chip style	L/W	L	ESL
2225	22/25	22	1,534 pH
2220	22/20	22	1,680 pH
1825	18/25	18	1,175 pH
1812	18/12	18	1,425 pH

consists of several ceramic capacitors in parallel, which results in a smaller ESL. Therefore, the parasitic inductance due to ESL of DC bus decoupling capacitor can be ignored. Note however, if converters use other types of decoupling capacitors or other types of decoupling schemes, their ESL may be larger and may have to be accounted for. The gate drive decoupling capacitor and its ESL may also have limited impact, since the capacitor chip size used in gate drive is much smaller than that in DC bus due to the low-voltage rating and less clearance requirements.

In summary, the effort of parasitic management here focuses on the placement of power device, gate drive, and power stage board to reduce the effective length of device's leads in the switching loop as well as PCB layout design for the power stage and gate drive.

There are two basic rules to follow for the parasitic minimization: one is the P-cell and N-cell concept, and the other is magnetic field cancellation.

As discussed in Chapter 4, Section 4.1, the switching commutation always occurs between either upper switch and lower diode or lower switch and upper diode. Therefore, instead of placing the switch and its corresponding antiparallel diode as close as possible, it is preferred to locate the switch with its complementary freewheeling diode next to each other, so that the physical distance in the actual switching commutation loop is naturally minimized. This is shown in Figure 6.8 and called the P-cell and N-cell concept [16].

The P-cell and N-cell concept is a well-accepted layout principle, especially for Si IGBT-based power converters, since the additional antiparalleled diode has to be used in the phase-leg configuration. The situation for a WBG-based power converter is different. Considering the excellent performance of WBG device body diode (e.g., SiC MOSFET) or "diode-like behavior" (e.g., GaN HEMT), a dedicated SBD is not necessary to improve the dynamic performance. Specifically, for SiC MOSFETs, since the lifetime of minority carriers in SiC semiconductor is shorter than Si, the reverse recovery charge of SiC MOSFET's body diode is reduced. On the other hand, although the dedicated SiC SBD can mitigate reverse recovery, it also introduces additional junction capacitance, leading to extra switching loss. It is observed that, based on the latest generation Wolfspeed SiC MOSFETs, the reverse recovery charge introduced by a MOSFET's body diode is comparable to or less than the charge of SBD junction capacitance over a wide temperature range [17]. Accordingly, there is a trend to eliminate the usage of

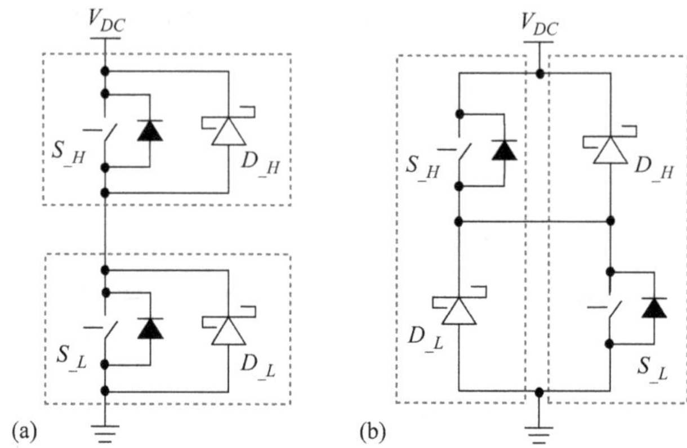

Figure 6.8 Parasitic minimization using P-cell and N-cell concept: (a) traditional approach; (b) P-cell and N-cell concept

SBDs in SiC-based power converters. For GaN HEMTs, its "diode-like behavior" does not experience reverse recovery; it is therefore not necessary to introduce dedicated an antiparallel SBD to improve the switching performance. In total, it is important to utilize P-cell and N-cell concept for the configuration where the dedicated antiparallel SBD is added. But it may become unnecessary for WBG-based design due to the diminished importance of having SBD.

The other principle for parasitic minimization is the magnetic field cancellation. The loop inductance is essentially determined by the loop area. Small loop area means less magnetic flux under a given current or magnetic field density and therefore lower equivalent parasitic inductance. Therefore, minimizing parasitic inductance, beyond minimizing the conductor length, is best achieved by canceling out stray magnetic fields through having a returning current path as close as possible, forming a minimized loop.

According to the layout design principle, understanding of physical contributors for parasitics, and basic concept of parasitics minimization discussed earlier, several case study examples are presented next to demonstrate the practical implementation of the layout design and parasitic management.

6.3 Case study

Device package significantly affects the layout design for a DPT board. Therefore, before the illustration of the case study, it is necessary to first give a brief overview of the package of state-of-the-art WBG devices. Then, devices based on two typical packages (TO-series for SiC and surface mount for GaN) are selected as examples to demonstrate the implementation of the parasitic management. Later, a special consideration of the dynamic current measurement and its impact on the DPT layout are discussed. Finally, the layout design considerations of the gate drive board are summarized.

6.3.1 Brief overview of WBG devices' package

Generally speaking, the state-of-the-art package of WBG devices can be categorized into three groups: SiC-based discrete package, SiC-based power module, and GaN-based discrete device. WBG packaging technology is currently an active research topic, and there are some other advanced packaging techniques under development, which are out of the scope of this discussion.

The TO-series package is popular for discrete SiC devices. As illustrated in Table 6.2, the state-of-the-art packages can be grouped from two aspects: (1) through-hole versus surface mount and (2) non-Kelvin connection versus Kelvin connection.

The primary difference between through-hole and surface-mount packages is the tradeoff between cooling and parasitics: a device with a through-hole package has better cooling capability, since its case can directly attach with the thermal management system via the thermal interface material, while the surface-mount device usually has to utilize the thermal via through PCB for heat dissipation, leading to large thermal resistance. However, a surface-mount device has short leads and less parasitics than that of the through-hole device. Also, the optimized layout design method can be utilized for the surface-mount device. As a result, switching loop parasitics for surface-mount device-based design is smaller.

Table 6.2 Summary of SiC discrete packages

	Non-Kelvin connection	**Kelvin connection**
Through-hole		
Surface mount		

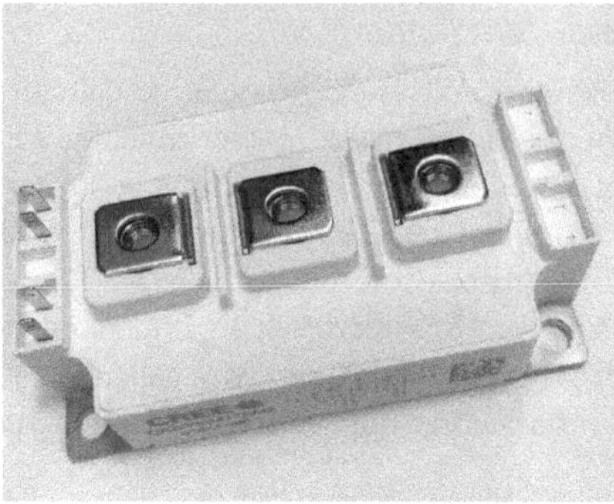
(a)

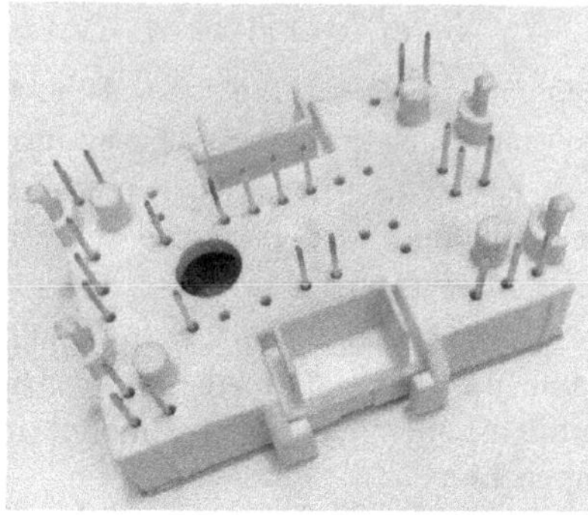
(b)

Figure 6.9 Examples of the package for SiC power modules: (a) 62 mm half bridge power module from Wolfspeed; (b) lightweight EASY package from Infineon

(a)

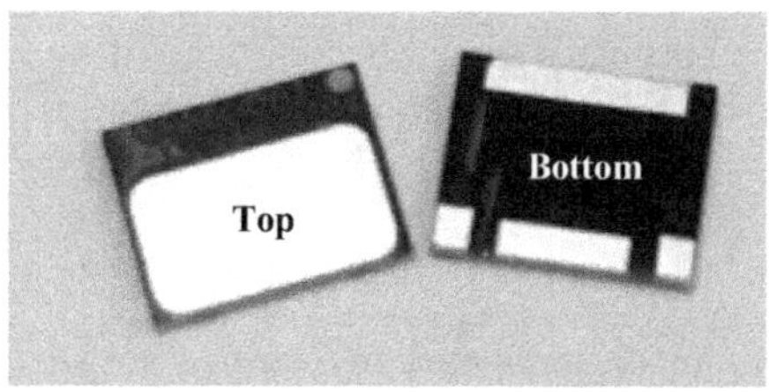

(b)

Figure 6.10 Examples of the package for GaN transistors: (a) GS 66508B from GaN Systems; (b) top-side cooled from GaN Systems

In comparison with the non-Kelvin connection, the key improvement for Kelvin connection is the separated source terminals for gate and power loops and its resultant less common source inductance. As discussed earlier, this allows the SiC device to have better switching performance and less switching loss.

The SiC power module is popular for high-current high-power applications. Instead of a single device, usually a basic circuit cell (e.g., phase-leg configuration) with multiple dies is integrated into the power module. Also, Kelvin connection is provided for common source inductance minimization and switching performance improvement. Furthermore, a clear separation between thermal and electrical interfaces makes the power converter using a power module easy to design and optimize. Several examples from the state-of-the-art SiC manufacturers are summarized in Figure 6.9.

Regarding the device package for GaN transistor, leadless surface-mount (e.g., chipscale, near-chipscale, DFN) is popular due to the high-speed switching capability and ultra-low parasitic requirement, while TO-series and other package styles are also offered. A Kelvin connection is typically included. Several examples from GaN Systems are listed in Figure 6.10.

6.3.2 Case study 1: TO-247 package SiC MOSFETs

Figure 6.11 displays an industry standard TO-247, 3-pin package, which is the most popular package for discrete SiC devices in today's market. According to the layout design principle, two aspects have to be taken into account: placement design for the effective lead length minimization and PCB layout design for the trace parasitics mitigation.

6.3.2.1 Placement design for effective lead length minimization

For power devices with TO-247 package, the conventional placement of power devices, gate drive, and power stage board in the phase-leg configuration is displayed in Figure 6.12. In this design, the lead associated with the source terminal of devices is involved in both gate loop and power loop, resulting in an inevitable common source inductance. To minimize this lead-induced common

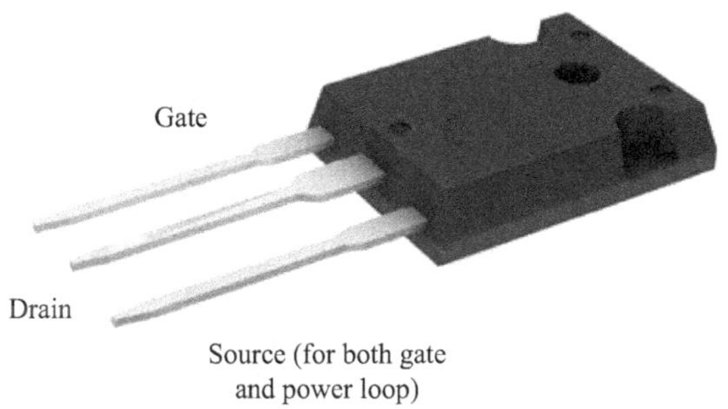

Figure 6.11 Standard TO-247 3-pin package

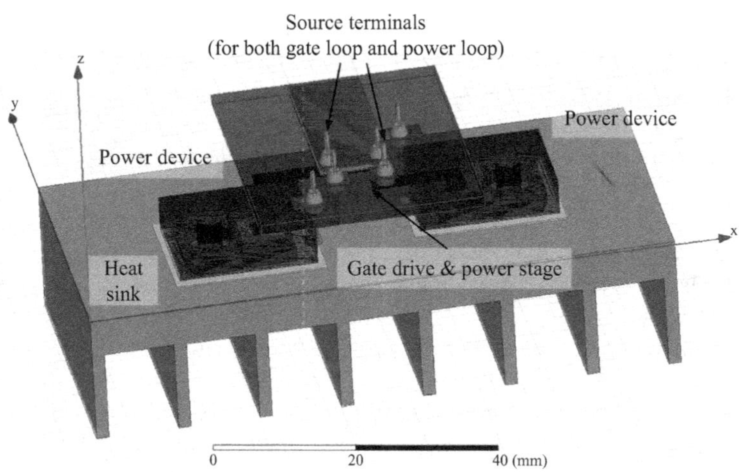

Figure 6.12 Conventional placement of power devices, gate drives, and power stage board in the phase-leg configuration

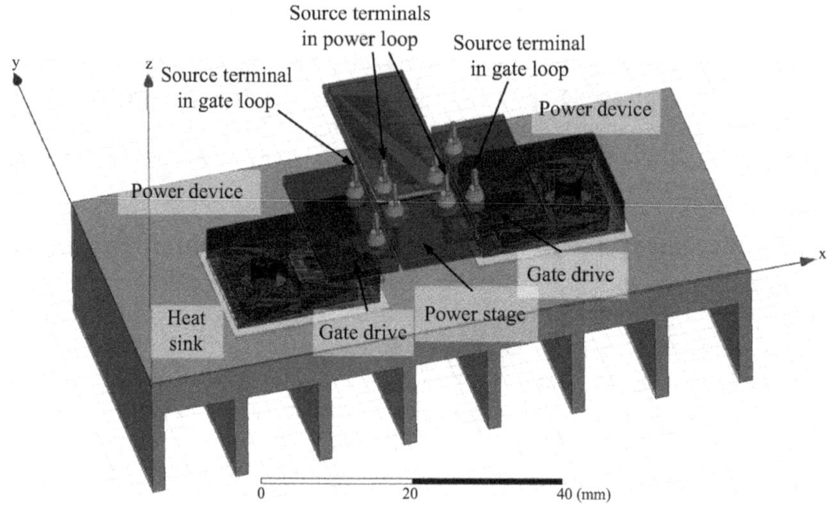

Figure 6.13 Placement of power devices, gate drives, and power stage board in the phase-leg configuration with Kelvin connection

source inductance, a Kelvin connection with separate source return path is recommended by power device manufacturers [18]. That is, a new source terminal is manually created by soldering a pin at the root of the device's source lead for the gate loop, as shown in Figure 6.13. Thereby, the source paths for the gate loop and power loop are separated externally, and then the common source inductance is reduced.

However, for the aforementioned two placements, the leads of the device are first bent, and then soldered into the PCB. No matter how close the PCB is placed on the top of power devices, at least 5 mm lead is necessary considering the height of the device, thickness of the PCB, as well as the size of components mounted on the bottom of the PCB. Practically, the lead-associated parasitics per millimeter is around 1 nH [7]. Hence, there will be around 10 nH additional parasitic inductance induced by leads of lower and upper switches in a phase-leg, which will greatly impact the parasitic ringing and over-voltage during the fast switching transient.

To further minimize the effective length of leads involved in the switching loop, a new placement of power devices, gate drives, and power stage board is highlighted in Figure 6.14. Separated from the power stage board, a gate drive daughter board is placed perpendicular to the power device and soldered at the root of device's gate and source leads, and then drain and source leads of the device directly attach the surface of the power stage board. Therefore, the effective lead lengths in both gate and power loop are greatly reduced. Also note that this new placement maintains the Kelvin connection, namely there is no extra common source inductance induced by the source lead of power devices. In total, minimum common source and power loop parasitic inductances can be achieved.

Based on Q3D Extractor, the gate loop, power loop, and common source inductances for different placement styles are simulated and summarized in

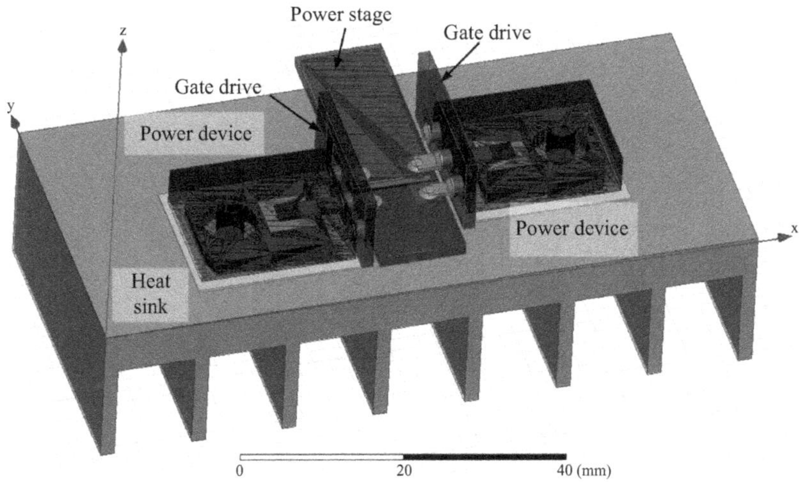

Figure 6.14 A parasitic minimization placement of power devices, gate drives, and power stage board in the phase-leg configuration

Table 6.3 Parasitic inductance dependence on different placements based on Q3D

Placement style	Gate loop inductance (nH)	Power loop inductance (nH)	Common source inductance (nH)
Conventional placement	20.4	23.9	6.8
Kelvin connection-based placement	17.7	23.8	4.5
Parasitic minimization placement	15.2	17.0	4.8

Table 6.3. As compared to the conventional and Kelvin connection-based placements, the new parasitic minimization placement enables the power loop inductance to decrease by up to 28.6%. Also, its common source inductance reduces by 29.4% when compared with that of the conventional placement and is almost identical to the optimal common source inductance achieved by the Kelvin connection-based placement.

To further compare the influence of different placement schemes on the switching performance, based on Wolfspeed second-generation 1200-V/36-A SiC MOSFETs, a voltage source converter is built with power stage board and gate drive daughter boards, as shown in Figure 6.15. Sharing with the same power devices, gate drives, and power stage board, Figures 6.16–6.18 display the voltage

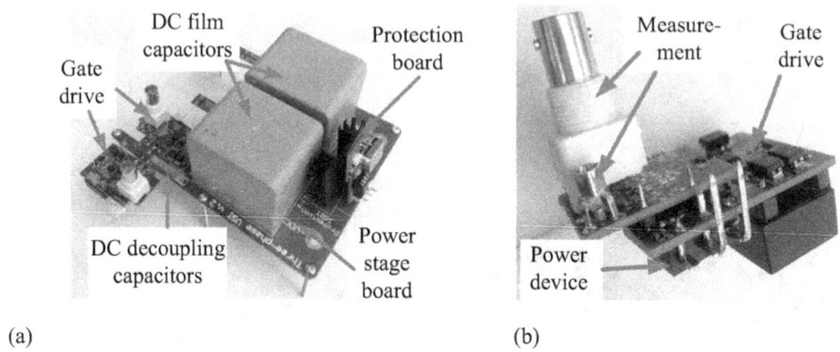

Figure 6.15 Hardware setup for different placement schemes comparison: (a) power stage with gate driver daughter boards; (b) gate drive daughter board

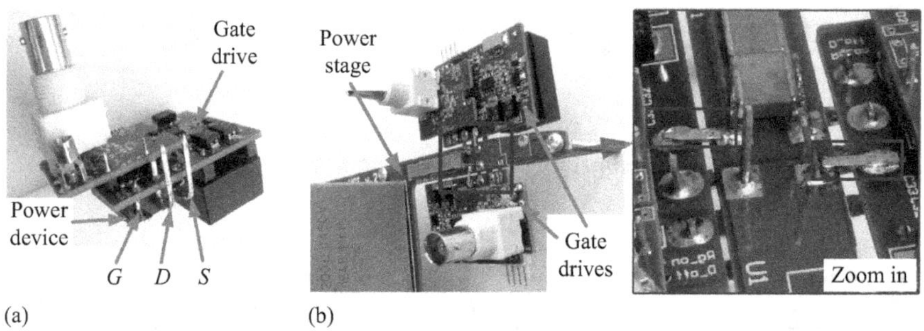

Figure 6.16 Phase-leg configuration built with the conventional placement: (a) gate drive versus power device; (b) power stage versus gate drives

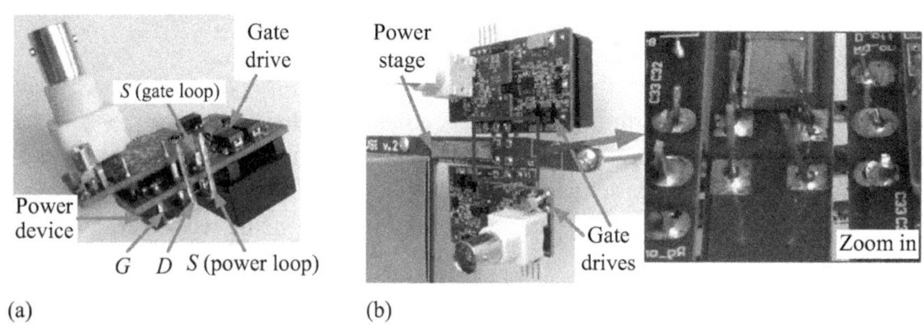

Figure 6.17 Phase-leg configuration built with the Kelvin connection-based placement: (a) gate drive versus power device; (b) power stage versus gate drives

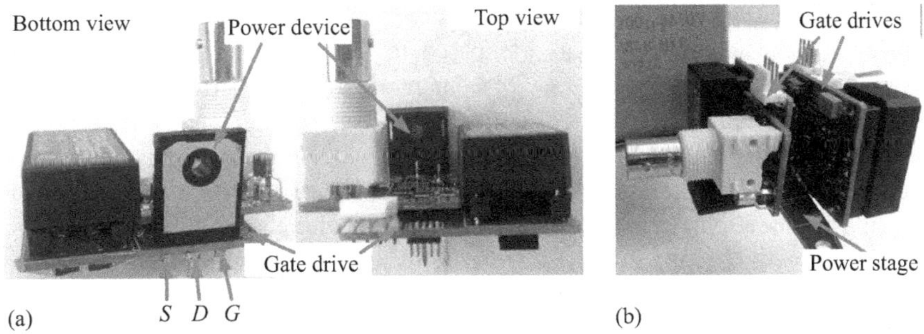

Figure 6.18 Phase-leg configuration built with the new parasitic minimization placement: (a) gate drive versus power device; (b) power stage versus gate drives

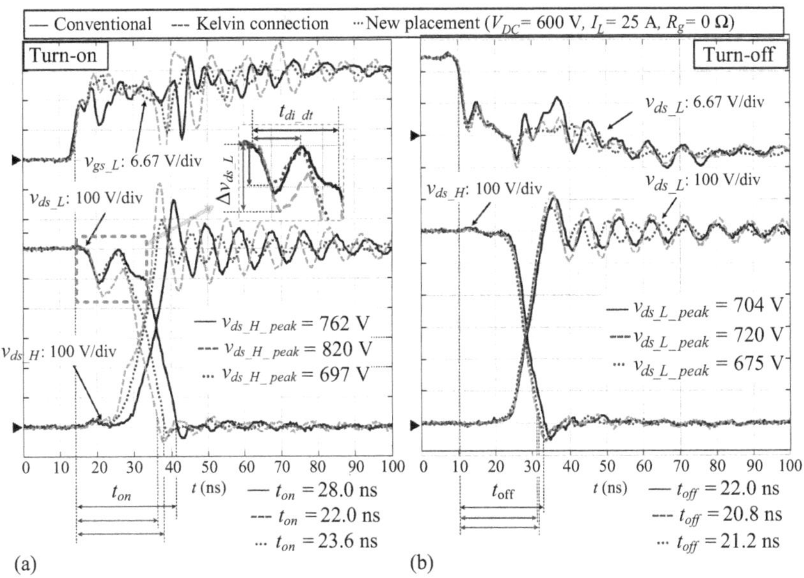

Figure 6.19 Switching waveform comparison under 600 V/25 A with 0-Ω external gate resistance: conventional and Kelvin connection-based placements versus new placement: (a) turn-on transient; (b) turn-off transient

source converters with different assemblies according to three placement styles discussed earlier. Therefore, switching behavior differences measured afterwards merely come from the different placement schemes.

Figure 6.19 shows the comparison waveforms of three different placement styles under the operating condition of 600 V/25 A with 0-Ω external gate resistance. The switching waveforms in Figure 6.19(a) show that during the turn-on transient of the lower switch, as compared to the conventional placement, the switching time by using the new placement decreases from 28.0 to 23.6 ns. Although the drain current cannot be directly measured in this hardware setup, the shorter interval from the moment when the drain-source voltage starts to drop to the moment at which the drain-source voltage falls rapidly (i.e., t_{di_dt} in Figure 6.19(a)) by using the new placement approach clearly indicates that the current rise time with the conventional placement is longer. This means that the primary contributor of the increased turn-on speed for the new approach is because of less common source inductance, which agrees with the Q3D simulation results in Table 6.3. Additionally, as compared to the Kelvin connection-based placement, it is observed that the voltage spike by using the new placement is significantly suppressed from 820 to 697 V with less severe parasitic ringing. This is due to the smaller power loop inductance achieved by shorter effective lead length involved in the switching loop.

Another interesting phenomenon observed in Figure 6.19(a) is that the turn-on time of Kelvin connection-based placement is shorter than that when the new parasitic minimization placement is adopted. The reason is because the snubber effect induced by power loop inductance during the fast *di/dt* transient makes a voltage dip (i.e., L_{loop} *di/dt*) across the drain-source terminals of the power device. Larger power loop inductance causes deeper voltage dip (see Δv_{ds_L} in Figure 6.19(a)), which means that it would be faster to let the drain-source voltage drop to zero during the subsequent *dv/dt* transient. In the end, the turn-on time by using the Kelvin connection-based placement becomes shorter.

Similar switching behaviors can be observed during the turn-off transient in Figure 6.19(b): the turn-off time with the parasitic minimization placement is shorter than that with the conventional approach; also among three placement schemes, the new parasitic minimization placement results in the lowest over-voltage with the least parasitic ringing.

More data in terms of voltage spike and switching time under different load currents during both turn-on and turn-off transients are shown in Figures 6.20 and 6.21. Additionally, Table 6.4 summarizes the highest over-voltages under the worst operating condition for three placement schemes. It shows that the new parasitic minimization placement offers the best tradeoff between the switching speed and the voltage stress among a wide operating range.

6.3.2.2 PCB layout design for trace parasitics minimization

In addition to the parasitic inductance as a result of device leads, PCB layout design is also critical. Since the impact of parasitic inductance in the power loop on the switching performance is more severe than that in the gate loop, in this section, the PCB layout design focuses on the power stage board.

As discussed in Section 6.2, there are two basic rules to follow when designing the power stage board for the parasitic minimization: one is P-cell and N-cell concept, and the other is magnetic field cancellation. By leveraging these

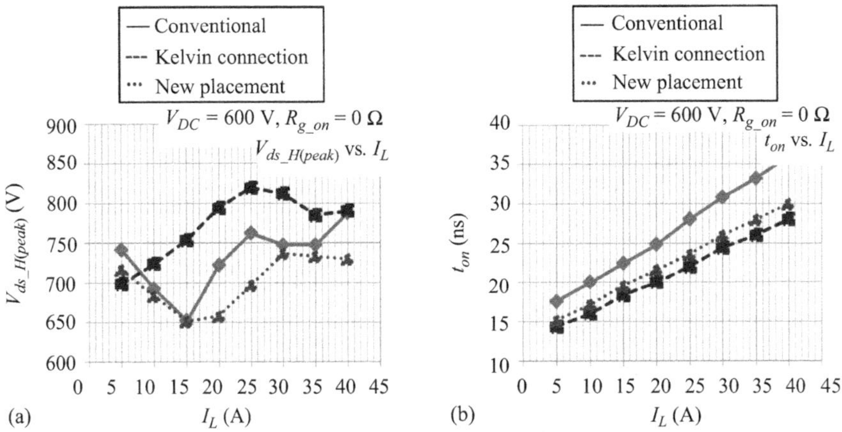

Figure 6.20 Comparison of voltage spike and switching time dependence on load current during the turn-on transient: (a) $V_{ds\_H(peak)}$ dependence on I_L; (b) t_{on} dependence on I_L

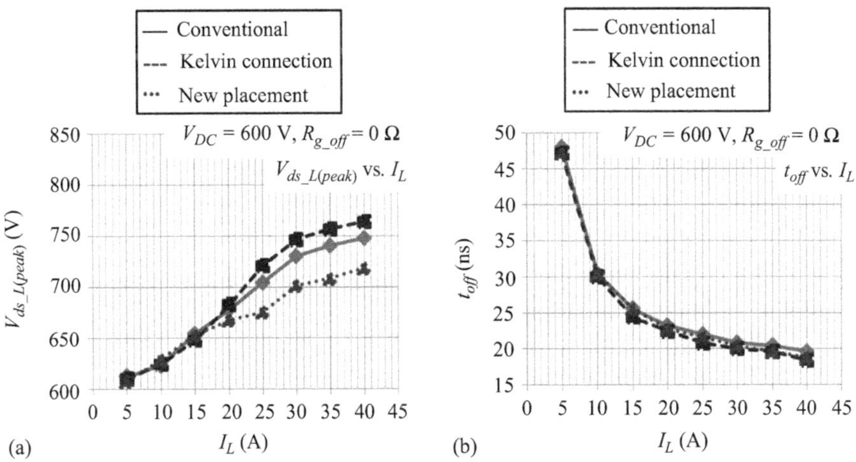

Figure 6.21 Comparison of voltage spike and switching time dependence on load current during the turn-off transient: (a) $V_{ds\_L(peak)}$ dependence on I_L; (b) t_{off} dependence on I_L

two PCB layout design principles together with the parasitic minimization placement scheme presented earlier, the interconnection design for the phase-leg configuration consisting of discrete SiC MOSFETs and Schottky diodes is developed, as shown in Figure 6.22. Figures 6.23 and 6.24 illustrate how the

Table 6.4 Summary of the highest over-voltages under the worst operating condition

Over-voltage	Conventional	Kelvin connection	New placement
Turn-on	764 V @ 600 V/5 A	821 V @ 600 V/25 A	731 V @ 600 V/5 A
Turn-off	743 V @ 40 A	768 V @ 600 V/40 A	718 V @ 600 V/40 A

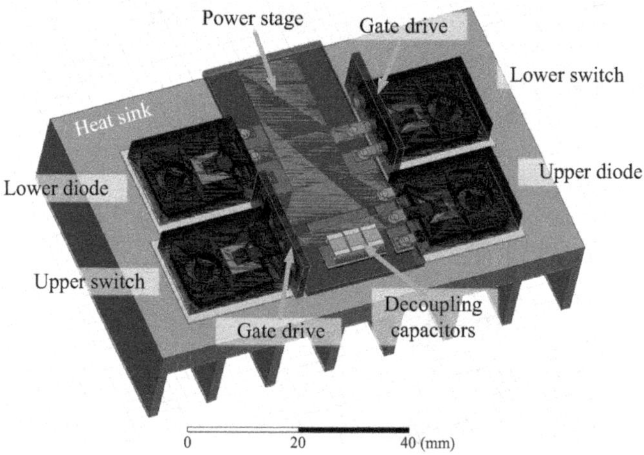

Figure 6.22 The new parasitic minimization placement and layout design in switch-diode pair-based phase-leg configuration

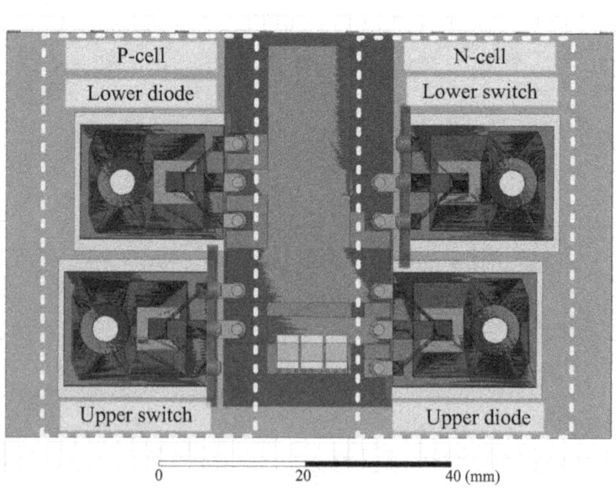

Figure 6.23 Power stage PCB layout design based on P-cell and N-cell concept

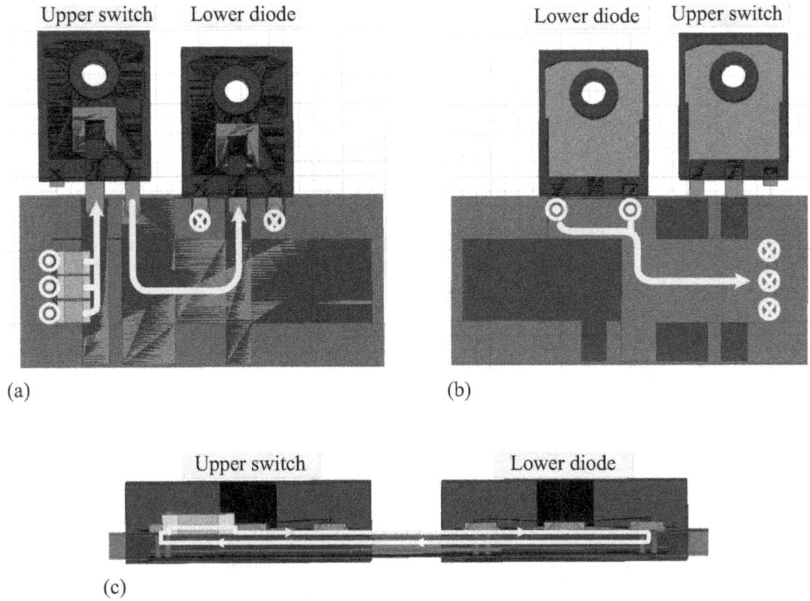

Figure 6.24 Power stage PCB layout design based on magnetic field cancellation (taking P-cell as an example): (a) top view; (b) bottom view; (c) lateral view

P-cell and N-cell concept and magnetic field cancellation are practically implemented in the PCB layout design.

First, as can be observed in Figure 6.23, the main power switching device and its complementary freewheeling diode are placed next to each other (i.e., P-cell and N-cell concept), such that the physical distance of PCB trace in the actual switching commutation loop is naturally minimized.

Second, as displayed in Figure 6.24, the PCB layout design utilizes the first inner layer as a power loop return path (Figure 6.24(b)). This return path is located directly underneath the top layer's power loop (Figure 6.24(a)), achieving the smallest physical loop size and hence near magnetic field cancellation. The lateral view (see Figure 6.24(c)) illustrates the switching current flowing path of creating a low-profile self-cancelling loop for the power loop inductance minimization.

In total, based on Q3D Extractor results, the power loop inductance due to PCB is 2.2 nH. As compared to 14.8 nH power loop inductance due to the device's package, the power stage PCB layout design is acceptable. Figure 6.25 illustrates the power loop inductance breakdown. It is observed that instead of the placement and layout design, the TO-247 package itself becomes the hurdle to further reduce the power loop inductance. In other words, the adopted placement and layout design are good enough for the switching loop parasitic minimization.

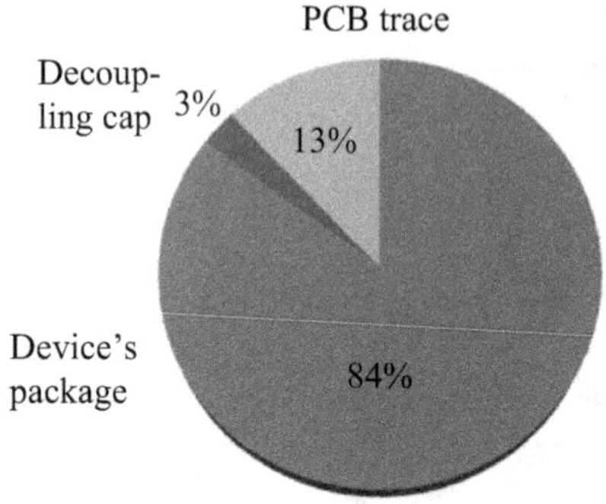

Figure 6.25 Power loop inductance breakdown

6.3.3 Case study 2: surface-mount WBG device

For surface-mount WBG devices, short leads (e.g., SiC MOSFETs with TO-268-2/TO-263-7) or leadless (e.g., chipscale or near-chipscale GaN HEMTs) package makes the minimization of effective lead length in the switching loop insignificant. Thereby, minimizing PCB layout resultant parasitics become more critical.

In general, two main schemes are utilized for the PCB layout of surface-mount devices. The two schemes are lateral loop layout design and vertical loop layout design. An example based on EPC GaN transistors in the phase-leg configuration is given next to highlight and compare these two layout concepts [19].

Figure 6.26(a) shows the PCB layout based on the lateral loop design. From the top view, it can be seen that DC capacitor and two transistors are located on the top layer and the power loop only travels the top layer. Based on the lateral loop design, the power loop area, which determines the parasitic inductance, is highly dependent on the footprint of DC capacitor and transistors as well as the clearance and creepage requirements. Usually, a relatively large parasitic inductance is introduced. But considering the single-layer-based power loop, there is no PCB trace overlap between layers, leading to negligible capacitive coupling and parasitic capacitance. Practically, if possible, the first inner layer can be utilized as a "shield layer" in which the opposite current resultant magnetic field is induced and counteracts the original power loop's magnetic field, leading to reduced power loop parasitics.

Figure 6.26(b) illustrates the PCB layout according to the vertical loop design. In this case study, DC capacitor and GaN transistors are placed on top and bottom layers separately. The power loop is oriented perpendicular to the board with vias from top to bottom. Magnetic field cancellation is achieved by the top and bottom traces themselves without the "shield layer." However, its effectiveness is determined by the PCB board thickness. Thinner board allows the smaller loop area, leading to the decreased parasitic inductance, but possibly with the penalty of higher cost of PCB fabrication. Also, the overlap of PCB traces between the top and bottom, which contributes to the better magnetic field cancellation, intensifies the capacitive coupling; as a result, excessive parasitic capacitance may be introduced.

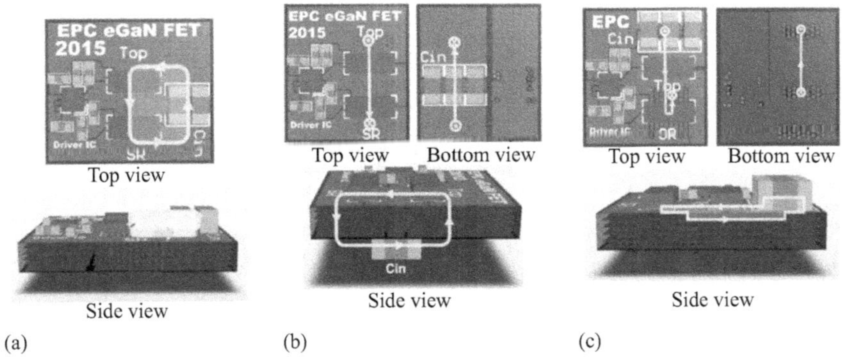

Figure 6.26 Power loop layout design with EPC GaN transistors in the phase-leg configuration [19]: (a) lateral power loop; (b) vertical power loop I; (c) vertical power loop II. © [2018] IEEE. Reprinted with permission from [19]

To further enhance the magnetic field self-cancellation, another vertical loop design-based PCB layout can be used. As shown in Figure 6.26(c), DC capacitor and two transistors are located on the top layer while the power loop travels perpendicular to the board with vias from the top to the first inner layer. With this, the smallest physical loop area with magnetic field self-cancellation leads to the smallest parasitic inductance. On the other hand, the capacitive coupling becomes more severe with the penalty of externally introducing additional equivalent junction capacitance.

In summary, vertical loop design generally provides better magnetic field cancellation without dedicated "shield layer," leading to smaller parasitic inductance but larger parasitic capacitance, as compared to the lateral loop design. However, considering the primary guideline for the DPT layout, that is, the layout design for device dynamic characterization should be easily leveraged for the future converter design, layout optimization for parasitic minimization has to keep the thermal management into account. This is especially significant for surface-mount WBG devices where the electrical and thermal interfaces are always coupled.

Taking the bottom-side cooled GaN HEMT from GaN Systems as an example, the source terminal for the power loop (i.e., pin 4 in Figure 6.27) is also the thermal pad for the heat dissipation [20]. The typical thermal management for surface-mount WBG devices is shown in Figure 6.28 [21]. Heat generated from the device die primarily flows down to the thermal pad and then transfers to the PCB. Within the PCB, the heat first spreads through the copper plane on the PCB top layer, then travels from the top copper to the bottom side of the PCB through the low thermal resistance path provided by thermal vias or other similar approaches (e.g., metal PCB). A heatsink is attached to the bottom copper plane via the thermal interface material and finally dissipates the heat to the ambient. Therefore, the existence of the vertical heat dissipation path (e.g., thermal vias) limits the electrical layout

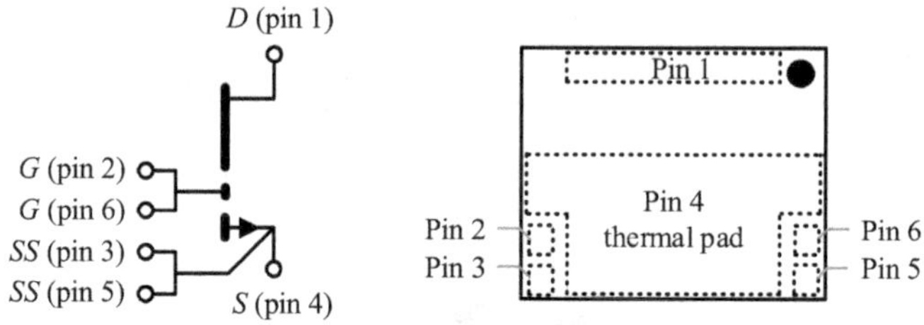

Figure 6.27 Circuit symbol and package outline of GaN Systems GS66516B bottom-side cooled GaN transistor [20]

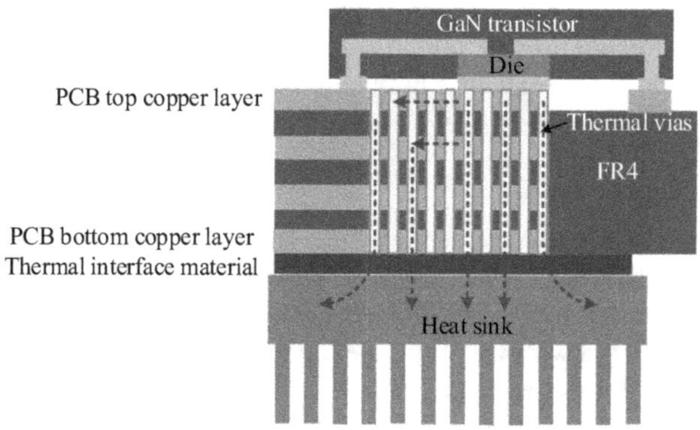

Figure 6.28 Typical thermal management for surface-mount WBG devices (taking bottom-side cooled devices from GaN Systems as an example) [21]

design of the PCB. Specifically, the vertical loop design cannot be utilized since the power loop return path is blocked due to the thermal path. Similarly, for the lateral loop design, the integration of the "shield layer" has to be sacrificed. In the end, the effectiveness of the magnetic field cancellation is adversely affected. Figure 6.29 illustrates a typical layout design based on the lateral loop concept with the consideration of the thermal management system [22]. Generally, this layout will introduce more parasitic inductance.

The conflict between thermal and electrical optimizations can be resolved thanks to the advent of the top-side cooled devices. Based on the package, the thermal pad is located on the top of the device while the electrical pins are all on the bottom side. As a result, thermal and electrical interfaces can be fully decoupled, leading to the minimization of the thermal impedance and parasitic inductance simultaneously. Figure 6.30 illustrates an example of the PCB layout design for the top-side cooled WBG devices considering the thermal management [22]. However, the assembly of heatsink is more challenging: traditional heatsink can be utilized if

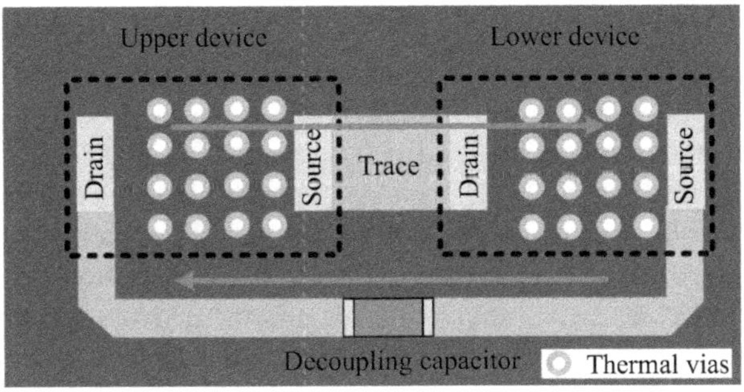

Figure 6.29 PCB layout design of bottom-side cooled WBG devices considering the thermal management [22]

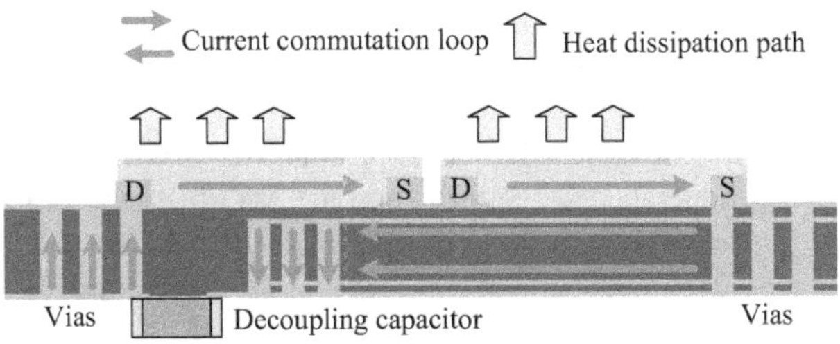

Figure 6.30 PCB layout design of top-side cooled WBG devices considering the thermal management [22]

Table 6.5 Summary of key characteristics dependence on different PCB layout design

Key characteristics	Lateral loop design	Vertical loop design
Minimized parasitic inductance	No	Yes
Minimized parasitic capacitance	Yes	No
Preferred for bottom-side cooled WBG devices	Yes	No
Preferred for top-side cooled WBG devices	No	Yes

no component on the top layer close to the devices is taller than the device itself; otherwise, a custom heatsink has to be designed.

Table 6.5 highlights key characteristics and their dependence on different PCB layout designs. In summary, taking the thermal management into account, for bottom-side cooled WBG devices, lateral loop design has to be employed; while with the top-side cooled WBG devices, vertical loop design is preferred.

6.3.4 With consideration of current measurement in DPT

In DPT, it is necessary to insert switching current measurement into the switching loop, which will affect the power loop layout. For the purpose of introducing the layout design with consideration of current measurement in DPT, the coaxial shunt (widely adopted for WBG DPT) is presented here, while the detailed consideration for switching current measurement will be discussed in Chapter 8. As illustrated in Figure 6.31, the coaxial shunt is a high bandwidth sense resistor installed between the source of the lower switch in a phase-leg and the negative terminal of the DC source. The main impact of the coaxial shunt on the PCB layout is its resultant insertion inductance due to (1) coaxial shunt itself and (2) the extra PCB traces for the interconnection of the shunt. As a user, we have to accept the insertion inductance introduced by the coaxial shunt body and focus on the optimization of coaxial shunt placement and PCB layout to minimize the parasitics generated by the additional connecting PCB traces.

Figure 6.32 illustrates the PCB layout based on the lateral power loop design. With the coaxial shunt included, two options can be utilized: one, as shown in Figure 6.32(b), is to place the coaxial shunt following the lateral loop. This approach is straightforward but will inevitably introduce larger power loop area and increased parasitic inductance. The other approach, as illustrated in Figure 6.32(c), is to create a local vertical loop between the lower switch and the coaxial shunt. With this, the magnetic field self-cancellation is able to minimize the parasitic inductance added by extra PCB traces for the coaxial shunt connection. As a

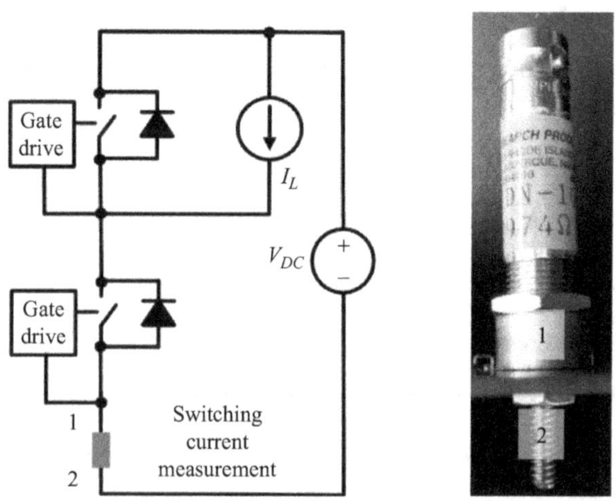

(a) (b)

Figure 6.31 Switching current measurement and physical implementation in DPT: (a) switching current measurement in DPT; (b) coaxial shunt mounted on DPT board

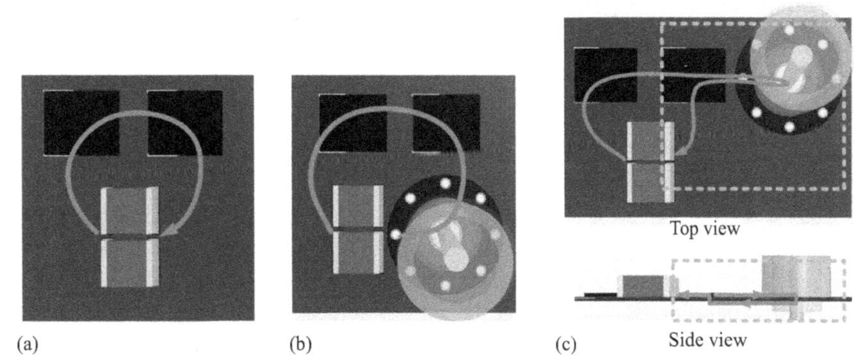

Figure 6.32 Lateral power loop with consideration of current measurement: (a) lateral power loop w/o coaxial shunt; (b) lateral power loop w/ coaxial shunt based on lateral power loop layout; (c) lateral power loop w/ coaxial shunt based on vertical power loop layout

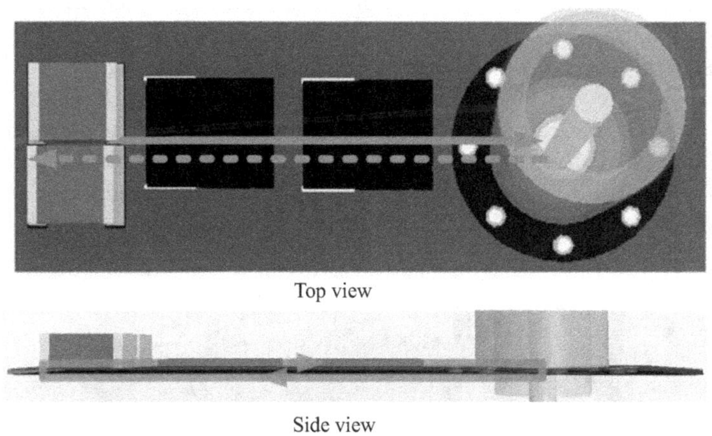

Figure 6.33 Vertical power loop with the consideration of current measurement

result, the adverse impact of the coaxial shunt and its resultant insertion inductance are limited.

For the PCB layout based on the vertical power loop design, the coaxial shunt can be seamlessly integrated with the existing vertical power loop with a slight extension of the power loop. As shown in Figure 6.33, although the power loop length increases due to the coaxial shunt, the limited loop area results in a small power loop parasitic inductance. However, in this case, the contribution of the shunt's own internal inductance may still be significant.

In total, it is preferred to employ the vertical loop layout concept for the placement of coaxial shunt along with its corresponding PCB layout, no matter whether the main power loop layout adopts the lateral loop layout concept or vertical loop layout concept.

6.3.5 Gate drive

As discussed in Chapter 5, the gate drive is the link between the world of control and the world of power in a system. Good isolation is critical to avoid interference between the power stage and signal stage for WBG devices with high-speed switching capability. In addition to the proper selection of isolation components, for example, sufficiently high CMTI capability and low coupling capacitance of signal isolator and isolated power supply of the gate drive, discussed in Chapter 5, the gate drive layout for these insolation components is also crucial. Therefore, the primary guideline for the gate drive layout is to mitigate the parasitics coupled between power stage and signal stage as much as possible. Practically, it is suggested to introduce a clear gap between the primary and secondary sides of isolation components, as illustrated in an example of the gate drive PCB layout for WBG devices in Figure 6.34.

The other consideration of the gate drive layout is to design a dense gate loop for the minimization of the gate loop parasitic inductance. It is noteworthy that a completed gate loop consists of gate drive decoupling capacitor, gate drive IC/buffer circuit, gate resistor, and device. Therefore, it is not sufficient to merely place the gate drive IC close to the device without considering the location of the decoupling capacitor. As shown in Figure 6.35, there are two separated gate loops: one for turn-on, and the other for turn-off. Similar to the power loop layout design principle, the magnetic field cancellation concept is still valid and

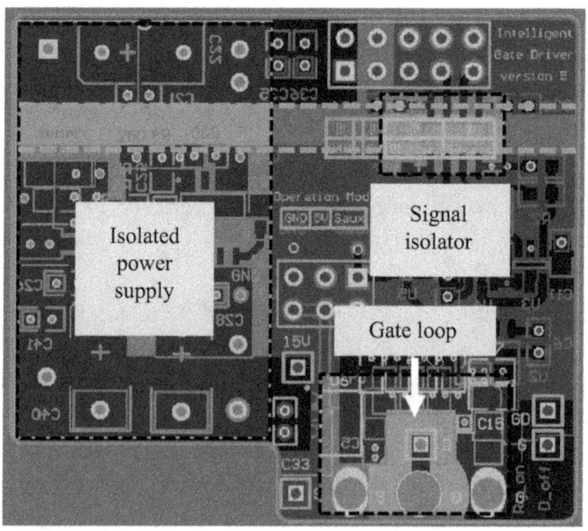

Figure 6.34 Consideration of gate drive layout design

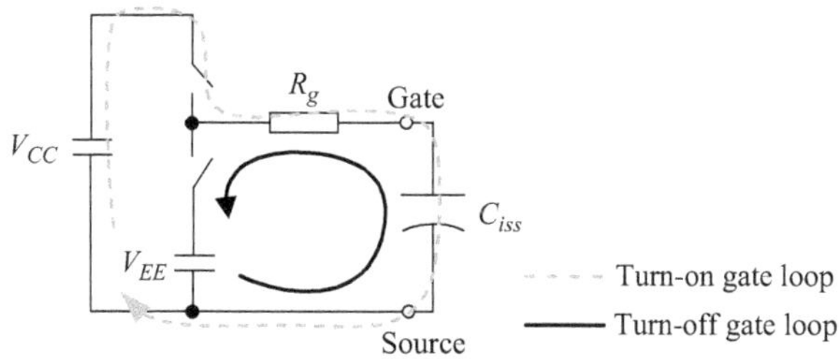

Figure 6.35 Gate loop layout for gate parasitic inductance minimization

important for the gate loop layout. Turn-on and turn-off gate loops have to be considered respectively to minimize the gate loop parasitics during both turn-on and turn-off transients.

6.4 Summary

1. The parasitic ringing due to fast switching speed is more sensitive to the main switching loop inductance than the gate loop inductance. The common source inductance and Miller capacitance work as negative feedback from the switching loop to the gate loop. Power loop parasitics severely impact the switching commutation and parasitic ringing.
2. The purpose of a DPT is to experimentally gather the current and voltage waveforms of the power devices to evaluate their switching behaviors in an actual power converter. Primary guideline for the DPT layout is the similarity between the DPT and the intended converter. Second, parasitics minimization should be considered.
3. Physically, parasitics are contributed by equivalent series inductance of decoupling capacitor, PCB trace, and bonding wire (or other bonding interface materials) and leads from package. Two aspects should be focused on for power stage design, including minimization of the effective length of the leads in the switching loop, if applicable (e.g., TO-series package), as well as the minimization of the PCB trace resultant parasitics.
4. For surface-mount WBG devices, usually there is a tradeoff between electrical and thermal optimization. Generally speaking, for bottom-side cooled devices, lateral power loop layout is suitable for thermal management (i.e., thermal vias beneath the thermal pad of the device) but with the penalty of increased power loop parasitics, while the vertical power loop layout is able to achieve small power loop parasitics but with the challenge of cooling system design. This issue can be mitigated if top-side cooled device is employed. But the assembly of heatsink is more challenging.
5. In DPT, it is necessary to insert a dynamic current measurement, typically coaxial shunt, into the switching loop. The main impact of the coaxial shunt on the PCB layout is its resultant insertion inductance due to (1) coaxial shunt

itself and (2) the extra PCB traces for the interconnection of coaxial shunt. As a user, we have to accept the insertion inductance introduced by the coaxial shunt body and focus on the optimization of coaxial shunt placement and PCB layout to minimize the parasitics generated by the additional connecting PCB traces. It is preferred to employ the vertical loop layout concept for the placement of coaxial shunt.

6. Two aspects should be focused on for gate drive layout design: (1) gap between primary and secondary side of isolation components for the mitigation of coupling capacitance and (2) dense gate loop layout for gate loop parasitic minimization. Also note that gate loop is not from the gate drive IC to the device, but from the gate drive decoupling capacitor to gate drive IC and then the device. Magnetic field cancellation concept can be used to minimize the gate loop parasitics.

References

[1] H. Zhu, A. R. Hefner, and J. S. Lai, "Characterization of power electronics system interconnect parasitics using time domain reflectometry," *IEEE Transactions on Power Electronics*, vol. 14, pp. 622–628, 1999.

[2] F. Merienne, J. Roudet, and J. L. Schanen, "Switching disturbance due to source inductance for a power MOSFET: analysis and solutions," in *27th Annual IEEE Power Electronics Specialists Conference*, Baveno, Italy, vol. 2, 1996, pp. 1743–1747.

[3] W. Teulings, J. L. Schanen, and J. Roudet, "MOSFET switching behaviour under influence of PCB stray inductance," in *Thirty-First IAS Annual Meeting Industry Applications Conference*, San Diego, CA, USA, vol. 3, 1996, pp. 1449–1453.

[4] Y. Xiao, H. Shah, T. P. Chow, and R. J. Gutmann, "Analytical modeling and experimental evaluation of interconnect parasitic inductance on MOSFET switching characteristics," in *Nineteenth Annual IEEE Applied Power Electronics Conference and Exposition*, Anaheim, CA, USA, 2004, pp. 516–521.

[5] Z. Chen, D. Boroyevich, and R. Burgos, "Experimental parametric study of the parasitic inductance influence on MOSFET switching characteristics,' in *Power Electronics Conference (IPEC), 2010 International*, Sapporo, Japan, 2010, pp. 164–169.

[6] Z. Chen, D. Boroyevich, and J. Li, "Behavioral comparison of Si and SiC power MOSFETs for high-frequency applications,' in *Applied Power Electronics Conference and Exposition (APEC)*, Long Beach, CA, USA, 2013, pp. 2453–2460.

[7] Z. Chen, "Characterization and modeling of high-switching-speed behavior of SiC active devices," M.S. Thesis, Electrical Engineering, Virginia Polytechnic Institute and State University, Blacksburg, 2009.

[8] http://epc-co.com/epc/.

[9] A. Lemmon, M. Mazzola, J. Gafford, and C. Parker, "Stability considerations for silicon carbide field-effect transistors," *IEEE Transactions on Power Electronics*, vol. 28, pp. 4453–4459, 2013.

[10] Rohm, SiC application note: SiC power devices and modules [Online]. Available: www.rohm.com/documents/11303/2861707/sic_app-note.pdf. Accessed in August 2014.
[11] R. W. Erickson and D. Maksimovic, *Fundamentals of Power Electronics*, Springer, Boston, MA, 2001.
[12] D. Grant, *Power MOSFETs: Theory and Applications*, Wiley, New York, 1989.
[13] L. Balogh, "Design and application guide for high speed MOSFET gate drive circuits," in *Power Supply Design Seminar*, 2001.
[14] J. Cain. Parasitic inductance of multilayer ceramic capacitors [Online]. Available: http://www.avx.com/.
[15] Datasheet: CCS020M12CM2 [Online]. Available: http://www.cree.com/.
[16] S. Li, L. M. Tolbert, F. Wang, and F. Z. Peng, "Stray inductance reduction of commutation loop in the P-cell and N-cell-based IGBT phase leg module," *IEEE Transactions on Power Electronics*, vol. 29, no. 7, pp. 3616–3624, July 2014.
[17] C. DiMarino, W. Zhang, N. Haryani, Q. Wang, R. Burgos, and D. Boroyevich, "A high-density, high-efficiency 1.2 kV SiC MOSFET module and gate drive circuit," in *IEEE Workshop on Wide Bandgap Power Devices and Applications*, 2016.
[18] B. Callanan. CREE CPWR AN-08. Application considerations for silicon carbide MOSFETs [Online]. Available: http://www.cree.com.
[19] D. Reusch and J. Strydom, "Understanding the effect of PCB layout on circuit performance in a high frequency gallium nitride based point of load converter," *IEEE Transactions on Power Electronics*, vol. 29, no. 4, pp. 2008–2015, 2014.
[20] GaN Systems GS66516B Datasheet. 2017 [Online]. Available: http://www.gansystems.com/gs66516b.php.
[21] GaN Systems, Application note: PCB thermal design guide for GaN enhancement mode power transistors, 2016.
[22] H. Li, X. Li, Z. Zhang, C. Yao, and J. Wang, "Design consideration of high power GaN inverter," *IEEE Workshop on Wide Bandgap Power Devices and Applications (WiPDA)*, Fayetteville, AR, 2016, pp. 23–29.

Chapter 7
Protection design for double pulse test

Today, WBG power devices are often expensive or sampled in limited quantities. In addition, intrinsic characteristics of WBG semiconductors enable WBG devices with very high voltage capability, for example, 10 kV SiC MOSFETs. Thus, due to the limited availability, high cost, and unique capability of devices, proper protection for WBG device DPT is important. The protection includes two categories: one is to protect DUT and DPT circuit, and the other is for the operator safety, especially when testing high-voltage WBG devices.

Figure 7.1 illustrates two potential failure modes of DUT in DPT. One is the shoot-through failure in switch/switch configuration once the upper switch accidentally turns on while the lower switch is on. In practice, this may be caused by a mistake in the double-pulse generation, or a false trigger of the gate drive of the upper switch due to noise during the fast switching transient of the lower switch. The other failure mode is load inductor saturation as a result of a long turn-on duration of the lower switch, due to the false generation of the double-pulse signal. Thereby, the impedance of the load inductor significantly reduces, leading to over-current through the lower switch and damage to the lower switch. This case can exist in both switch/switch and switch/diode configurations. In summary, short-circuit and over-current protection is critical for WBG-based dynamic characterization.

It should be noted that WBG devices and DPT circuit could have other fault conditions in addition to short-circuit and over-current, such as over-voltage and ground fault. However, these faults are less common during DPT. As a result, the focus of this chapter is on short-circuit and over-current fault detection and protection.

In addition, the special considerations of high-voltage WBG device DPT, including the safety consideration and protection schemes, are discussed with preferred practice summarized based on a 10 kV SiC MOSFET DPT.

7.1 Overview of state-of-the-art protection schemes for WBG devices

The requirement of fast short-circuit and over-current protection for WBG devices has been shown to be much stricter than conventional Si devices. This can be explained by the much smaller die size of WBG devices, resulting in reduced thermal capacitance, and the potential for a more rapid spike in junction

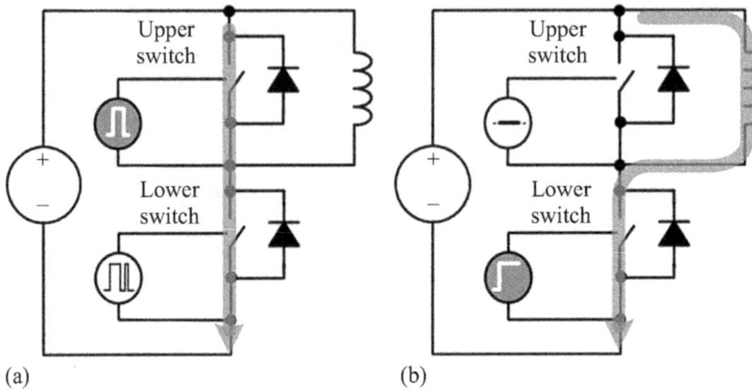

Figure 7.1 Potential failure modes for DPT: (a) shoot-through failure; (b) load inductor saturation

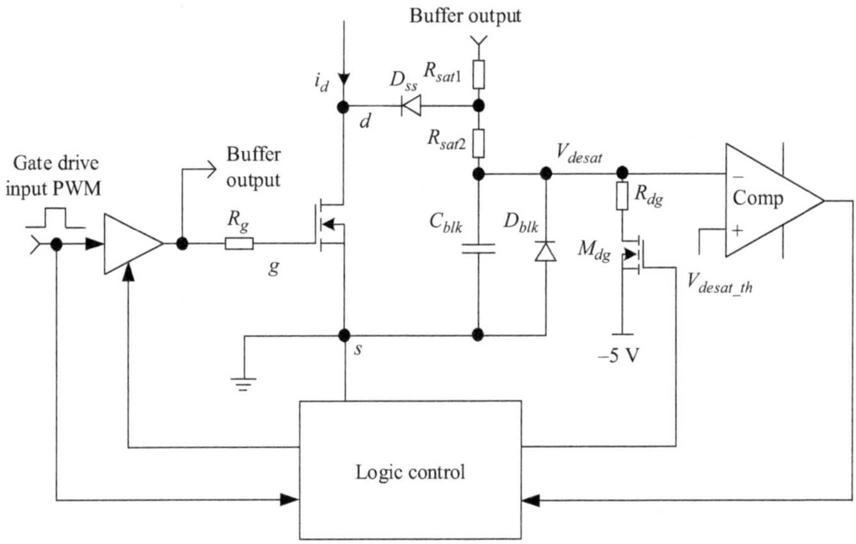

Figure 7.2 Schematic of a desaturation overcurrent protection scheme [2]

temperature [1, 2]. In comparison with the Si devices, which generally guarantee at least 10 μs short-circuit withstand time (SCWT), 1 μs is the typical time for SiC devices [3]. The situation is even worse for GaN transistors. Recent literature has shown that a healthy GaN HEMT can be destroyed by a short-circuit pulse as brief as 400–600 ns at 400 V [4–12], and in some cases even only 200 ns [13–15]. This short-circuit capability increases significantly with lower bus voltages as a result of lower leakage current and loss after the device is turned off. In any case, it is desirable to turn off the devices as quickly as possible under short-circuit conditions, as sustained short-circuit conditions will likely lead to over-current and over-temperature, causing device degradation.

Perhaps the most common short-circuit protection scheme is desaturation protection, as shown in Figure 7.2 [2]. Desaturation protection requires sensing of the power terminal voltage (drain-source voltage for a MOSFET or collector-emitter voltage for an IGBT). It triggers the gate drive to turn off, when this voltage indicates that the device has left its normal working region (linear region for a MOSFET or saturation region for a bipolar device like IGBT), and entered the abnormal high current region (saturation region for an MOSFET or active region for a bipolar device). In fact, the desaturation protection gets its name from its use in bipolar devices. However, WBG devices covered in this book mainly refer to the majority carrier devices like SiC MOSFET and GaN HEMT. The corresponding terminologies will be used when discussing desaturation and other protection schemes.

Once properly designed, desaturation protection has been shown to function well for both SiC and GaN [2, 16–18] with a total response time of $\sim$200 ns. Desaturation protection scheme has several limitations and shortcomings, especially with regard to its potential use in WBG DPT protection. First, a diode must be used to protect the sensing circuit from the much higher drain-source voltage, when the device is in off state, which reduces the response time and adds capacitive loading to the output capacitance of the device. Second, the over-current threshold is set by the saturation current level corresponding to the selected gate drive voltage and cannot be selected independently. Third, with the given gate voltage, the saturation current is a strong function of junction temperature, and in general, for GaN HEMT, it drops as temperature increases while the opposite trend can be observed for SiC MOSFET. As a result, a fixed setting of the desaturation protection cannot predictably respond to over-current faults of WBG devices in a wide range of operating temperatures. More importantly, desaturation protection relies on having sufficient knowledge on the device to be protected, which is a challenge for DPT.

Figure 7.3 shows another over-current protection scheme, which was developed for Si IGBTs and also applied to SiC MOSFETs [2, 19]. This scheme is based on sensing the voltage between the power source and Kelvin source terminals of the device. The voltage across the parasitic inductance L_s between them represents $L_s di_d/dt$, where i_d is the device drain current. By passing the voltage signal through an analog integrator, i_d can then be sensed by a comparator for over-current fault detection. This scheme relies on a sufficient parasitic inductance existing between the source and Kelvin source terminals to enhance the sensitivity. With near-chipscale GaN FETs or low parasitic packaging based SiC devices and modules, the inductance may not be adequate to implement this scheme. This issue can be resolved by using an external Rogowski coil to pick up the switching loop *di/dt* with the controllable coil inductance to guarantee the measurement sensitivity. Usually, a custom pick-up coil integrated with PCB is preferred to fit with the package of WBG devices, and in the meantime, to improve the density of the whole measurement system. This concept has been utilized for SiC power module and GaN transistors [20–22] for high-speed dynamic current measurement, and short-circuit and over-current protection. The limitation of this approach is that the custom pick-up coil design depends on the footprint of the specific device; as a result, the generality of the approach is lost.

A high bandwidth, small footprint switching current sensor can also be used in over-current detection and protection scheme for WBG devices, which practically

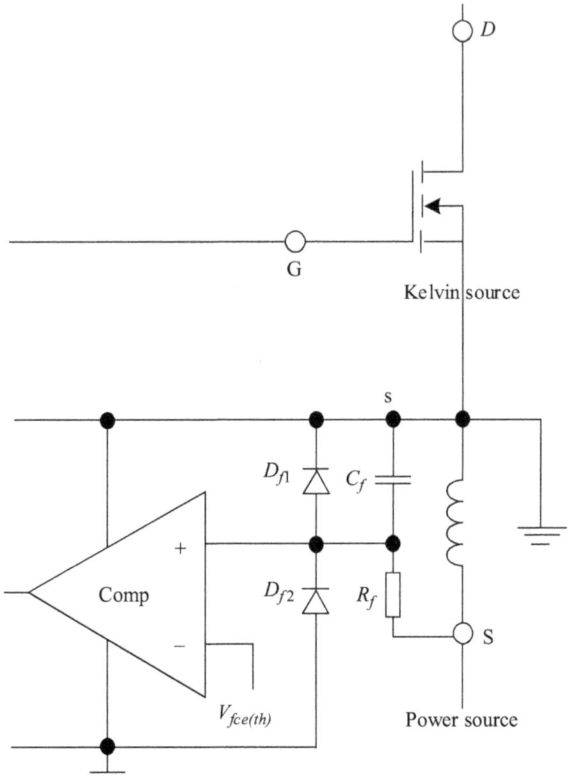

Figure 7.3 Schematic of a di/dt integration overcurrent protection scheme [2]

can be implemented by the shunt resistance or Si sensing MOSFET embedded with WBG devices at the die or package level [23, 24]. One approach using a monolithically embedded shunt resistance to measure current for GaN FETs was proposed in [24]. This solution obviously requires the fabrication of custom semiconductor dies, which is not applicable in many cases and may also be costly. Similar solutions have been used in the past with discrete shunt resistors. The challenge for implementing this scheme includes extra losses, increased switching loop inductance, isolation concerns, and noise immunity issues [1].

7.2 Solid-state circuit breaker

Several over-current detection and protection schemes discussed in the previous section either require prior knowledge of the WBG devices under test, or their implementation will interfere with the device switching performance, or both. Considering the limited knowledge of DUT before the dynamic characterization, it is preferable to employ a universal protection method, which is independent of the specific characteristics and package of the DUT. Therefore, in this section, solid-state circuit breaker (SSCB), as a practical and universal solution, is introduced to

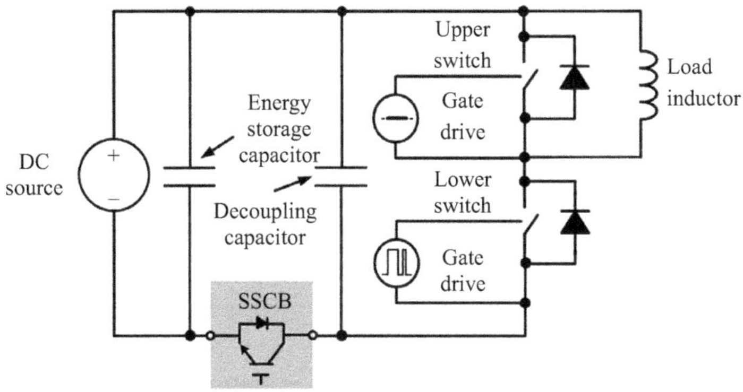

Figure 7.4 DPT circuit with protection board highlighted

protect the DUT and DPT circuit against short-circuit/over-current faults. The SSCB operation principle, circuit implementation, and design consideration, as well as a design example, are presented.

SSCB provides a universal protection method, even without knowing the specific characteristics of WBG devices under test [2, 25]. As illustrated in Figure 7.4, the normally-on SSCB is inserted between the bulk energy storage capacitor and the rest of the DPT circuit. Under the normal operating condition, the SSCB stays on. In the case of short-circuit and over-current faults, the SSCB will be tripped to isolate the DPT from the energy source, including the bulk energy storage capacitor and DC source. As a result, the DPT is protected. For WBG-based DPT, the unique challenge for SSCB design is the ultra-fast response time required, since generally speaking the short-circuit withstand capability of WBG devices is limited. Also, the high noise immunity capability is critical due to the high-speed switching.

As can be observed in Figure 7.4, a decoupling capacitor is placed after the SSCB and close to DUTs. This can help minimize power loop parasitics, especially for the extra insertion inductance introduced by SSCB. The drawback of this arrangement is that the energy stored in the decoupling capacitor will be dissipated through DUTs under the fault condition. Usually, the decoupling capacitance is small (10s–100s nF), thus the stored energy should be much lower than the short-circuit critical energy of DUTs. But it is still worth noticing that the selection of decoupling capacitance just needs to follow the design criteria described in Chapter 4. Oversizing of decoupling capacitance does not offer additional benefits for voltage spike minimization during the fast switching, while it may worsen the survivability of DUTs under the fault condition.

7.2.1 Operation principle

SSCB operation relies on the well-known desaturation protection of Si IGBTs. The basic idea is to utilize the output characteristic of the IGBT to detect the short-circuit and over-current event, and then clear the fault by turning off the IGBT.

Figure 7.5 illustrates a typical output characteristic (v_{CE}–i_C curve) of IGBT dependence on gate-emitter voltage v_{GE}. Under the normal operating point (taking

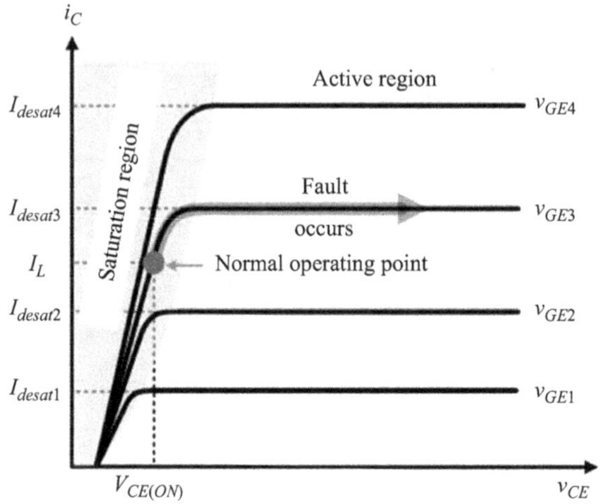

Figure 7.5 Typical output characteristics (V–I curve) of IGBT under different gate voltage

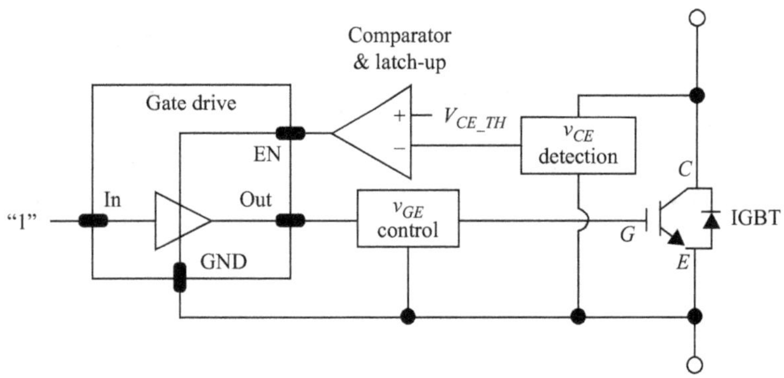

Figure 7.6 Normally on SSCB block diagram

v_{GE3} in Figure 7.5 as an example), the IGBT turns on and operates in the saturation region, where with the given load current I_L, on-state collector-emitter voltage $V_{CE(on)}$ is low. Under the fault condition, the short-circuit current forces the IGBT to enter the active region from the saturation region and is clamped to the desaturation current I_{desat}, which depends on v_{GE}. In the meantime, collector–emitter voltage v_{CE} quickly rises. Therefore, the protection current is limited to I_{desat} and can be tuned by v_{GE} according to the IGBT *V–I* curve. Also, v_{CE} is a highly sensitive indicator for over-current detection and can be used to trip the short-circuit fault.

According to the aforementioned principle, Figure 7.6 illustrates the block diagram of a normally-on SSCB for a DPT, primarily including gate drive, v_{GE} control unit, v_{CE} detection unit, and comparator and latch-up unit. Under the normal operating condition, the gate drive logic input is always controlled to be high to keep the

IGBT on. Proper v_{GE} is set according to the desired protection current by the v_{GE} control unit. v_{CE} is online monitored by the v_{CE} detection unit and compared with the preset v_{CE} protection threshold V_{CETH}, which is practically set to enable the desaturation protection under the fault condition, when v_{CE} rises to the boundary between the saturation region and the active region of the IGBT. The output of the comparator and latch-up circuit is connected with the Enable/Disable (EN) pin of the gate drive to turn off the IGBT in case of a fault. Also, some of the commercially available gate drive integrated circuits (ICs) have an embedded comparator and latch-up unit to simplify the needed peripheral circuits for the desaturation protection.

7.2.2 Circuit implementation and design consideration

Figure 7.7 gives an example circuit of a SSCB based on an Infineon IR2127 gate drive IC. Since IR2127 integrates comparator and latch-up circuit, SSCB circuit consists of three functional blocks: gate drive, v_{GE} control unit, and v_{CE} detection unit. Detailed circuit implementation and design consideration are discussed in the following sections.

7.2.2.1 Gate drive IC

Electrical connection of the IR2127 gate drive IC is illustrated in Figure 7.7. "VCC" and "Vb" are power supply pins for gate drive IC and desaturation circuit. "COM" and "Vs" pins are connected with IGBT emitter terminal as the ground of the gate drive IC and desaturation circuit. As the logic input, "IN" pin is always set to be high. Practically, it can be connected with "VCC" and "Vb" pins. "Fault" pin reports the status of desaturation protection. "HO" is the gate drive output pin, in this specific case, always outputting supply voltage V_{CC} under the normal condition. "CS" pin is the v_{CE} sense input, and it is also connected to a built-in comparator and latch-up circuit to shut down the gate drive output "HO" when the input voltage of "CS" pin exceeds its preset threshold.

7.2.2.2 v_{GE} control unit

The v_{GE} control unit consists of a resistor divider R_{GE1} and R_{GE2}, turn-off gate resistor R_{g_off} and diode D_{g_off}, and Zener diode D_Z, as shown in Figure 7.7.

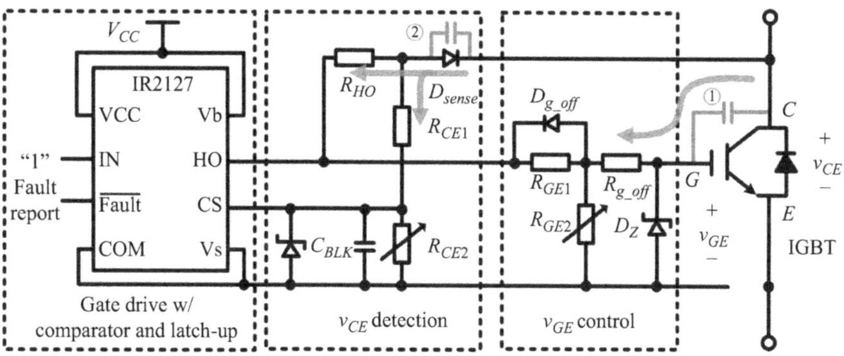

Figure 7.7 SSCB circuit

Specifically, resistor divider R_{GE1} and R_{GE2} is responsible for tuning the actual gate voltage v_{GE} of the IGBT, based on:

$$v_{GE} = \frac{R_{GE2}}{R_{GE1} + R_{GE2}} V_{CC}. \tag{7.1}$$

Practically, a trimmer is used for either R_{GE1} or R_{GE2} to control v_{GE} conveniently. IGBT input capacitance is the load of the resistor divider, and its resultant loading effect in the steady state is negligible. Then R_{GE1} and R_{GE2} with large resistances can be selected (usually in a range of 100s–1000s Ω) to minimize the power loss dissipated by R_{GE1} and R_{GE2} under the normal condition.

Due to the use of relatively large R_{GE1}, a simple turn-off gate resistor network consisting of R_{g_off} and D_{g_off} is introduced to reduce the turn-off gate resistance and increase the turn-off speed under the fault condition. In this network, antiparalleled D_{g_off} bypasses R_{GE1}, allowing R_{g_off} to dominate the turn-off gate impedance. Also note R_{g_off} cannot be extremely small; otherwise, a collector-emitter voltage spike may damage the IGBT during the fast turn-off transient after the shoot-through current is cleared. Practically, 10s Ω of R_{g_off} is preferred.

Additionally, under the fault condition, IGBT v_{CE} rises rapidly due to first the short-circuit current, and then the turn-off of the IGBT. This high dv_{CE}/dt introduces displacement current via gate-collector capacitance (i.e., Miller capacitance) of the IGBT flowing through the gate drive, leading to a spurious positive gate voltage higher than v_{GE} under the normal condition. In the end, the shoot-through current (i.e., saturation current of the IGBT) becomes larger. To mitigate the impact of this Miller effect-induced noise current on the gate voltage, a Zener diode D_Z is introduced across the gate-emitter of the IGBT. The breakdown voltage of the selected D_Z only needs to be slightly higher than v_{GE} under the normal condition. Also, D_Z should physically be placed as close to gate-emitter terminals of the IGBT as possible. Another design consideration for spurious positive gate voltage mitigation is to select an IGBT with small internal gate resistance. Since D_Z can only clamp the voltage across the external gate-emitter terminals of the IGBT, the smaller internal gate resistance allows the internal and external v_{GE} to be closer.

7.2.2.3 v_{CE} detection unit

The v_{CE} detection unit consists of the sensing diode D_{SAT}, charging resistor R_{CHG}, v_{CE}-related resistor divider R_{CE1} and R_{CE2}, and blanking capacitor C_{BLK}, as shown in Figure 7.7.

Specifically, under the normal condition when IGBT is on or under the fault condition before IGBT is turned off, as illustrated in Figure 7.8(a), the gate drive "HO" pin outputs V_{CC}. It offers the charging current via R_{CHG}, making D_{SAT} forward biased. As a result, the voltage across the resistor divider v_{RD} follows the voltage forward drop of the sensing diode v_{f_DSAT} plus v_{CE}, yielding:

$$v_{RD} = v_{f\_DSAT} + v_{CE} \tag{7.2}$$

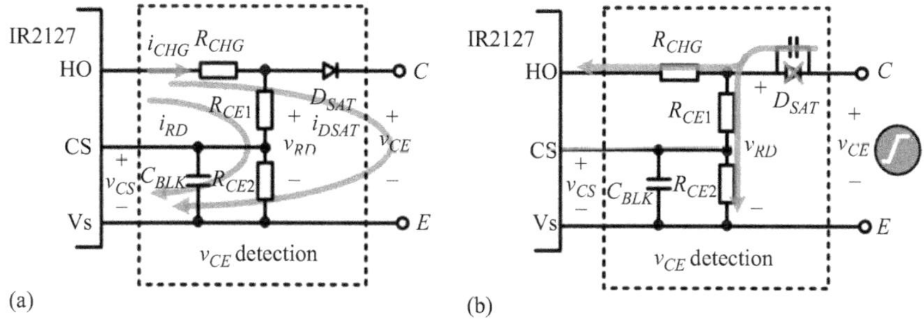

Figure 7.8 Operation principle of v_{CE} detection circuit: (a) under the normal condition or fault condition before IGBT is tripped; (b) under the fault condition when IGBT is tripped

Charging current i_{CHG} is calculated as

$$i_{CHG} = \frac{v_{CC} - v_{RD}}{R_{CHG}} = \frac{v_{CC} - v_{f\_DSAT} - v_{CE}}{R_{CHG}} \tag{7.3}$$

Charging current flowing through the resistor divider i_{RD} is given by

$$i_{RD} = \frac{v_{RD}}{R_{CE1} + R_{CE2}} = \frac{v_{f\_DSAT} + v_{CE}}{R_{CE1} + R_{CE2}} \tag{7.4}$$

Accordingly, current through sensing diode i_{DSAT} is expressed as

$$i_{DSAT} = i_{CHG} - i_{RD} = \frac{v_{CC} - v_{f\_DSAT} - v_{CE}}{R_{CHG}} - \frac{v_{f\_DSAT} + v_{CE}}{R_{CE1} + R_{CE2}} \tag{7.5}$$

Maximum current of the sensing diode $i_{DSAT(max)}$ is determined by the lowest v_{CE} when there is no current flowing through the IGBT. In DPT, this is almost always the case. Accordingly:

$$i_{DSAT(max)} = \frac{v_{CC} - v_{f\_DSAT} - v_{CE,knee}}{R_{CHG}} - \frac{v_{f\_DSAT} + v_{CE,knee}}{R_{CE1} + R_{CE2}} \tag{7.6}$$

where $v_{CE,knee}$ is the on-state voltage of IGBT under no load condition.

Voltage across "CS" pin is determined by the resistor divider R_{CE1} and R_{CE2}, shown as

$$v_{CS} = \frac{R_{CE2}}{R_{CE1} + R_{CE2}} \times v_{RD} = \frac{R_{CE2}}{R_{CE1} + R_{CE2}} \times \left(v_{f\_DSAT} + v_{CE}\right) \tag{7.7}$$

According to the operation principle of the desaturation protection discussed earlier, once v_{CE} approaches the boundary between IGBT saturation region and active region v_{CETH}, v_{CS} should be equal to "CS" pin threshold v_{CSTH} preset by the gate drive IC, then the IGBT is tripped. Thus, according to (7.7), the ratio of the resistor divider R_{CE1} and R_{CE2} is designed as

$$\frac{R_{CE2}}{R_{CE1}} = \frac{v_{CSTH}}{v_{f\_DSAT} + v_{CETH} - v_{CSTH}} \tag{7.8}$$

Additionally, when v_{CE} reaches v_{CETH}, i_{DSAT} is preferred to be controlled close to zero, since v_{CE} starts to increase dramatically when the gate drive shuts down. Instead of forward biased to sense v_{CE}, D_{SAT} is about to be reverse biased to block v_{CE}. Therefore, it is preferred to design R_{CHG}, R_{CE1}, and R_{CE2} so as to set i_{DSAT} just slightly larger than zero when v_{CE} reaches v_{CETH}. Also note i_{DSAT} cannot be zero before v_{CE} approaches v_{CETH}, since if D_{SAT} turns off earlier, v_{RD} stops following the increase of v_{CE}; as a result, v_{CS} cannot reach v_{CSTH} to turn off the IGBT. Thus, according to (7.5), another design criterion is given:

$$i_{DSAT} = i_{CHG} - i_{RD} = k_{DSAT} \times i_{DSAT(max)}, \quad \text{when } v_{CE} = v_{CETH} \tag{7.9}$$

where k_{DSAT} is a coefficient to limit the current through D_{SAT} when IGBT is turned off. Practically, it can be set to 5%–10%.

Combining (7.6)–(7.9) yields

$$\frac{R_{CE1}}{R_{CHG}} = \frac{v_{fDSAT} + v_{CETH} - v_{CSTH}}{v_{fDSAT} + v_{CETH}} \times \frac{\left(v_{fDSAT} + v_{CETH}\right) - k_{DSAT} \times \left(v_{fDSAT} + v_{CE,knee}\right)}{(1 - k_{DSAT}) \times v_{CC} - \left(v_{fDSAT} + v_{CETH}\right) + k_{DSAT} \times \left(v_{fDSAT} + v_{CE,knee}\right)} \tag{7.10}$$

$$\frac{R_{CE2}}{R_{CHG}} = \frac{v_{CSTH}}{v_{fDSAT} + v_{CETH}} \times \frac{\left(v_{fDSAT} + v_{CETH}\right) - k_{DSAT} \times \left(v_{fDSAT} + v_{CE,knee}\right)}{(1 - k_{DSAT}) \times v_{CC} - \left(v_{fDSAT} + v_{CETH}\right) + k_{DSAT} \times \left(v_{fDSAT} + v_{CE,knee}\right)} \tag{7.11}$$

Similar to the resistor divider R_{GE1} and R_{GE2} of the v_{GE} control unit, a trimmer is suggested for either R_{CE1} or R_{CE2}.

Selection of D_{SAT} is critical for the performance of the v_{CE} detection unit and SSCB. First, the breakdown voltage of D_{SAT} must be higher than the maximum operating voltage of DPT circuit, and its current rating should be at least $i_{DSAT(max)}$ calculated by (7.6). More importantly, the selected D_{SAT} should be either a super-fast recovery diode or a Schottky diode to mitigate the noise introduced by the reverse recovery when D_{SAT} turns off. Also, junction capacitance of D_{SAT} is preferred to be as small as possible, since displacement current will be induced by the *dv*/*dt* generated when the IGBT is turned off. (see Figure 7.8(b)). Less junction capacitance of D_{SAT} is beneficial for displacement current mitigation and noise immunity enhancement of the SSCB circuit. Practically, several low-voltage fast recovery or Schottky diodes in series can be employed to reduce the equivalent junction capacitance of D_{SAT}.

Additionally, it is recommended to add a blanking capacitor C_{BLK} across the gate drive "CS" pin for noise immunity of the SSCB circuit. C_{BLK} should be selected much larger than the junction capacitance of D_{SAT} to mitigate the impact of D_{SAT} displacement and recovery current on v_{CS}, yielding:

$$C_{BLK} > k_{Cj\_DSAT} \times C_{j\_DSAT} \tag{7.12}$$

where C_{j_DSAT} is the junction capacitance of D_{SAT}, and k_{Cj_DSAT} is a coefficient for C_{BLK} determination to mitigate the adverse impact of D_{SAT}. Practically it is selected to be greater than 50.

Also, blanking time introduced by C_{BLK} should be limited to achieve the fast response of short-circuit protection. Thus, the time constant τ_{vcs} for v_{CS}, determined by the network consisting of R_{CE1}, R_{CE2}, and C_{BLK} in parallel, as expressed in (7.13), should be adequately small to allow the protection response time shorter than t_{SCWT} of the DUT:

$$\tau_{vcs} = \left(\frac{R_{CE1} \times R_{CE2}}{R_{CE1} + R_{CE2}} \right) \times C_{BLK} < SCWT \tag{7.13}$$

According to the reported short-circuit withstand capability of the state-of-the-art SiC devices, protection response time is desired to be less than 1 μs [2]. The requirement for GaN transistors is even more challenging and still an active research topic. According to the recent data reported in [9], SCWT for GaN HEMTs is highly dependent on operating conditions and can be as short as 100s of ns. In summary, the selection of τ_{vcs} depends on the specific DUT.

In the end, combining (7.10)–(7.13), the value of R_{CHG}, R_{CE1}, R_{CE2}, and C_{BLK} can be calculated. Also note that a Zener diode is usually added in parallel with C_{BLK} to clamp the "CS" pin voltage within its maximum rating. Regarding the physical layout, C_{BLK} and/or Zener diode should be placed in close proximity to "CS" pin. Also, the junction capacitance of the Zener diode should be considered as part of the C_{BLK}.

7.2.3 Test setup and procedure

Once the SSCB is designed and built, the performance has to be experimentally verified before it is utilized for DPT. Thus, a test setup needs to be devised. Figure 7.9 illustrates an example circuit for SSCB test, which is similar to the testing circuit for DPT shown in Figure 7.4. However, DUTs are replaced by another SSCB to purposefully incur a shoot-through failure to evaluate the performance of SSCB under test.

A second SSCB can be leveraged to act as the "SSCB assisting test" with several gate drive circuit modifications:

1. Instead of a constant high logic control, one short pulse is sent as the input of gate drive "IN" pin to create the fault. The duration of this pulse should be set slightly longer than the expected short-circuit protection response time of SSCB under test, but much shorter than 10 μs, the typical SCWT of Si IGBT.

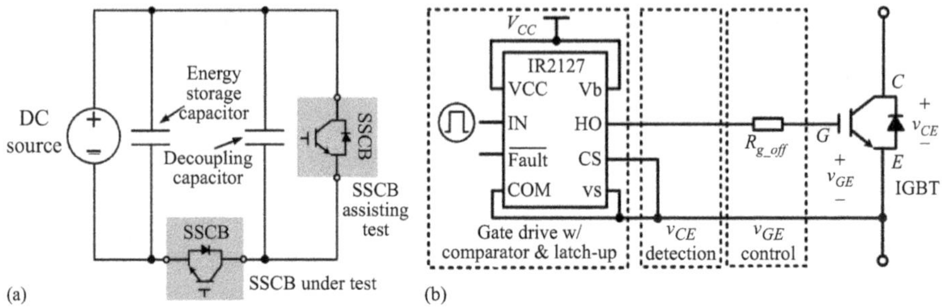

Figure 7.9 Testing setup for SSCB performance evaluation: (a) testing circuit for SSCB; (b) gate drive for SSCB assisting test

Table 7.1 Summary of DUT

Supplier	**Part number**	**Description**	**Max testing voltage/current**
Wolfspeed	C3M0065090D	900 V/36 A SiC MOSFET	450 V/35 A
GaN systems	GS66508B-E01-MR	650 V/30 A GaN transistor	300 V/30 A

Thus, in case the SSCB under test is not tripped under the fault condition, SSCB assisting test can be regarded as the secondary protection to clear the fault and protect the testing setup.

2. The components related to v_{CE} detection unit do not need to be populated. Gate drive "CS" pin is shorted with ground to disable the desaturation protection of the SSCB assisting test.
3. For v_{GE} control unit, R_{g_off} is kept as the gate resistance, D_{g_off} is shorted, and the rest of the components do not need to be soldered. In the end, just the basic gate driving function is employed for the SSCB assisting test. The detailed circuit is depicted in Figure 7.9(b).

During the test, two important parameters need attention: one is the short-circuit protection response time, and the other is the over-voltage across the IGBT of SSCB under test when the shoot-through current is being cleared. Based on the short-circuit withstand capability of WBG devices [2], it is suggested to ensure the response time less than 500 ns. If the v_{CE} over-voltage approaches the breakdown voltage of IGBT, R_{g_off} should be tuned larger to slow down the turn-off speed.

7.2.4 Case study

As discussed earlier, SSCB is a universal circuit for DPT protection. In the case study, assume one SSCB circuit is designed for two different WBG devices, including a SiC MOSFET from Wolfspeed and a GaN HEMT from GaN Systems. Table 7.1 summarizes the basic information and desired testing conditions of these two DUTs.

Table 7.2 Summary of IGBT

Vendor/model	**Microsemi/APT**
Part number	APT75GN120LG
Description	1,200 V/200 A

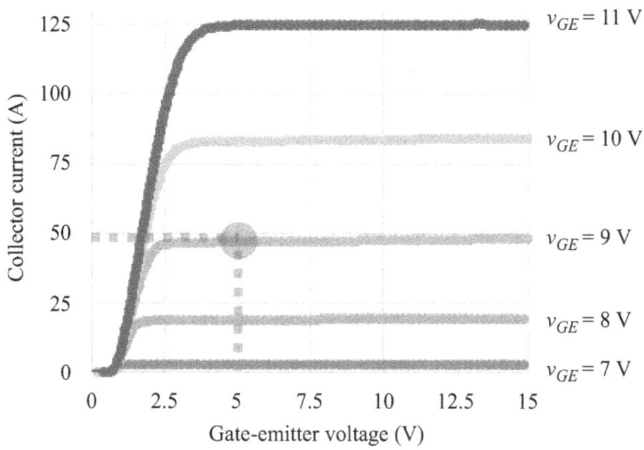

Figure 7.10 Output characteristics of 1,200 V/200 A APT75GN120LG IGBT

7.2.4.1 IGBT selection

Step no. 1: Determine protection current I_{SSCB} and maximum operating voltage V_{SSCB} according to the maximum operating voltages and currents of DUTs. Since two devices are included, the worst scenario has to be considered.

$$V_{SSCB} = 450\ \text{V}$$
$$I_{SSCB} = 50\ \text{A} > 35\ \text{A}$$

Step no. 2: Select IGBT with sufficient voltage and current capability. Since IGBT is used for protection, relatively large voltage and current margins are adopted. In this case study, 1,200 V/200 A APT75GN120LG IGBT from Microsemi is employed, as summarized in Table 7.2.

Step no. 3: Determine the v_{GE}, v_{CETH}, and $v_{CE,knee}$ based on the output characteristics of the selected IGBT, as illustrated in Figure 7.10.

$$v_{GE} = 9\ \text{V}@\ I_{SSCB} = 50\ \text{A}$$
$$v_{CETH} = 5\ \text{V}$$
$$v_{CE,knee} = 0.5\ \text{V}$$

7.2.4.2 Gate drive IC selection

Following the aforementioned design and implementation, Infineon IR2127 gate drive IC is used in this case study with the detailed information summarized in Table 7.3.

Table 7.3 Summary of gate drive IC

Vendor	International rectifier	Model number	IR 21271
Description	Current sensing single channel drive with built-in comparator and latch-up circuit for desaturation protection		
v_{CSTH}	1.8 V		
V_{CC}	15 V is selected in this case study, provided by an isolated power supply MEJ1S0515SC from Murata Power Solutions Inc.		

Table 7.4 Summary of v_{GE} control unit

	Vendor/model	Description
R_{GE1}	Yageo, RC0805FR-07300RL	300 Ω chip resistor
R_{GE2}	Vishay Sfernice, TS63Y502KR10	5 kΩ 0.25W trimmer (tune R_{GE2} to 3.175 kΩ)
D_{g_off}	Nexperia USA Inc., PMEG3010BEA,115	30 V, 1 A, Schottky diode
R_{g_off}	Stackpole Electronics Inc., RMCF0805JT20R0	20 Ω chip resistor
D_Z	Diodes Incorporate, BZT52C9V1-7-F	9.1 V, 500 mW, Zener diode

7.2.4.3 Design of v_{GE} control unit

Step no. 1: Select resistor divider R_{GE1} and R_{GE2} according to (7.1) where $v_{GE} = 9$ V and $V_{CC} = 15$ V.

$$\frac{R_{GE2}}{R_{GE1} + R_{GE2}} = \frac{v_{GE}}{V_{CC}} = \frac{9}{15}$$

Assume $R_{GE1} = 300\,\Omega$, then $R_{GE2} = 3.175\,\text{k}\Omega$. The selection of physical resistors for R_{GE1} and R_{GE2} is summarized in Table 7.4.

Step no. 2: Select turn-off gate resistor R_{g_off} and diode D_{g_off} according to the design criteria mentioned earlier. The selection of physical resistor for R_{g_off} and diode D_{g_off} is summarized in Table 7.4.

Step no. 3: Select Zener diode D_Z based on the desired v_{GE} of IGBT. The selection of D_Z is summarized in Table 7.4.

7.2.4.4 Design of v_{CE} detection unit

Step no. 1: Select sensing diode D_{SAT} considering breakdown voltage and reverse recovery characteristics. The selection of D_{SAT} is summarized in Table 7.5 with the forward drop voltage v_{f_DSAT} and junction capacitance C_{j_DSAT} highlighted next.

$$v_{f\_DSAT} = 0.5\text{ V}$$

$$C_{j\_DSAT} = 20\text{ pF}$$

Table 7.5 Summary of v_{CE} detection unit

	Vendor/model	Description
D_{SAT}	Micro Commercial Co., MUR1100-TP	1 kV/1 A superfast recovery diode with 20 pF junction capacitance
C_{BLK}	KEMET, C0805C152J5GACTU	1,500 PF 50 V C0G/NP0 ceramic capacitor
R_{CE2}	Vishay Sfernice, TS63Y501KR10	500 Ω 0.25 W trimmer, (tune R_{CE2} to 334 Ω)
R_{CE1}	TE, RN73C2A681RBTDF	681 Ω, 1/10 W, chip resistor
R_{CHG}	Stackpole, RNCP0805FTD1K65	1.65 kΩ, 1/4W, chip resistor

Step no. 2: Select blanking capacitor C_{BLK} according to (7.12) where the coefficient k_{Cj_DSAT} is selected to be 75. The physical selection of C_{BLK} is summarized in Table 7.5.

$$C_{BLK} = k_{Cj\_DSAT} \times C_{j\_DSAT} = 75 \times 20\,\text{pF} = 1500\,\text{pF}$$

Step no. 3: Select charging resistor R_{CHG} and resistor divider R_{CE1} and R_{CE2} according to (7.10), (7.11), and (7.13) where k_{DSAT} is selected to be 5% and τ_{vcs} is selected to be 500 ns. The values of R_{CHG}, R_{CE1}, and R_{CE2} are listed next with their physical selections summarized in Table 7.5.

$$R_{CHG} = 1.65\,\text{k}\Omega$$

$$R_{CE1} = 681\,\Omega$$

$$R_{CH2} = 334\,\Omega$$

7.2.4.5 Testing

Based on the aforementioned design, a prototype is built to evaluate the performance of SSCB board, as shown in Figure 7.11. Two SSCB boards are assembled: one works as the SSCB board under test with the short-circuit protection functionality enabled, while the other is controlled as the SSCB assisting test to generate the fault by purposefully turning on its associated IGBT with protection functionality disabled. Following the test principle illustrated in Figure 7.9, the SSCB board is tested under the DC bus voltage of 450 V (for Wolfspeed SiC MOSFET) and 300 V (for GaN Systems GaN HEMTs), respectively. Four key waveforms are measured to evaluate the performance of the designed SSCB, including v_{GE}, v_{CE}, v_{CS}, and shoot-through current I_{SC}. Figure 7.12 illustrates the measurement setup for SSCB testing and summarizes probes used in the case study to measure these four key parameters. Also, tip adapters are used to ensure the accuracy of the high-speed dynamic measurement.

As can be observed in Figure 7.13, for DC bus voltages of 300 and 450 V, the short-circuit response time, that is, the duration from the moment when I_{SC} approaches the protection current threshold to the moment when v_{GE} starts to decrease, is around 580 ns, which is close to the design value of 500 ns and

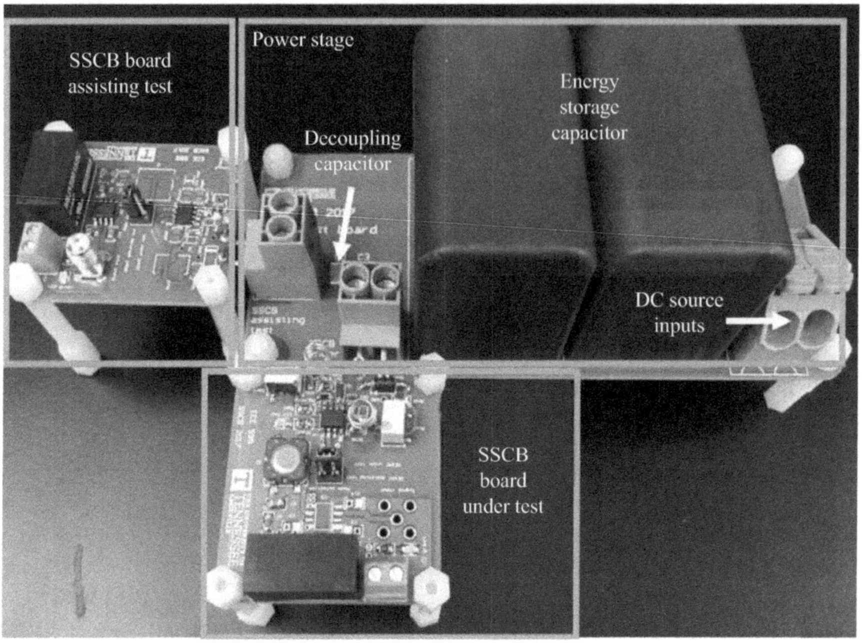

Figure 7.11 SSCB testing setup

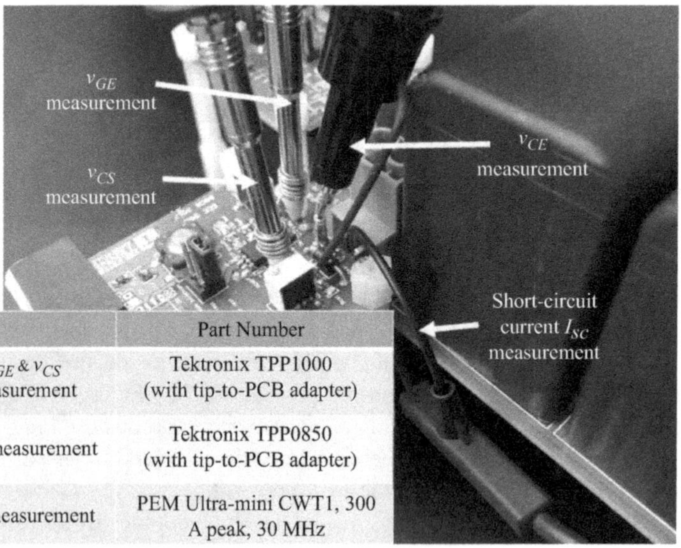

	Part Number
v_{GE} & v_{CS} measurement	Tektronix TPP1000 (with tip-to-PCB adapter)
v_{CE} measurement	Tektronix TPP0850 (with tip-to-PCB adapter)
I_{SC} measurement	PEM Ultra-mini CWT1, 300 A peak, 30 MHz

Figure 7.12 Measurement of SSCB testing

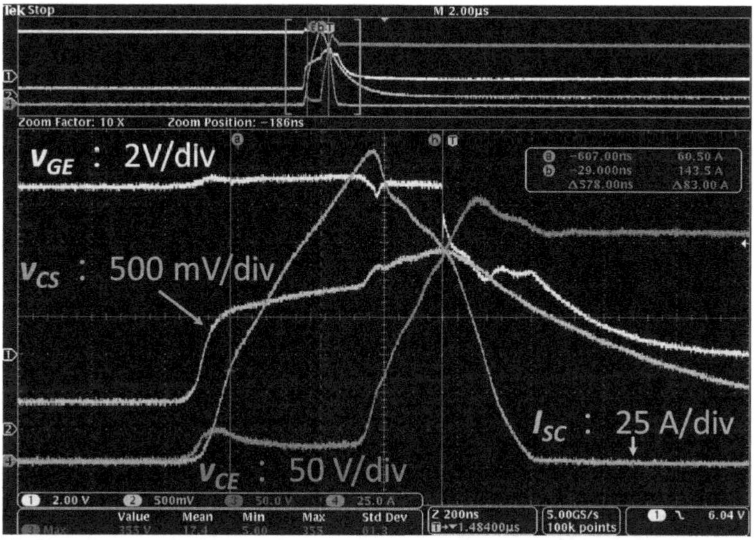

(a)

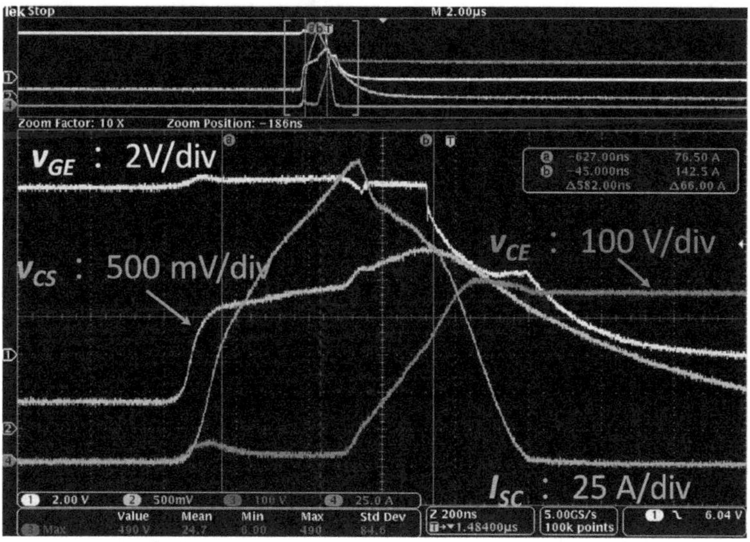

(b)

Figure 7.13 SSCB testing waveforms: (a) DC bus voltage of 300 V; (b) DC bus voltage of 450 V

sufficiently fast. Also, during the fault before the protection is tripped, thanks to the Zener diode D_Z across the gate-emitter terminals of the IGBT, v_{GE} stays at 9 V stably without positive spurious variation. Besides, due to C_{BLK}, the v_{cs} waveform is clean. Furthermore, when the shoot-through current is cleared, the v_{CE} voltage overshoot is only around 50 V higher than the DC bus voltage. Considering the

1,200 V breakdown voltage of the selected IGBT, the v_{CE} safety margin is enough. In summary, the designed SSCB is able to protect the DUT under the fault condition with fast response time, high noise immunity, and low-voltage spike.

7.3 Consideration for high-voltage WBG device DPT

Considering that the breakdown field strength of WBG is much higher than traditional Si, WBG power semiconductors show superior performance, especially in medium- and high-voltage levels (>3.3 kV). In recent years, high-voltage SiC devices (>3.3 kV) have been rapidly developed including JBS and PiN diodes, JFETs, MOSFETs, BJTs, and IGBTs. These devices have potential applications in medium- and high-voltage power converters, for example, medium-voltage motor drives and power grid applications. It is important to characterize these emerging devices. Compared to the lower voltage power semiconductor (<3.3 kV) test, the safety issue and protection scheme should be seriously taken into account.

7.3.1 Safety consideration

Several safety considerations are listed next:

1. A dedicated room is used for high-voltage test. Warning signs and barriers should be placed in the test area. Live terminals should be shielded to prevent any contact with high voltage.
2. Instrumentation should be used to detect the test voltages and currents constantly.
3. High-voltage and high-power tests should be performed and supervised by qualified personnel. The test should be performed by at least two persons.
4. The required grounding point should be solidly connected to the ground. The connection should be easily verified by the operator visually.
5. The ground stick for discharge should be available. The test objectives should be deenergized by the ground stick before any manual operation.
6. All the equipment in the power room should have remote control including power supply, DUT, oscilloscope, etc. During the test, the operation of the equipment can be remotely controlled from the control room.
7. The remote control should be implemented using a voltage isolation technique, e.g., fiber optic cable.
8. Power supply should shut off after the DC capacitor bank is charged to the desired voltage before the DPT starts, in order to avoid constant power input during the DPT under fault.
9. All high-voltage generating equipment should have a single obvious control to switch the equipment off under emergency conditions.

7.3.2 Protection scheme

The DPT platform for the high-voltage power semiconductors should have protections including:

1. Over-current protection for DUT's gate drive

   The gate drive of DUT should have over-current protection and the function should be validated. A gate drive example of 10 kV SiC MOSFET with its function test is illustrated in Figure 7.14.

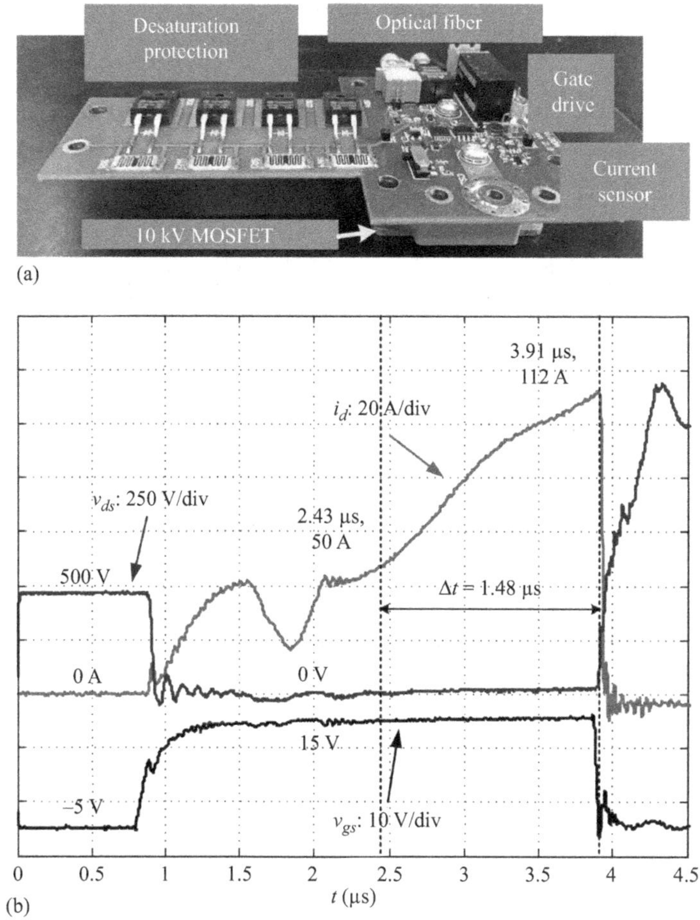

Figure 7.14 Over-current protection of 10 kV SiC MOSFETs: (a) gate drive of 10 kV SiC MOSFET; (b) over-current protection waveform

2. High-voltage SSCB

   High-voltage SSCB is required (1) in DPT test and (2) short-circuit test. In DPT test, the SSCB protects the DUT from the short-circuit. The saturation current of SSCB should be lower than that of DUT. In short-circuit test, the SSCB can protect the rest of the platform if the DUT fails in its short-circuit test. The saturation current of SSCB is higher than that of DUT. The high-voltage SSCBs are commonly developed with mature Si devices. Figure 7.15 shows a SSCB based on 4 kV Si IGBT. With three in series connection, the SSCB can be used as the protection of the DPT platform of 10 kV SiC MOSFET.
3. Current limitation of DC power supply

   The DC power supply with current limitation needs to be selected. If the fault happens in DPT platform, the over-current may also occur at the input of the DC power supply. The current limitation of the DC power supply is important

Figure 7.15 SSCB using 4 kV Si IGBT

in two aspects: (1) protect the DC power supply from over-current failure; and (2) prevent cascaded failure of the DPT platform caused by the power surge from the DC power supply to the DPT platform.

7.4 Summary

1. This chapter focuses on short-circuit and over-current detection and protection for the DUT and its DPT circuit. The popular detection and protection schemes used in power converters include desaturation scheme, *di/dt* sensing such as Rogowski coil-based scheme, and direct current sensing such as shunt resistor scheme. None of these schemes are ideal for WBG DPT protection, either due to their dependence on device characteristics or their added impact on device switching performance.
2. As a universal protection method without need for the specific characteristics of WBG devices under test, SSCB is recommended for protecting the short-circuit and over-current fault for DPT.
3. A normally-on SSCB is inserted between the bulk energy storage capacitor and the rest of DPT circuit. The decoupling capacitor is placed between the SSCB and DUT, and close to DUT, in order to minimize power loop parasitics, especially the extra insertion inductance introduced by SSCB.
4. SSCB relies on desaturation protection of IGBT and consists of gate drive, v_{GE} control unit, v_{CE} detection unit, and comparator and latch-up unit (which is usually embedded into the gate drive IC). For WBG-based DPT, the unique challenge for SSCB design is ultra-fast response time and high noise immunity capability. The key design criteria are highlighted next.

(a) Zener diode D_Z across the gate-emitter terminals of IGBT along with the small internal gate resistance of IGBT are effective approaches to mitigate the Miller effect and its resultant spurious gate voltage and higher shoot-through current when the fault occurs.
(b) Charging resistor R_{CHG} enables v_{CE} detection unit to follow the variation of IGBT v_{CE} under normal condition and should be designed along with resistor divider R_{CE1} and R_{CE2} to control the current through sensing diode D_{SAT} close to zero when IGBT is turned off. Sensing diode D_{SAT} should be selected to be either fast recovery diode or Schottky diode to mitigate the noise introduced by reverse recovery when the D_{SAT} turns off. Also, D_{SAT} with small junction capacitance is preferred.
(c) Blanking capacitor C_{BLK} is recommended to be added across the gate drive "CS" pin for the noise immunity of SSCB circuit. Its value should be much larger than the junction capacitance of sensing diode D_{SAT}.

5. High-voltage WBG devices (>3.3 kV) are the key enabler for applications in medium- and high-voltage power converters, for example, medium-voltage motor drives and power grid applications. It is important to characterize these emerging devices. Compared to the lower voltage device (<3.3 kV) DPT, the safety issue and protection scheme should be seriously taken into account.

References

[1] A. Maerz, T. Bertelshofer, R. Horff, and M. M. Bakran, "Requirements of short-circuit detection methods and turn-off for wide band gap semiconductors," in *Proceedings of the International Conference on Integrated Power Electronics Systems (CIPS)*, Nuremberg, Germany, 2016, pp. 1–6.

[2] Z. Wang, X. Shi, Y. Xue, L.M. Tolbert, F. Wang, and B.J. Blalock, "Design and performance evaluation of overcurrent protection schemes for silicon carbide (SiC) power MOSFETs," *IEEE Transactions on Industrial Electronics*, vol. 61, no. 10, pp. 5570–5581, 2014.

[3] Z. Wang, X. Shi, L. M. Tolbert, *et al.*, "Temperature-dependent short-circuit capability of silicon carbide power MOSFETs," *IEEE Transactions on Power Electronics*, vol. 31, no. 2, pp. 1555–1566, 2016.

[4] N. Badawi, A. E. Awwad, and S. Dieckerhoff, "Robustness in short-circuit mode: benchmarking of 600V GaN HEMTs with power Si and SiC MOSFETs," in *Proceedings of the IEEE Energy Conversion Congress and Exposition (ECCE),* Milwaukee, WI, USA, 2016, pp. 1–7.

[5] M. Fernández, X. Perpiñà, J. Roig-Guitart, *et al.*, "P-GaN HEMTs drain and gate current analysis under short-circuit," *IEEE Electron Device Letters*, vol. 38, no. 4, pp. 505–508, 2017.

[6] M. Fernández, X. Perpiñà, J. Roig-Guitart, *et al.*, "Short-circuit study in medium-voltage GaN cascodes, p-GaN HEMTs, and GaN MISHEMTs," *IEEE Transactions on Industrial Electronics*, vol. 64, no. 11, pp. 9012–9022, 2017.

[7] M. Fernández, X. Perpiñá, M. Vellvehi, *et al.*, "Short-circuit capability in p-GaN HEMTs and GaN MISHEMTs," in *Proceedings of the International Symposium on Power Semiconductor Devices and IC's (ISPSD)*, Sapporo, Japan, 2017, pp. 455–458.

[8] H. Li, X. Li, X. Wang, *et al.*, "E-mode GaN HEMT short circuit robustness and degradation," in *Proceedings of the IEEE Energy Conversion Congress and Exposition (ECCE)*, Cincinnati, OH, 2017.

[9] X. Li, "Short circuit capability and degradation mechanism analysis of e-mode GaN HEMT," M.S. thesis, Graduate Program in Electrical and Computer Science, The Ohio State University, Columbus, OH, 2017.

[10] T. Oeder, A. Castellazzi, and M. Pfost, "Experimental study of the short-circuit performance for a 600V normally-off p-gate GaN HEMT," in *Proceedings of the International Symposium on Power Semiconductor Devices and IC's (ISPSD),* Sapporo, Japan, 2017, pp. 211–214.

[11] W. B. Ray, M. Kim, A. Bilbao, J. A. Schrock, and S. B. Bayne, "Analysis on repetitive pulsed overcurrent operation of GaN power transistors," in *Proceedings of the Workshop on Wide Bandgap Power Devices and Applications (WiPDA)*, Fayetteville, AR, USA, 2016, pp. 353–356.

[12] L. Yu, S. Araujo, D. Pappis, and P. Zacharias, "Short-circuit capability: benchmarking SiC and GaN devices with Si-based technologies," in *Proceedings of the PCIM Europe*, Nuremberg, Germany, 2015, pp. 1–5.

[13] M. Landel, C. Gautier, D. Labrousse, and S. Lefebvre, "Experimental study of the short-circuit robustness of 600V E-mode GaN transistors," *Microelectronics Reliability*, vol. 64, no. Supplement C, pp. 560–565, 2016.

[14] M. Landel, C. Gautier, D. Labrousse, S. Levebvre, F. Zaki, and Z. Khatir, "Dispersion of electrical characteristics and short-circuit robustness of 600 V emode GaN transistors," in *Proceedings of the PCIM Europe*, Nuremberg, Germany, 2017, pp. 1–9.

[15] D. Pappis, K. Goebel, and P. Zacharias, "Short circuit capability of 650 V normally off GaN E-HEMT and MOSFET-HEMT cascode," in *Proceedings of the International Exhibition and Conference for Power Electronics, Intelligent Motion, Renewable Energy and Energy Management*, Nuremberg, Germany, 2017, pp. 1–8.

[16] B. Huang, Y. Li, T. Q. Zheng, and Y. Zhang, "Design of overcurrent protection circuit for GaN HEMT," in *Proceedings of the IEEE Energy Conversion Congress and Exposition (ECCE)*, Pittsburgh, PA, USA, 2014, pp. 2844–2848.

[17] H. Li, X. Li, Z. Zhang, C. Yao, and J. Wang, "Design consideration of high power GaN inverter," in *Proceedings of the Workshop on Wide Bandgap Power Devices and Applications (WiPDA)*, Fayetteville, AR, USA, 2016, pp. 23–29.

[18] P. Skarolek and J. Lettl, "Current limiting driver for GaN half-bridge," in *Proceedings of the International Conference on Applied Electronics (AE)*, Pilsen, Czech Republic, 2017, pp. 1–4.

[19] Z. Wang, X. Shi, L. M. Tolbert, F. Wang, and B. J. Blalock, "A di/dt feedback-based active gate driver for smart switching and fast overcurrent protection of IGBT modules," *IEEE Transactions on Power Electronics*, vol. 29, no. 7, pp. 3720–3732, 2014.

[20] J. Wang, Z. Shen, C. DiMarino, R. Burgos, and D. Boroyevich, "Gate driver design for 1.7kV SiC MOSFET module with Rogowski current sensor for shortcircuit protection," in *IEEE Applied Power Electronics Conference and Exposition (APEC)*, Long Beach, CA, 2016, pp. 516–523.

[21] S. Ji, S. Zheng, Z. Zhang, F. Wang, and L. Tolbert, "Protection and temperature-dependent switching characterization of latest generation 10 kV SiC MOSFETs," in *IEEE Applied Power Electronics Conference and Exposition (APEC)*, Tampa, FL, USA, 2016, pp. 783–788.

[22] K. Wang, X. Yang, H. Li, L. Wang, and P. Jain, "A high-bandwidth integrated current measurement for detecting switching current of fast GaN devices," *IEEE Transactions on Power Electronics*, vol. PP, no. 99, pp. 1–1, 2017.

[23] P. Liu, L. Zhang, A. Q. Huang, S. Guo, and Y. Lei, "High bandwidth current sensing of SiC MOSFET with a Si current mirror," *2016 IEEE 4th Workshop on Wide Bandgap Power Devices and Applications (WiPDA)*, Fayetteville, AR, 2016, pp. 200–203.

[24] J. Roberts, G. Klowak, D. Chen, and A. Mizan, "Drive and protection methods for very high current lateral GaN power transistors," in *Proceedings of the IEEE Applied Power Electronics Conference and Exposition (APEC)*, Charlotte, NC, USA, 2015, pp. 3128–3131.

[25] R. Lai, F. Wang, R. Burgos, D. Boroyevich, D. Zhang, and P. Ning, "A shoot-through protection scheme for converters built with SiC JFETs," *IEEE Transactions on Industry Applications*, vol. 46, no. 6, pp. 2495–2500, 2010.

Chapter 8

Measurement and data processing for dynamic characterization

This chapter will describe the techniques for measurement and post-processing of dynamic characterization data collected using the methodology described in Chapters 4–7. Measurement is a crucial consideration in performing an effective DPT. The oscilloscope, voltage probes, current sensors, and their connection to the test circuit must be carefully selected and designed.

Following the test, the data processing can be implemented with any computation tool or programming language. This chapter will discuss the generalized algorithms and methodologies as well as some example code and post-processing results implemented in MATLAB®. These data processing techniques allow the experimenter to analyze the dynamic behavior of the device under test and calculate key device characteristics.

8.1 Key considerations and challenges in dynamic characterization measurement

The three primary tools in measurement of a DPT are an oscilloscope, voltage probes, and current probes. The most crucial measurements are drain voltage (v_{ds}) and drain current (i_d). Optionally, the gate voltage (v_{gs}) and load inductor current (I_L) can be helpful additions to the measurement setup. The v_{ds} and i_d waveforms are necessary to calculate the switching energies (E_{on} and E_{off}) as well as voltage and current slew rates, overshoots, and rise/fall times. The v_{gs} waveform gives additional details such as the delay times and gate voltage overshoot. The load current is primarily used for calibrating the attenuation of the drain current sensor, and it is also helpful for configuring and debugging the DPT setup.

The greatest challenge in designing the DPT measurement system is capturing the fast switching transients with sufficient detail. This requires high-resolution data capture in both the vertical (voltage, current) and horizontal (time) dimensions. The vertical resolution is determined by the oscilloscope's analog-to-digital converter (ADC) bit length (e.g., eight-bit) as well as the vertical scaling on each channel. The oscilloscope must also be well-calibrated for the probe channel to compensate for the impedance of the measurement loop. Section 8.2 will detail these considerations [1–3].

The horizontal resolution is more complex. In order to capture the high-frequency detail of a nanosecond-scale transient, the oscilloscope and probe must both have a high enough bandwidth. Second, the scope's sampling rate must be high enough. Third, the measurement loop of the probe must be properly designed to limit common-mode (CM) ringing. Finally, once the time-scale resolution of all measurement channels is set, the measured signals must be deskewed to avoid time misalignment due to mismatched propagation delays between the various probes. Section 8.3 will detail these considerations [4].

8.2 Oscilloscope selection and setup

The first measurement tool to consider in a DPT is the oscilloscope, which captures time-domain waveforms of the switching transients. Oscilloscopes are available from several manufacturers, with a variety of options. Some of the key features of a DPT oscilloscope include the bandwidth and sampling rate, which determine the fastest switching transient that can be accurately measured. The primary frequency component of a transient voltage/current measurement can be estimated from the rise time of the signal being measured, according to

$$f_{eq} = \frac{0.35}{t_r} \tag{8.1}$$

or similarly for a fall transient using t_f. It is interesting to note that this frequency does not depend on the voltage/current slew rate, but rather the duration of the ramp. In this way, a WBG semiconductor DPT may require a similarly high bandwidth regardless of the voltage or current level being tested. In reality, the bandwidth of the oscilloscope and the sampling rate must be much higher than this equivalent frequency to capture sufficient detail of the waveform. Typical WBG transients are on the order of nanoseconds. For example, to capture a 1 ns transient requires a bandwidth of at least 1 GHz and a sampling rate of at least 2 GHz. A lower bandwidth will result in filtering of the high-frequency components of the signal, and a lower sampling rate will result in aliasing errors. Higher bandwidth and sampling rates will improve the detail of the waveform capture, which is critical in a DPT setup [5,6].

In addition to the time axis, the vertical resolution is also limited by the oscilloscope's ADC and the maximum scale of the screen. For example, an eight-bit oscilloscope with a vertical axis set to $\pm 1{,}000$ V can only digitize the captured data with a resolution of 7.8 V, according to

$$resolution = \frac{Max\ scale - Min\ scale}{2^{bits}}. \tag{8.2}$$

This means that the vertical scale for each channel on the oscilloscope should be fine-tuned to maximize this resolution for each waveform capture. In fact, for some switching transients, the turn-on and turn-off transients must be captured separately in order to measure the turn-off transient accurately. This is particularly true for light load conditions, when the peak drain current can be several times higher in the turn-on transient than in turn-off. The final consideration for oscilloscopes is the availability of probes that are compatible with the selected system, which will be discussed next.

8.3 Voltage probe selection and setup

Each oscilloscope can be used with a number of compatible voltage measurement probes. These probes may be differential, active single-ended, or passive. Each probe has a specified attenuation factor, voltage rating, and bandwidth. For the typically low-voltage v_{gs} measurement, there are many options for high-bandwidth measurement in the desired voltage range. However, the drain voltage measurement can be much more challenging. Few probes are available with sufficiently high voltage and bandwidth. Table 8.1 lists a few examples of potential v_{ds} measurement probes and their relative performance specifications. The bandwidth can be used to estimate the rise/fall time that can be measured accurately with this probe. For most WBG devices, the two differential probes listed do not have sufficient bandwidth to capture the waveform accurately, and passive probes must be used [7,8].

Passive probes require additional steps to ensure accurate waveform capture. Perhaps the most important consideration with passive voltage probes is the ground lead inductance. Many oscilloscope probes come pre-fitted with alligator clips and grounding wires. These standard probe connections are not suitable for DPT measurement. The inductance introduced by these long leads causes a resonance with the input capacitance of the probe, resulting in severe ringing and incorrect rise/fall time measurements. To solve this problem, the DPT board must be outfitted with probe tip adapters that minimize the measurement loop area enclosed by the probe tip and ground connection. Minimizing this area also reduces the measurement loop inductance and eliminates the excessive inductance introduced by the long grounding wire. Figure 8.1 shows the equivalent circuit of this resonant circuit, and Figure 8.2 shows several examples of probe tip

Table 8.1 Examples of voltage probes suitable for drain voltage measurement

Types	**Models**	**Maximum voltage (V)**	**Bandwidth (MHz)**	**Rise time (ns)***	**Pros**	**Cons**
Differential probes	THDP0200 TMDP0200	±1,600 ±750	200 200	8.75 8.75	Galvanic isolation	Limited bandwidth, long leads
Passive probes	P5150 TPP0850	2,500 2,500	500 800	3.50 2.19	High bandwidth	No galvanic isolation

*Calculated to include at least 5× the equivalent frequency from Eq. (8.1).

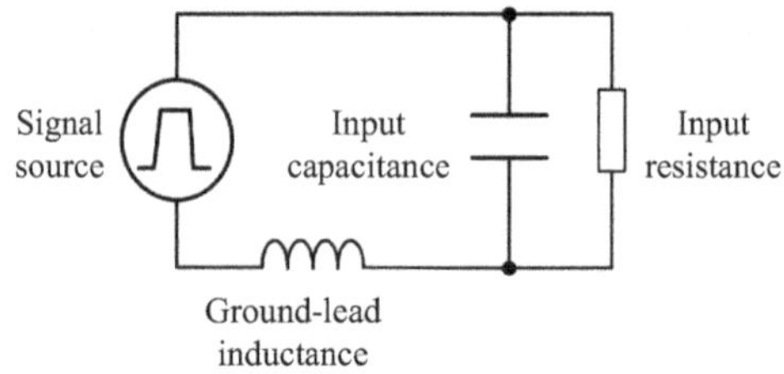

Figure 8.1 *Equivalent circuit of probe measurement loop with added inductance from a long ground lead*

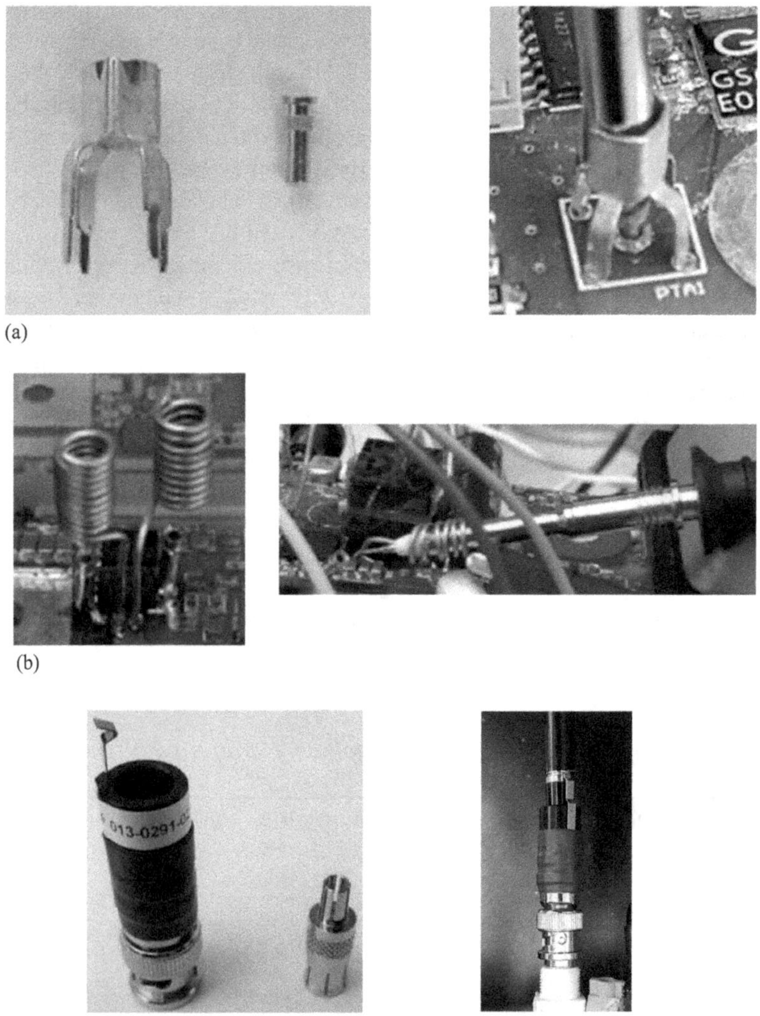

Figure 8.2 *Oscilloscope probe tip adapter options: (a) four-pronged "lunar lander" style; (b) hand-wound "paper-clip" style; (c) tip-to-BNC adapters for two different Tektronix probes*

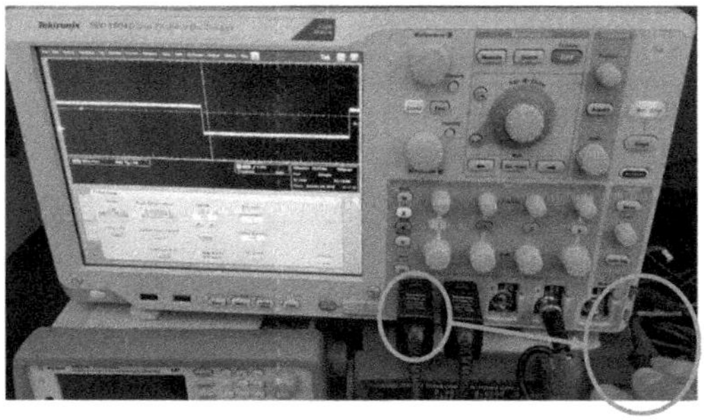

Figure 8.3 Voltage probe channel compensation/calibration in a Tektronix oscilloscope

adapters. The four-pronged "lunar lander" adapter provides solid mechanical connection and ultra-low inductance. However, its footprint may be too large for some board designs, and the pin spacing may not have sufficient clearance for all voltage levels. The second option, the "paper-clip" adapter, only requires two vias in the PCB, with spacing as required for clearance at the targeted voltage range. This option can be hand-made from solid-core wire or a discrete through-hole resistor lead, and it may provide suitable mechanical support for the probe, depending on the gauge of wire used. The third option can be more expensive than the rest, but it is sometimes preferred for higher voltage applications. This tip-to-BNC adapters has a small measurement loop area and provides a very sturdy mechanical connection.

Once the probe and oscilloscope are selected, the final step in voltage probe setup is the compensation of the measurement circuit in the oscilloscope. This procedure is typically only needed for passive probes, and it varies for different oscilloscopes. It usually involves connecting the probe to a square wave input and adjusting the circuit parameters until it captures an accurate rise time. Over-compensation or under-compensation can result in very poor waveform accuracy, so this procedure must be followed whenever a new scope or probe is introduced to the test setup. Figure 8.3 shows the procedure for one example oscilloscope.

8.4 Current sensor selection and setup

In addition to the voltages (v_{gs} and v_{ds}), current measurement is also desirable in a DPT. Table 8.2 lists some examples of current probes typically used in a DPT, and Figure 8.4 shows some of these examples as well [9–12]. For the load inductor current I_L, the split-core current probe is typically the most convenient option because size and bandwidth requirements are not as significant as for drain current. In the case of i_d measurement, typically only a current shunt provides sufficient bandwidth. This shunt cannot simply be a surface-mount resistor, because the series inductance of its package will severely limit its measurement bandwidth. An array of resistors in parallel may minimize this concern if effectively designed. One popular option is a coaxial current-viewing shunt resistor with four leads,

Table 8.2 Examples of current probes suitable for drain current measurement

Types/models	Current type	Bandwidth (MHz)	Rise time (ns)	Pros	Cons
Coaxial shunt SSDN-10	DC/AC	2,000	0.88	High bandwidth, high accuracy	No galvanic isolation, not suitable for continuous operation
Current transformer Pearson 2877	AC	200	8.75	High bandwidth, galvanic isolation	Not suitable for high current or continuous operation due to saturation
Split core current probe TCP0030A	DC/AC	120	14.58	Galvanic isolation, continuous operation	Low accuracy, limited bandwidth
Rogowski Coil CWT Mini	AC	17	102.9	Galvanic isolation	Low bandwidth and accuracy

including two terminals for the current flow and two separate terminals for Kelvin measurement of the sensing resistor. The Kelvin measurement terminals must be connected internally in such a way that the package inductance is not included. Figure 8.5 shows the results of a test to verify the improved accuracy of a shunt current measurement compared with other measurement techniques. The switching energy of a synchronous switch in a DPT is expected to sum to zero for adjacent turn-on and turn-off transients. In this way, the accuracy of the current measurement can be extrapolated from the error in switching energy measurement from zero. The higher bandwidth shunt measurement produced the most accurate current measurement in this SiC example.

While the coaxial shunt shown in Figure 8.4(b) enables accurate drain current measurement, it also adds a significant insertion inductance into the circuit. Some of this inductance can be avoided through effective DPT board layout as discussed in the previous chapters. However, the shunt case itself inserts several nH of inductance into the power loop. Figure 8.6 shows various connection techniques for a shunt from T&M Research, and the approximate insertion inductance resulting from each. This added inductance changes the behavior of the DPT circuit and of the device under test, especially with the near-chipscale packages commonly used for GaN FETs. In some cases, adding a shunt to the power loop can increase its total inductance by a factor of ten or more. One alternative concept is the surface-mount coaxial shunt resistor (SMCSR), which has been shown to have significantly lower insertion inductance than the regular coaxial shunts.

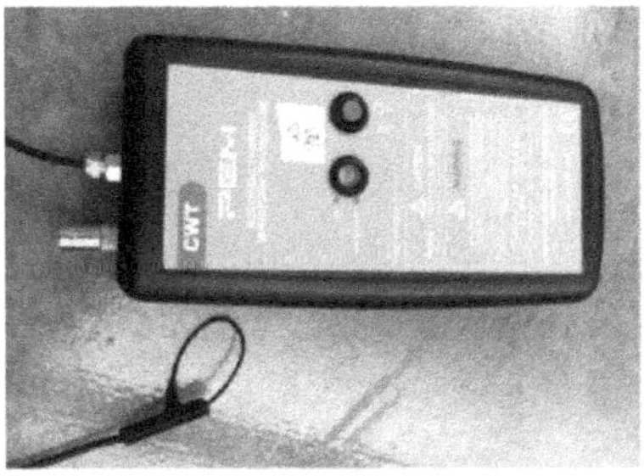

(a)

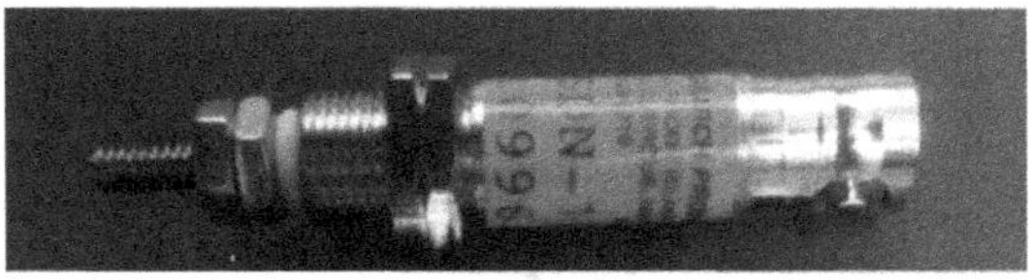

(b)

(c)

Figure 8.4 Typical current measurement options for a DPT, including (a) Rogowski coil, (b) coaxial shunt resistor, and (c) Pearson current transformer probe

8.5 Time-alignment (deskew) of measurements

After the probes have been selected for all voltage and current measurements, all of the oscilloscope channels must be aligned in time to a shared reference point. Each probe has a slightly different propagation delay, and the time misalignment of two channels can cause large errors in the calculation of switching energy and delay times. Figure 8.7 shows the effect of time misalignment between the drain current and voltage on the calculated switching loss.

For the v_{gs} and v_{ds} probes, the procedure to "deskew" the channels is relatively straightforward. The key to deskewing two channels is to identify a sharp and distinct waveform event that both probes can measure simultaneously. Using one of the two probes as a reference, the deskew setting on that oscilloscope channel can be adjusted until both probes' measurements match. The best way to perform this measurement is to connect both probes to the v_{gs} or v_{ds} measurement point on the

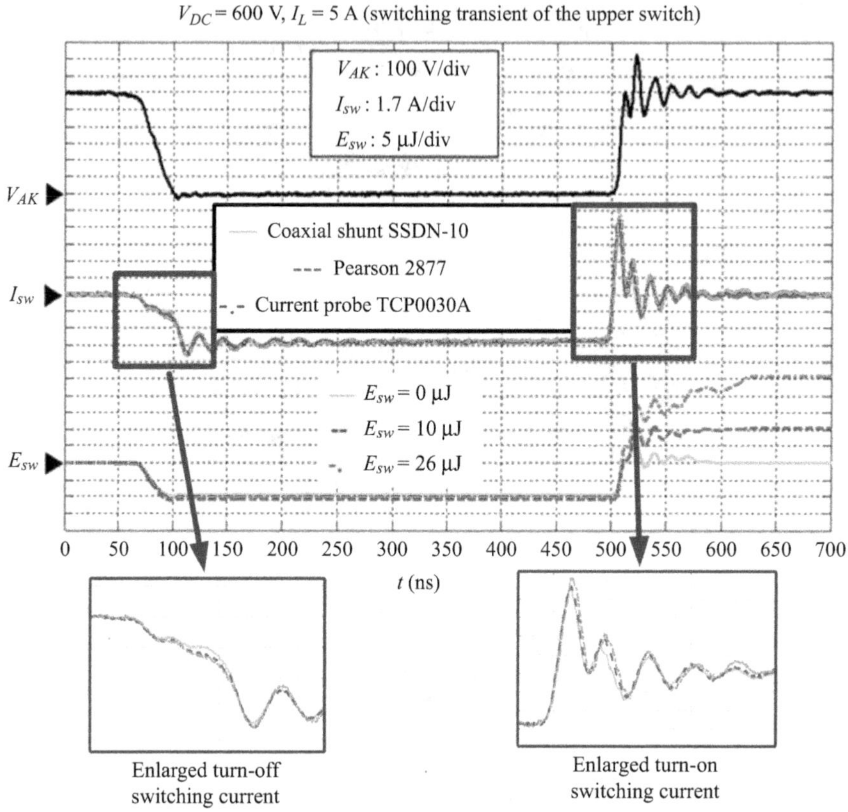

Figure 8.5 Comparison of the accuracy of several current measurement techniques

DPT circuit, which likely produces a sharper rise than most benchtop function generators. Deskewing the i_d measurement presents a different challenge. A voltage and current probe cannot both measure the same signal, since they are fundamentally different types of measurement. Although the scope is measuring the voltage on a current shunt, the shunt itself adds some amount of delay to the measurement, and this must also be deskewed. Therefore, the procedure for deskewing a current measurement may require a special fixture or test in some cases. One convenient workaround is to harness the inherent relationship between i_d and v_{ds} in a DPT circuit. In a fast switching transient, some events can be expected to occur simultaneously in both waveforms, and the current channel can be deskewed using one of these waveform features. However, this procedure is more effectively performed during the post-processing stage than in the measurement setup, and will be discussed in more detail later in this chapter.

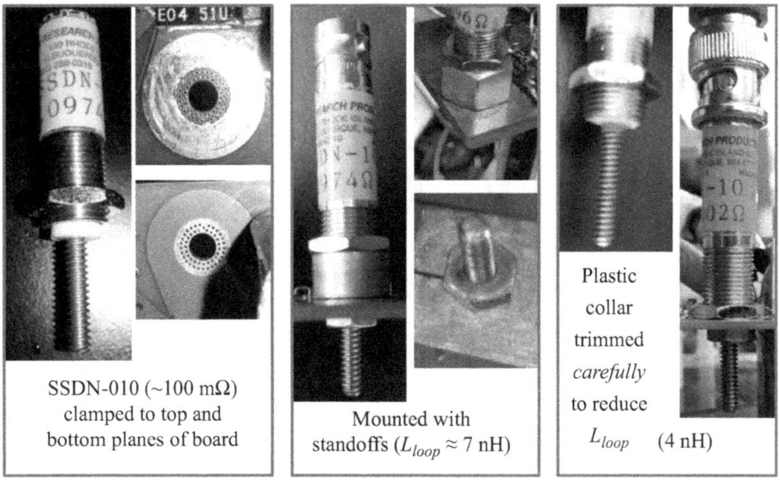

Figure 8.6 Connection and resulting insertion inductance of a coaxial current-viewing shunt resistor

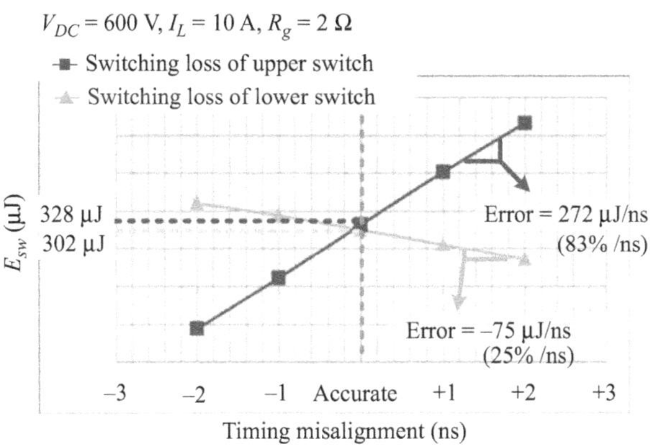

Figure 8.7 Example of the sensitivity of loss calculation to time-misalignment in current measurement

8.6 Consideration of CM ringing

As previously mentioned, passive probes are preferred in a DPT measurement setup because of the higher bandwidth typically available. However, this introduces a ground point at the source terminal of the low-side switch. The remainder of the DPT circuit should be isolated from ground, but each isolation point has a small amount of parasitic capacitance. These isolation capacitances, particularly in the

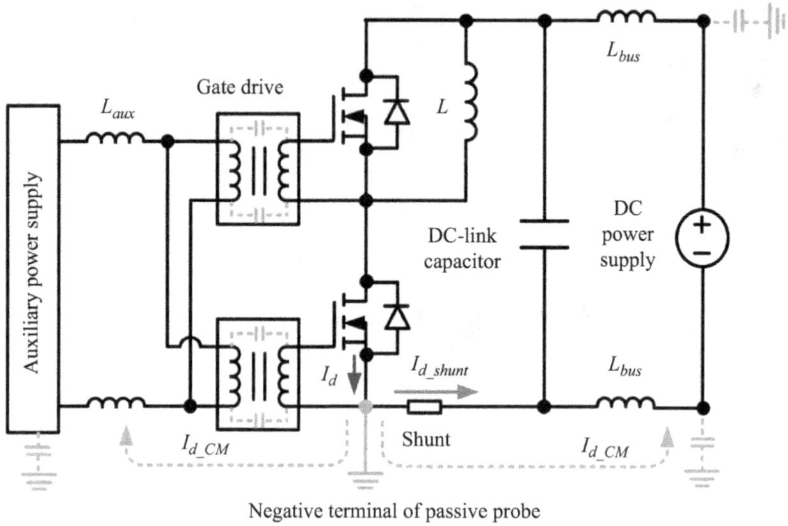

Figure 8.8 Equivalent circuit of a DPT showing the cause of common-mode ringing in parasitic impedance paths

high-side gate drive circuit and all isolated power supplies, resonate with the parasitic inductances in the circuit. This CM ringing shows up in oscilloscope measurements immediately following any high-speed switching transient. CM ringing is especially problematic in high-voltage, light-load measurements, when the magnitude of the ringing can be even higher than the magnitude of the actual drain current. Figure 8.8 shows an equivalent circuit of a DPT circuit, indicating the primary causes of CM ringing. Several solutions have been proposed to mitigate this error, such as floating the oscilloscope to eliminate the ground loop, and adding choke inductance to all loops in and out of the DPT board. Floating the scope can be dangerous for the user and the equipment and is typically not recommended. Rather, adding clip-on ferrite cores to the probe cables, power supply cables, and signal wires can be a simple and effective solution to this problem. Figure 8.9 shows the potential severity of CM ringing in a drain current measurement, and the improvement seen by implementing these solutions. Figure 8.10 shows a DPT setup with clip-on ferrite cores used as chokes on all wires between the DPT circuit and other equipment.

8.7 Goals of dynamic data processing

The key characteristics that are generally desired from DPT results are typically the switching loss (E_{on}, E_{off}), voltage overshoots ($V_{ds,pk}$, $V_{gs,pk}$), slew rates (dv_{ds}/dt, di_d/dt), and switching interval durations ($t_{d,on}$, t_{cr}, t_{vf}, $t_{d,off}$, t_{vr}). These values can be obtained directly from an oscilloscope using math functions and cursors, without further computational post-processing. However, implementing these functions in

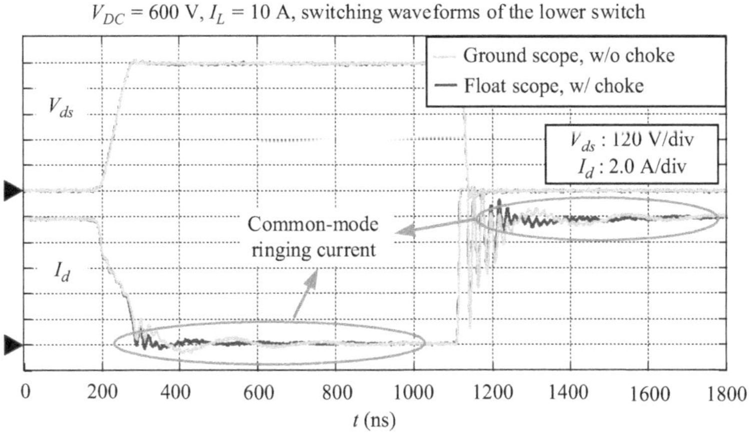

Figure 8.9 Example of the low-frequency oscillation in a DPT due to common-mode ringing

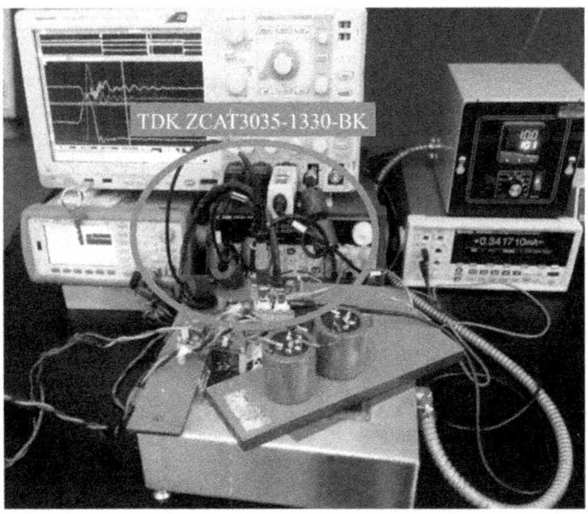

Figure 8.10 Addition of clip-on ferrite cores for common-mode inductance on all ground-connected loops

post-processing has several advantages. First, the methodology used for calculation can be standardized for application to all data points, which reduces the variable human error that may cause discrepancies when comparing results from different test conditions. Second, additional functions can be implemented in software once the data has been imported.

The flow chart for computational post-processing is shown in Figure 8.11, including several subroutines that will be discussed later in this chapter. The input for this data processing is primarily the measurement data from DPT results, tested at whatever operating conditions are of interest. This typically includes several DC

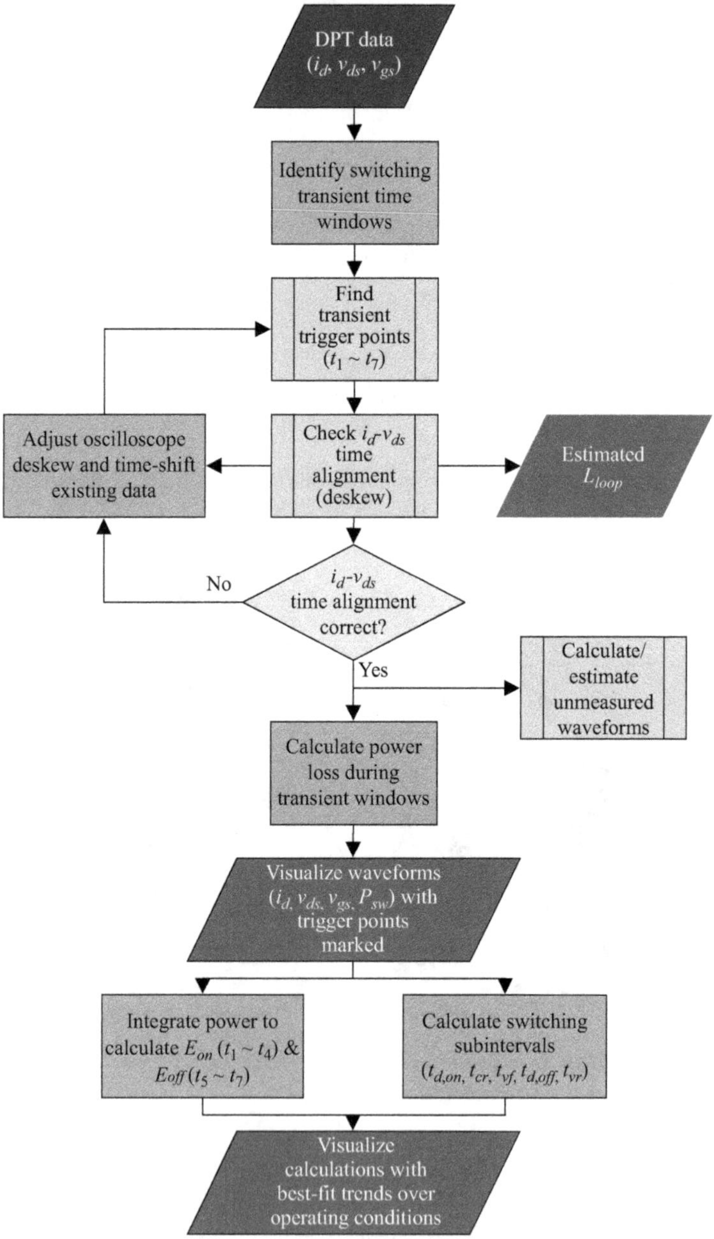

Figure 8.11 Flow chart for computational post-processing of DPT data. Different shapes are used to indicate processes, subroutines, decisions, and data input/output

voltages, load currents, and junction temperatures. Each DPT result data set includes at least one turn-on transient and one turn-off transient. The two pulses in a DPT actually produce two turn-ons and two turn-offs, although sometimes only a small subset of this data is saved to conserve storage space. The most useful data is the first turn-off transient (at the end of the first long pulse) and the second turn-on transient (at the beginning of the second short pulse), as well as a desired margin before and after these transients. If more data is captured and saved, then the first step in data processing is identifying these transients within the saved data. Once they are identified, the time range for each transient can be parsed out based on the midpoint of the transient (typically when $v_{ds} = V_{dc}/2$) and the desired window of time before and after this midpoint. The turn-on transient is processed to calculate the energy loss (E_{on}), switching subinterval durations ($t_{d,on}$, t_{cr}, t_{vf}), and slew rates (di_d/dt, dv_{ds}/dt). The waveforms are also visualized to help interpret the results and analyze features of interest. Likewise, the turn-off transients are processed to calculate the energy loss (E_{off}), switching subinterval durations ($t_{d,off}$, t_{vr}), voltage overshoots ($V_{ds,pk}$, $V_{gs,pk}$), and slew rates (di_d/dt, dv_{ds}/dt). To ensure that the i_d and v_{ds} measurements are accurately deskewed, the turn-on waveforms can be checked for time alignment. Conveniently, this process also gives an estimate of the power loop inductance (L_{loop}). A further post-processing option is to use the voltage-dependent junction capacitance curves (C_{oss}, C_{iss}, C_{rss}) to calculate waveforms for variables that were not measured. These variables include the internal output capacitance current (i_{Coss}), internal channel current (i_{chan}), gate current (i_g), and synchronous switch drain voltage ($v_{ds,sync}$). If the parasitic packaging inductances (L_d, L_s) and internal gate resistance ($R_{g,int}$) are known, then the v_{ds} and v_{gs} measurements can also be compensated to obtain the internal voltages on the drain and source contacts of the die.

The remaining sections of this chapter will detail the methodology for each of these post-processing steps, focusing on two examples: a SiC MOSFET and an enhancement-mode GaN FET. The same methodology can apply to any other fast-switching device. An implementation of this procedure using MATLAB script is included for reference in Appendix B.

8.8 Identification of switching transient subintervals

The first step in post-processing is identifying the time indexes that correspond to the turn-on and turn-off transients from within the saved data. This can easily be accomplished by searching for trigger points based on the expected behavior during each transient. Figure 8.12 shows a flow chart for identifying the turn-on and turn-off transient intervals, as well as further subintervals within each transient.

The turn-on transient can be identified based on v_{ds} dropping to half of the DC bus voltage, $V_{dc}/2$. Depending on the magnitude of noise and ringing, there may be repeated instances of v_{ds} crossing $V_{dc}/2$ with a negative slope. In this case, the first instance can typically be used. The turn-on transient window can then be selected based on this trigger point, with the desired time margin before and after to show

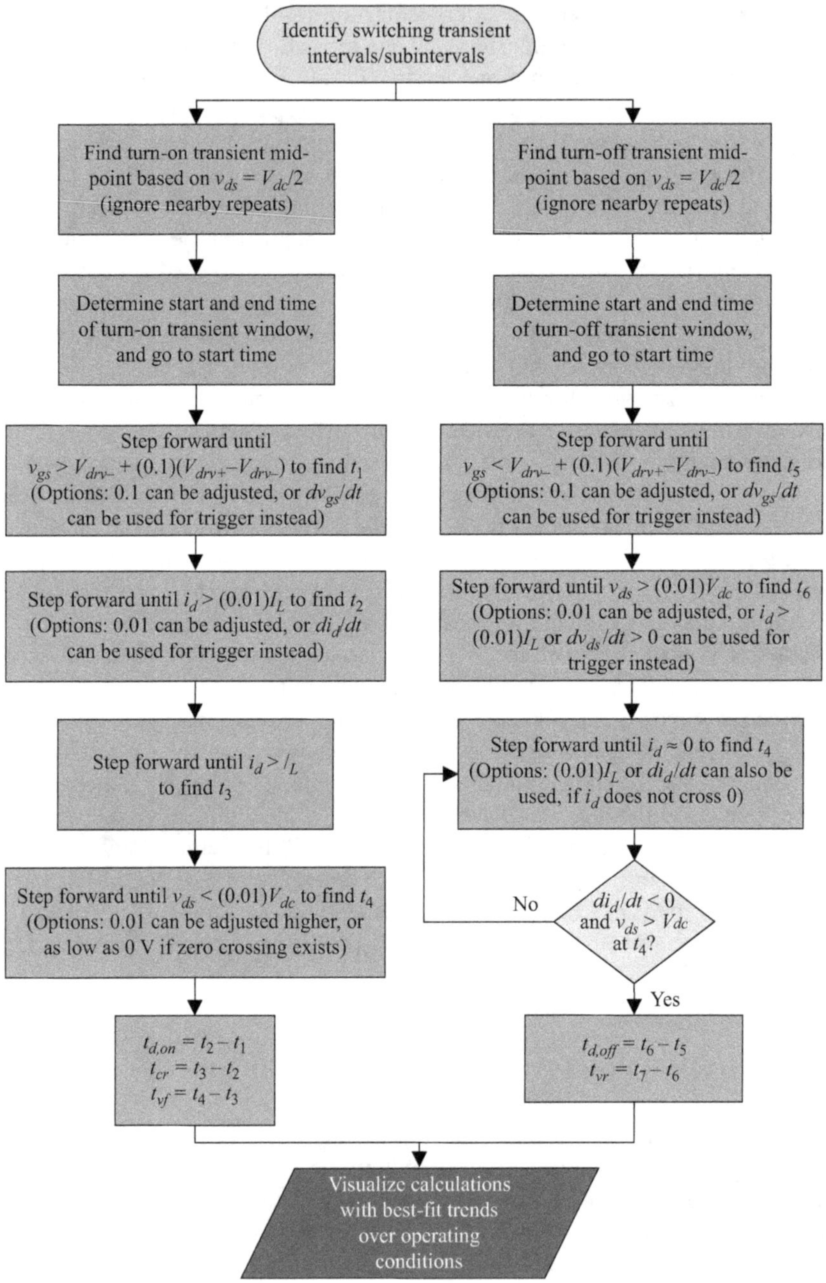

Figure 8.12 Flow chart for identifying trigger points between switching subintervals in DPT data. Different shapes are used to indicate processes, decisions, and data input/output

the desired information in waveform visualization. Stepping forward through the data from the start of the transient window, the start of the turn-on delay time can be triggered based on the gate voltage, either based on exceeding some predetermined percentage of its voltage swing (e.g., 0.01 V_{drv+}) or based on its slope becoming significantly positive. This time can be marked as t_1. Continuing to step forward, the next trigger point t_2 is similarly found when a positive change occurs in drain current. Next, t_3 is triggered by the drain current equaling or exceeding the load inductor current (i.e., $i_d \geq I_L$). Finally, the end of the turn-on transient t_4 is triggered by a zero crossing in v_{ds}. Typically, the drain voltage drops below 0 V at the end of the transient and resonates around it due to the parasitic inductance within the package ($L_d + L_s$). However, in cases where the v_{ds} measurement does not cross zero at the end of the turn-on transient, t_4 can also be triggered based on a change in the slope dv_{ds}/dt from negative to positive. The subinterval durations and overall turn-on transient duration can now be calculated from these times, according to

$$t_{d,on} = t_2 - t_1 \tag{8.3}$$

$$t_{cr} = t_3 - t_2 \tag{8.4}$$

$$t_{vf} = t_4 - t_3 \tag{8.5}$$

$$t_{on} = t_4 - t_1 \tag{8.6}$$

where $t_{d,on}$ is the turn-on delay time, t_{cr} is the current rise time, t_{vf} is the voltage fall time, and t_{on} is the turn-on transient time.

Figure 8.13 shows example turn-on transient waveforms for a SiC MOSFET and GaN FET, with the subintervals described earlier denoted by dashed lines. The instantaneous power loss of the device is also shown here, which is simply the product of i_d and v_{ds}. Determining switching energy loss from this power will be explained later in this chapter. Figure 8.14 shows subinterval trends for the GaN example as a function of load current and bus voltage. As indicated, the current rise time of WBG devices typically shows a strong positive correlation with load current, while the delay time and voltage fall time show a weaker relationship. Neither curve indicates a noticeable trend with operating bus voltage. Next, Figure 8.15 continues the analysis of the DPT waveforms with trends of subinterval times over junction temperature and current. The SiC results indicate a more noticeable positive correlation between switching time and load current, as well as a weak negative correlation between switching time and temperature. This is partly due to the threshold shift in SiC MOSFETs with temperature, although this effect is somewhat counteracted by the negative temperature coefficient of transconductance. In these example GaN results, the negative temperature coefficient of transconductance also produces a weak correlation between temperature and switching time. This correlation is more noticeable in some GaN FETs than in others.

Next, the turn-off transient can be identified based on v_{ds} rising to half of the DC bus voltage, $V_{dc}/2$. Depending on the magnitude of noise and ringing, there may be repeated instances of v_{ds} crossing $V_{dc}/2$ with a positive slope. In this case, the

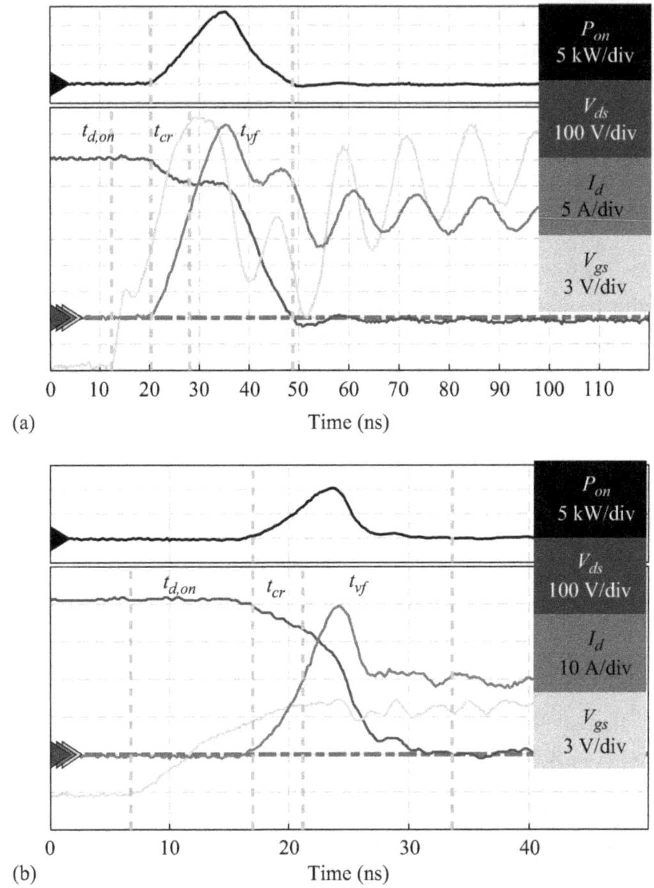

Figure 8.13 Example turn-on transient waveforms with subintervals denoted by dashed lined using DPT data for (a) a SiC MOSFET and (b) a GaN FET

first instance can typically be used. The turn-off transient window can then be selected based on this trigger point, with the desired time margin before and after to show the desired information in waveform visualization. The margins for turn-on and turn-off transient windows will likely be different because these two transient times vary significantly from each other in most operating conditions. Stepping forward through the data from the start of the transient window, the start of the turn-off delay time can be triggered based on the gate voltage, either based on falling below some predetermined percentage of its voltage swing (e.g., 0.9 V_{drv+}) or based on its slope becoming significantly negative. This time can be marked as t_5. Continuing to step forward, the next trigger point t_6 is found when a

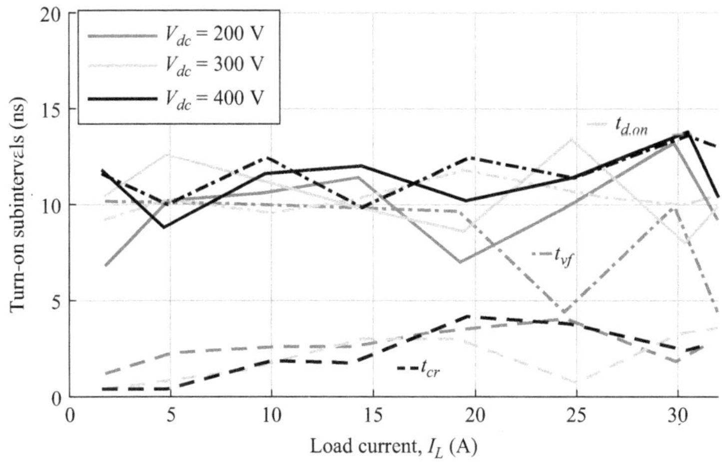

Figure 8.14 Example of turn-on transient subinterval trends over current and voltage using DPT data for a GaN FET

positive change occurs in v_{ds}. Next, t_7 is triggered by the drain current crossing 0 A with a negative slope. As mentioned previously in the turn-on transient analysis, triggering on a zero crossing can be problematic near the end of the switching transient. In some extremely light-load cases, the i_d measurement may not cross 0 A, and a slope polarity reversal can instead be used to detect the end of the turn-off transient. In some heavy-load cases, the i_d measurement may cross 0 A one or more times before the transient is actually complete. This can cause significant error in subinterval durations as well as switching energy loss. Therefore, the first zero crossing of i_d is not necessarily the correct trigger for t_7. At the moment when the turn-off transient is complete, the drain current should be crossing zero with a negative slope, and v_{ds} should be near its peak overshoot. Therefore, the trigger for t_7 can be refined by verifying that di_d/dt is negative and v_{ds} is greater than V_{dc} at this moment. If either of these is not true, then the next i_d zero crossing can be detected by continuing to step forward through the data, until one is found where $di_d/dt < 0$ and $v_{ds} > V_{dc}$.

The subinterval durations and overall turn-on transient duration can now be calculated from these times, according to

$$t_{d,off} = t_6 - t_5 \tag{8.7}$$

$$t_{vr} = t_7 - t_6 \tag{8.8}$$

$$t_{off} = t_7 - t_5 \tag{8.9}$$

where $t_{d,off}$ is the turn-on delay time, t_{vr} is the voltage rise time, and t_{off} is the turn-off transient time.

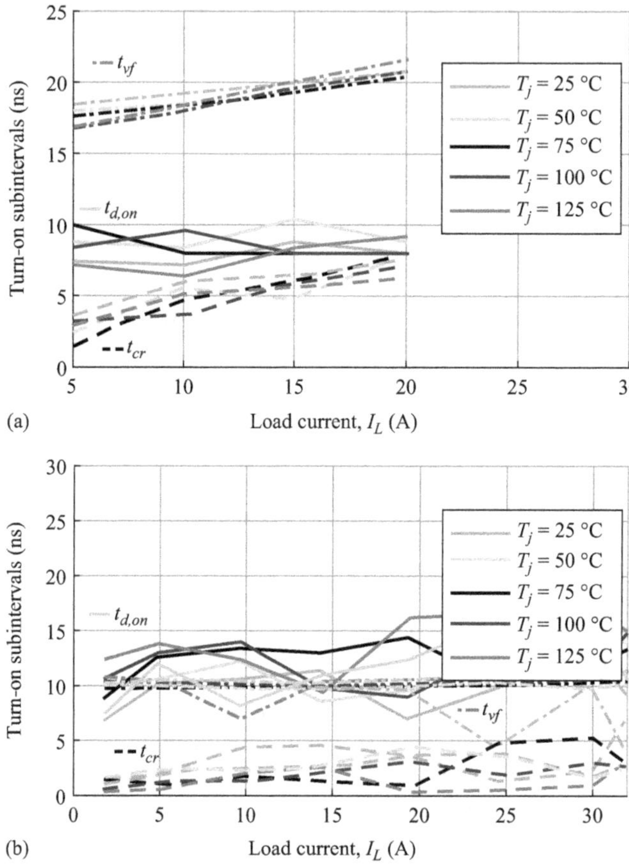

Figure 8.15 Example of turn-on transient subinterval trends over temperature using DPT data for (a) a SiC MOSFET and (b) a GaN FET

Figure 8.16 shows example turn-off transient waveforms for a SiC MOSFET and a GaN FET, with these subintervals denoted by dashed lines. Figure 8.17 shows subinterval trends for the GaN example as a function of load current and bus voltage. As indicated, the turn-off time of WBG devices typically shows a strong inverse relationship with load current. There is also a positive correlation with DC bus voltage due to the higher output capacitance charge to displace (Q_{oss}). The delay time does not vary much with voltage or current. Next, Figure 8.18 shows subinterval trends for both SiC and GaN examples over temperature and current. Here, the SiC MOSFET is more impacted by temperature than the GaN example, with a noticeably positive temperature coefficient in the delay time. This is due to the threshold voltage shift in the example SiC MOSFET. These trends vary between different SiC and GaN devices, depending on the temperature dependence of their threshold voltage and transconductance.

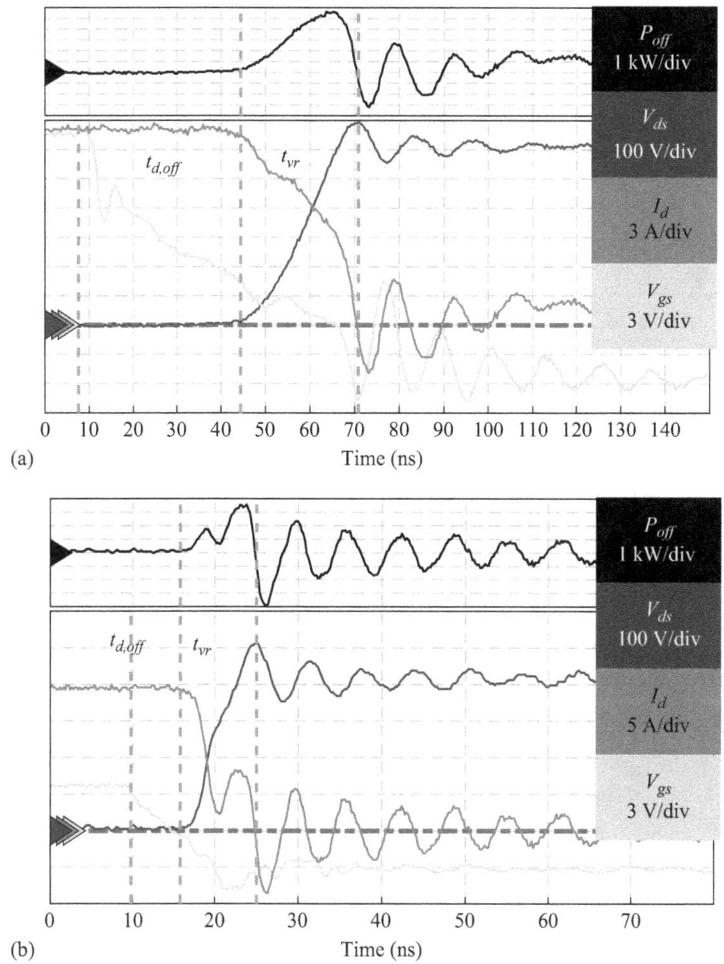

Figure 8.16 Example turn-off transient waveforms with subintervals denoted by dashed lined using DPT data for (a) a SiC MOSFET and (b) a GaN FET

8.9 Calculation of *I–V* misalignment for deskew adjustment

Accurate post-processing of the data requires that the measurements be time-aligned very precisely by adjusting the oscilloscope's deskew settings for each channel. The measurement equipment setup typically includes some procedures for deskew adjustment. However, the switching energy loss calculations are particularly sensitive to i_d–v_{ds} misalignment, even within fractions of a nanosecond. For this reason, the best practice is to verify the alignment of these two measurements as part of the post-processing software analysis by comparing them at a time when the

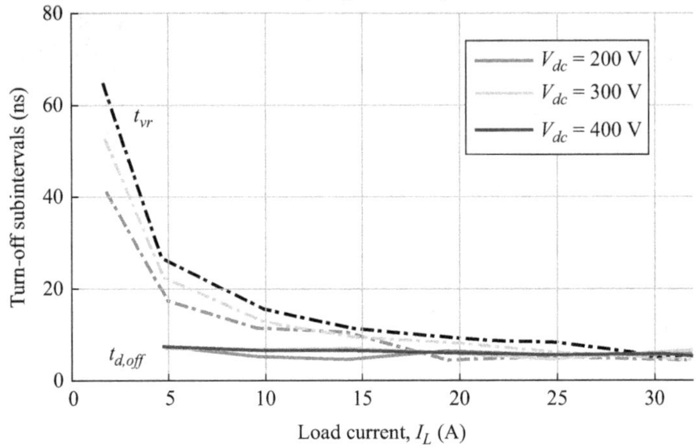

Figure 8.17 Example of turn-off transient subinterval trends over current and voltage using DPT data for a GaN FET

relationship between the two variables is well-defined and easily calculated. This procedure can also help the experimenter to determine realistic boundaries for *I–V* alignment tolerance, as well as the sensitivity of E_{on} and E_{off} calculations to potential misalignment error.

At any time during the switching transient, the drain voltage can be modeled as

$$v_{ds} = V_{dc} - v_{ds,comp} - i_d R_{shunt} - L_{loop}\frac{di_d}{dt} \tag{8.10}$$

where L_{loop} is the power loop inductance, R_{shunt} is the resistance of the coaxial current-viewing shunt resistor, and $v_{ds,comp}$ is the drain voltage of the complementary/synchronous switch. During most of the turn-on and turn-off transients, the relationship between i_d and v_{ds} is complicated by the changing voltage $v_{ds,comp}$. However, during the turn-on current rise time t_{cr}, $v_{ds,comp}$ is clamped by the body diode, antiparallel diode, or DLB mechanism to a known value. In this case:

$$v_{ds} = (V_{dc} + V_{bd}) - i_d R_{shunt} - L_{loop}\frac{di_d}{dt} \tag{8.11}$$

where V_{bd} is the forward voltage drop of the diode mechanism. In this relationship, the only nonconstant variables are i_d and v_{ds}. Therefore, v_{ds} can easily be calculated as a function of time using the measured i_d if the values R_{shunt} and L_{loop} are known. Comparing the calculated and measured v_{ds} waveforms during t_{cr} clearly indicates whether the two measurements are well time-aligned. Figure 8.19 shows a flow chart for this procedure.

This technique can be very effective in deskewing fast-switching transient data when properly implemented. Correctly calculating di_d/dt and estimating loop inductance are two key challenges that must be addressed in this procedure.

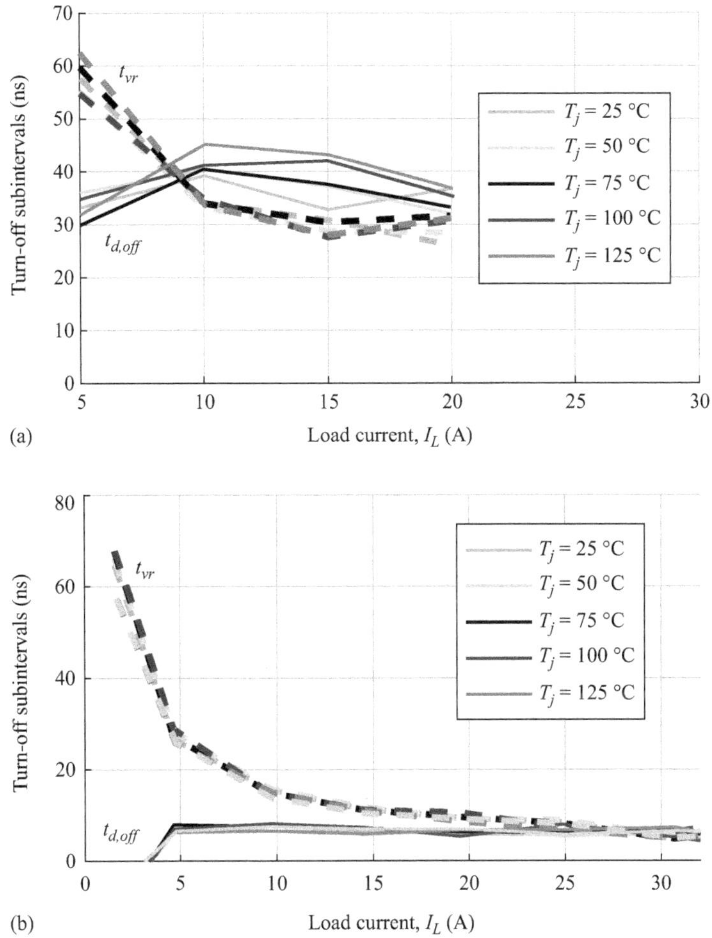

Figure 8.18 Example of turn-off transient subinterval trends over temperature using DPT data for (a) a SiC MOSFET and (b) a GaN FET

To calculate the current slew rate without introducing a time offset, the slope can be calculated at each time index according to

$$\frac{di_d}{dt}(t) = \frac{i_d(t + \Delta t) - i_d(t - \Delta t)}{2\Delta t} \tag{8.12}$$

where Δt is an integer multiple of the sampling period t_{sample}. This calculation produces a slope that is centered at each time index. Choosing a higher filtering period Δt reduces the noise in the result, but it also filters out the higher frequency components in the switching transient. Therefore, this filtering period should be selected as low as possible, without allowing the amplitude of noise to dominate the calculation.

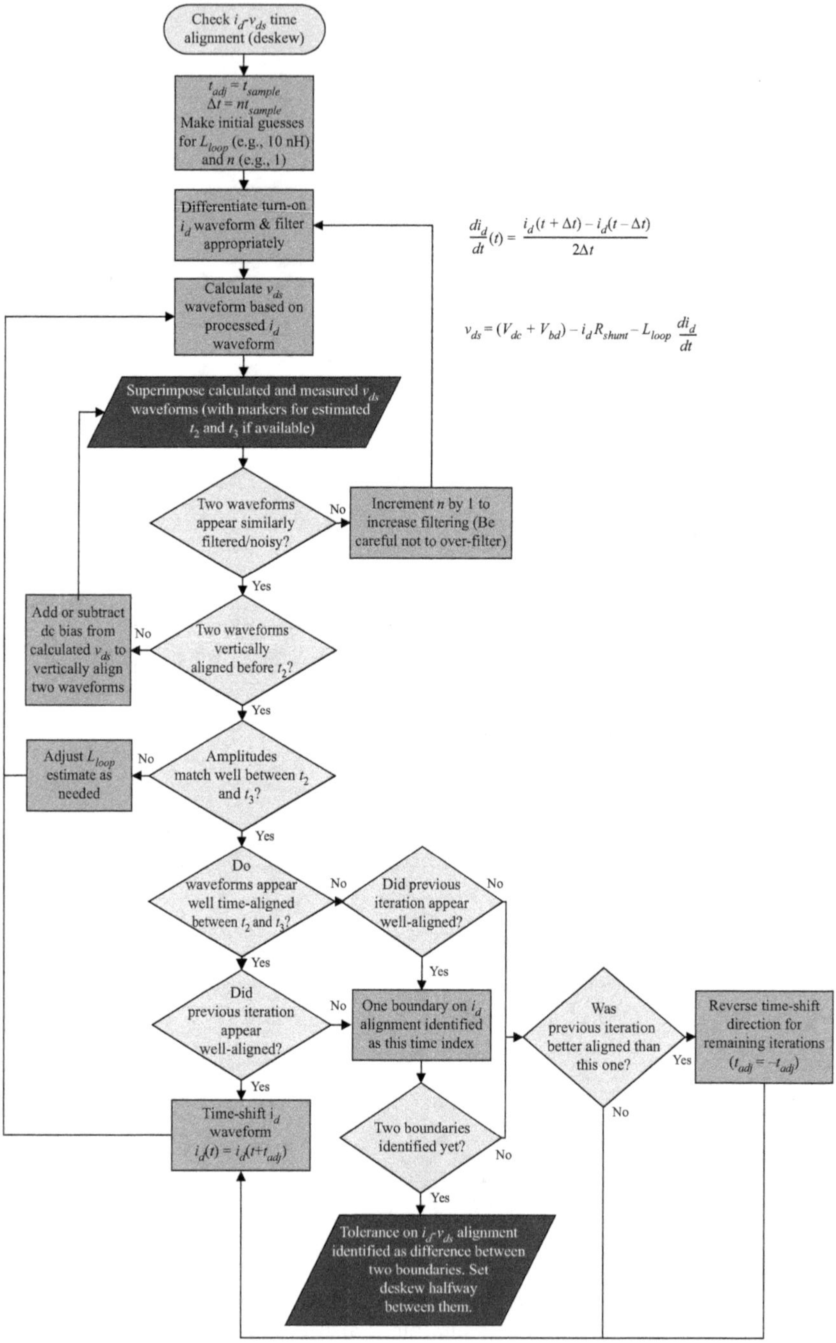

Figure 8.19 Flow chart for computational I–V alignment of DPT data for deskew adjustment. Different shapes are used to indicate processes, decisions, and data input/output

Next, the loop inductance must be estimated. An approximate value for L_{loop} may already be known for the design under test, which can be verified with this deskew procedure. In fact, L_{loop} can be found from this procedure even if no previous estimate has been made, without impacting the *I–V* alignment. The vertical amplitude match of the voltage drop during t_{cr} corresponds to the L_{loop} estimate. The horizontal offset match between the two compared waveforms indicates the *I–V* misalignment. The calculated and measured v_{ds} waveforms can be superimposed graphically to visually assess how well they match. Once the i_d data has been time-shifted to correct misalignment, and the L_{loop} has been estimated accurately, the two waveforms should be nearly identical.

The graphical alignment of two waveform plots is limited by the subjectivity of the visual analysis, as well as the selected degree of noise filtering. This subjectivity and its potential error can be characterized by time-shifting the i_d data to intentionally misalign the measurements. The time-shift is gradually increased until the two v_{ds} waveforms no longer appear well-aligned. Then, the data is time-shifted in the opposite direction until the two v_{ds} waveforms again appear misaligned. This establishes a tolerance on the accuracy of the *I–V* alignment, and the midpoint between these two tolerance boundaries should be selected as the correct amount of deskew required. With the data intentionally misaligned to each tolerance boundary, the switching energy loss can be calculated to show the potential impact of misalignment on the final E_{on} and E_{off} calculations.

This graphical *I–V* alignment technique is demonstrated in Figure 8.20 using the example of a GaN FET. It is helpful to identify the current rise-time interval before beginning this procedure, but the subinterval triggering procedure from the previous section must be repeated after any deskew adjustment is made. It is also important to remember that deskew is a fixed experimental parameter and does not change with operating conditions. Once a deskew setting is chosen based on *I–V* alignment verification, this setting should not be changed unless the experimental test setup is modified. Therefore, the accuracy of deskew selection can be further improved by testing multiple data points across a range of operating conditions. The duration of t_{cr} increases with higher load current, making this a particularly effective test condition to use for the procedure.

8.10 Calculation of switching energy loss

The instantaneous power loss during the switching transient is easily calculated from the product of i_d and v_{ds} measurements, as shown in Figures 8.13 and 8.16. The switching energy losses are simply the integral of this power loss over the transient interval, as given by

$$E_{on} = \int_{t_1}^{t_4} i_d v_{ds} dt \tag{8.13}$$

$$E_{off} = \int_{t_5}^{t_7} i_d v_{ds} dt. \tag{8.14}$$

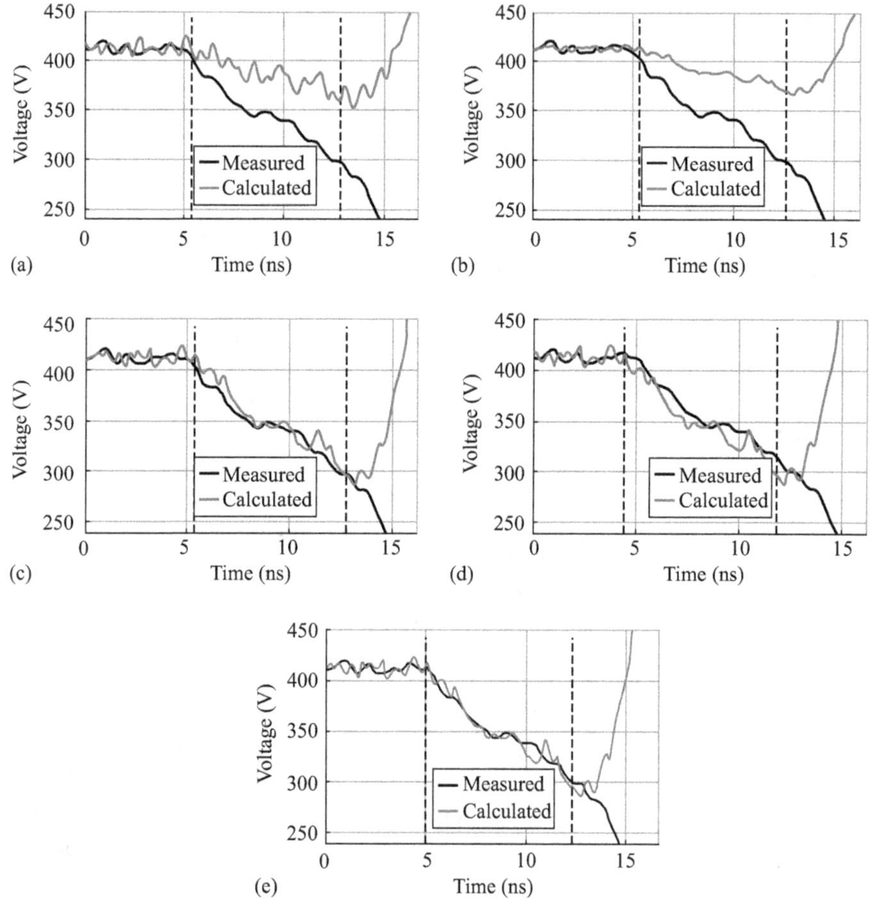

Figure 8.20 Example waveforms showing computational I–V alignment steps of DPT data for deskew adjustment using example GaN FET DPT data: (a) filter frequency too low; (b) loop amplitude too low; (c) i_d skewed to the right; (d) i_d skewed to the left; (e) final selected deskew setting

Figure 8.21 shows an example trend of the E_{on} calculations for the GaN example over various DC bus voltage and load current conditions. This trend indicates a positive trend with voltage and current. Next, Figure 8.22 shows the trends in E_{on} for the SiC and GaN examples with temperature and load current. Both demonstrate a positive trend with temperature. It is worth noting that SiC MOSFETs will sometimes demonstrate a slightly negative temperature coefficient in E_{on} because of the threshold voltage shift discussed previously. However, in this example, the reverse recovery charge in the high-side SiC MOSFET's body diode has a strong positive temperature coefficient, which increases the turn-on loss at higher temperature. The GaN example also has a

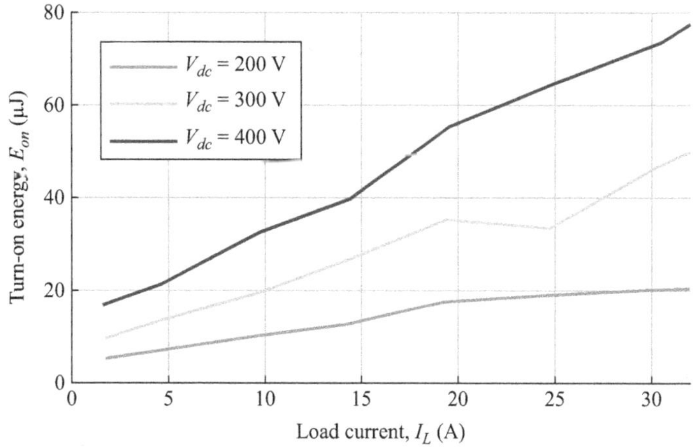

Figure 8.21 Example of turn-on switching loss trends versus current and voltage using DPT data for a GaN FET

weak positive temperature coefficient, although the strength of this trend varies between GaN FETs depending on the temperature dependence of their transconductance.

Figure 8.23 shows the trend in turn-off loss for the GaN example with operating voltage and current. This calculation is typically flat over current, although it has a positive correlation at lower voltage, higher current, or with slower gate drives. In all cases, the trend with voltage is positive, as there is more output charge (Q_{oss}) to displace. Next, Figure 8.24 shows the trends in E_{off} for the SiC and GaN examples over temperature and current. Here, the SiC example has a much more noticeable positive trend with current because this example uses a slower gate drive circuit than the GaN example. The SiC example also shows a slight positive temperature coefficient, whereas the GaN example has a slight negative coefficient. This is often true for these device types (although not always) because of the threshold voltage shift in SiC and transconductance curve in GaN.

As shown in these examples, the E_{off} trend appears nearly constant over current at light-load conditions. With WBG devices, the gate voltage can fall quickly enough during turn-off transients to avoid any Miller plateau and subsequent resistive channel losses. Therefore, the only measured E_{off} is the energy stored in the active switch C_{oss}. This energy is not actually lost during the turn-off transient and is in fact lost resistively during the following hard turn-on transient. In the case of resonant soft-switching, no hard turn-on transient occurs, so that stored energy is recovered rather than lost. The actual energy lost during hard turn-off and turn-on transients can therefore be estimated by compensating the calculations as in

$$E_{on,compensated} = \int_{t_2}^{t_4} i_d v_{ds} dt + E_{off,0A} \tag{8.15}$$

$$E_{off,compensated} = \int_{t_5}^{t_7} i_d v_{ds} dt - E_{off,0A}. \tag{8.16}$$

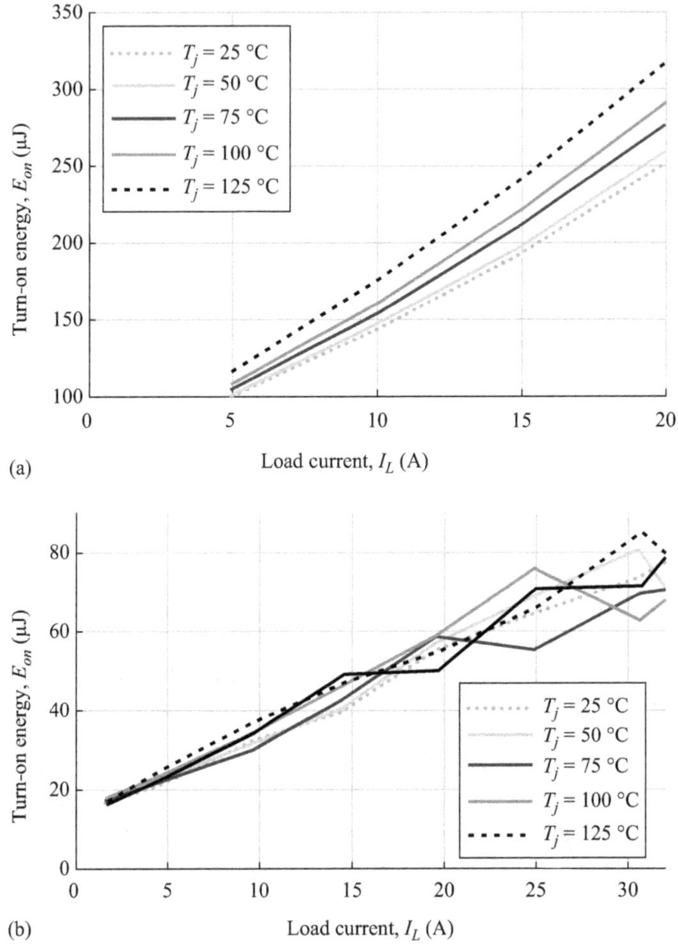

Figure 8.22 Example of turn-on switching loss trends versus temperature using DPT data for (a) a SiC MOSFET and (b) a GaN FET

These compensated switching energy losses are sometimes reported rather than the directly measured values, especially in characterizations intended to inform resonant soft-switching converter design. The no-load convergence points of each energy loss trend can also be calculated from the C_{oss}–v_{ds} curve as a function of DC bus voltage, as given by

$$E_{on,0A} = \int_0^{V_{dc}} \left[C_{oss}(v) + C_{stray,sync}\right] \times (V_{dc} - v)dv \tag{8.17}$$

$$E_{off,0A} = \int_0^{V_{dc}} \left[C_{oss}(v) + C_{stray,active}\right] \times v\, dv. \tag{8.18}$$

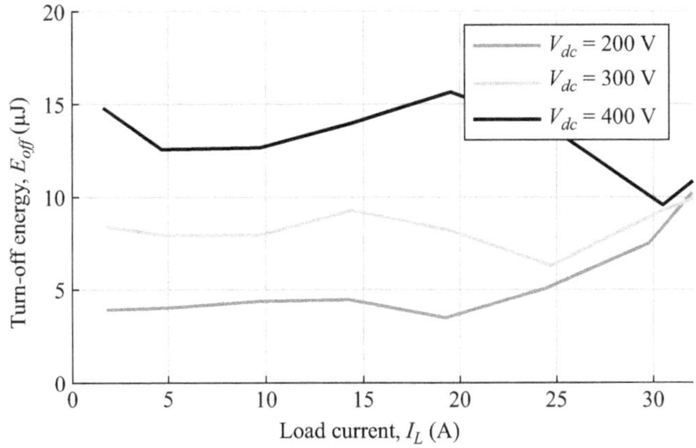

Figure 8.23 Example of turn-off switching loss trends versus current and voltage using DPT data for a GaN FET

where $C_{stray,sync}$ and $C_{stray,active}$ are the stray capacitances in parallel with the synchronous switch and active switch, respectively. The stray capacitance is often higher for the low-side switch because of isolation leakage capacitance at the switch node, so these two values may not be the same. Although it is difficult to measure stray capacitances, they can be determined by fitting (8.17) and (8.18) to experimental loss trends at one bus voltage. Then, the same equations can be applied to determine the no-load convergence point at other bus voltages.

8.11 Consideration of ringing and DC bias

The boundaries of the integrals in (8.13) and (8.14) can be widened to include the ringing that occurs at the end of turn-on and turn-off transients. However, this is not recommended for most cases. The energy encompassed by this ringing typically sums to zero, but capturing only a portion of the resonant energy will introduce error.

The integral must end at a negatively sloped zero crossing (of v_{ds} in turn-on and i_d in turn-off) to avoid accidentally capturing some of the energy resonating between the loop inductance (L_{loop}) and the output capacitance (C_{oss}) of the off-state device. This ringing is most prominent in v_{ds} after the turn-off transient, and in i_d after the turn-on transient. If the integral is extended past the point where the ringing is damped out, this can also be accurate. However, there is often a significant low-frequency resonance in both i_d and v_{ds} measurements immediately after the transient, which appears as a DC bias error relative to the transient window. Attempting to include this ringing energy in the calculation can cause significant error by integrating the DC bias, which is worsened by the subjective

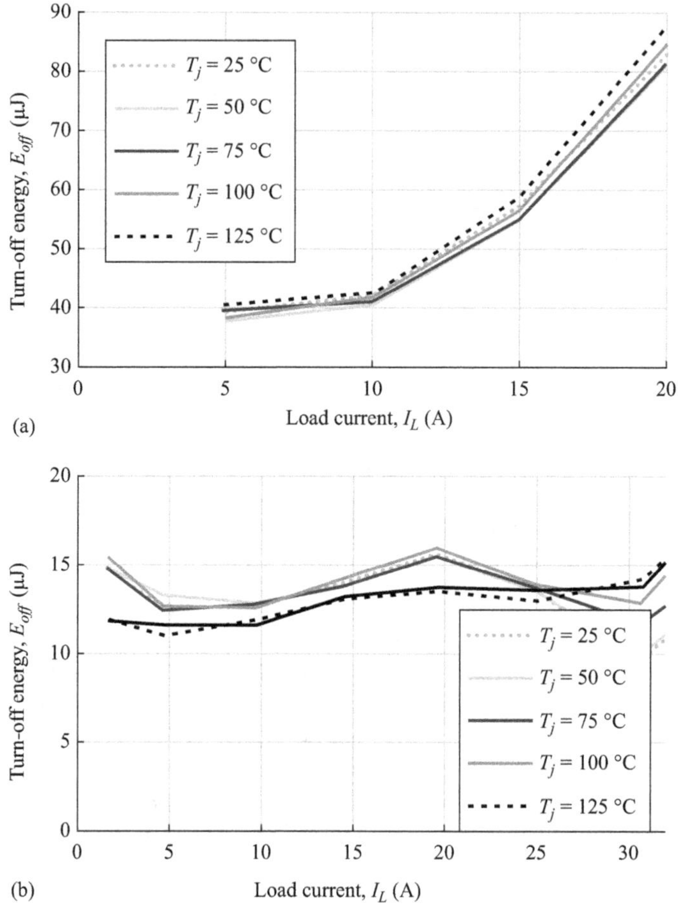

Figure 8.24 Example of turn-off switching loss trends versus temperature using DPT data for (a) a SiC MOSFET and (b) a GaN FET

selection of the integral duration. Ending the integral at the zero crossing eliminates this subjective error and minimizes the impact of any remaining DC bias error.

DC bias error compensation is a common practice in DPT post-processing. The goal is to improve the accuracy of calculated losses, but in fact this bias compensation can easily introduce new error. Figure 8.25 demonstrates the impact of attempting to "correct" an apparent DC bias error in the current measurement of a GaN FET. As pictured, the turn-off transient in the 1 A condition appears to show a DC bias error, because i_d resonates around a nonzero current after the transient has ended. Figure 8.25(b) shows an attempted bias compensation, re-centering the i_d waveform around 0 A after the transient completes. However, this creates a "dip"

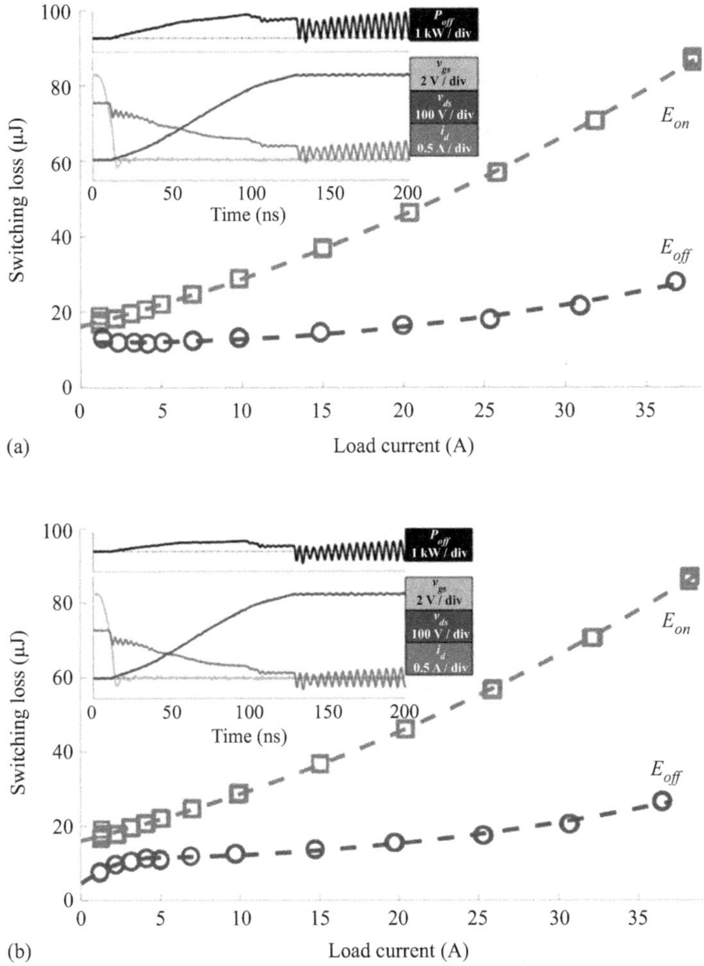

Figure 8.25 Impact of current shunt DC bias adjustment on the accuracy of a turn-off switching loss trend for an example GaN FET: (a) uncompensated loss trend, with inset showing 1 A turn-off transient; (b) compensated loss trend, with inset showing 1 A turn-off transient with DC bias adjustment applied

in the E_{off} trend as load current approaches zero. The error being adjusted here is not actually a DC bias at all, but a low-order CM resonance that occurs in many test setups and conditions (previously explained in Section 8.6). However, this bias is not significant at higher load conditions, because the integration is performed over the turn-off time, which is an inverse function of load current. At the very light load condition, this low-order resonance becomes much more significant. The most accurate treatment for such measurement error is simply to consider it a boundary condition for testing. When the current is low enough that this CM ringing introduces noticeable error to the turn-off loss, the test setup cannot accurately be used to calculate turn-off loss at that condition. Compensation in post-processing leads to further error. CM ringing can be reduced with the techniques previously mentioned in Section 8.6, which can improve the test setup's ability to accurately measure turn-off transients at lower load current. In most cases, ending the turn-off loss integral at the first zero crossing (according to the methodology presented in Figure 8.12) prevents this CM ringing from causing any significant error even in relatively light load conditions.

8.12 Alternative method for switching energy loss calculation

The conventional loss calculation methodology has several limitations when applied to WBG device characterization, some of which have already been explained in this chapter. However, there is an alternative approach to switching energy loss calculations that overcomes two of these limitations. The first is sensitivity to *I–V* misalignment, which can be quite severe, as previously mentioned. The second limitation is its sole focus on energy loss in the active switch. In fact, the switching transients in the synchronous switch are expected to be lossless, with the total combined energy during a turn-on and turn-off transient summing to zero. However, this assumption may be invalid in some cases (e.g., when cross-talk occurs, as discussed later in Chapter 9). The alternative loss calculation method relies on the conservation of energy in the circuit, which dictates that

$$E_{sw} = \int_{t_a}^{t_b} \left[V_{dc} i_{d,LS} - I_L v_{ds,HS}\right] dt \tag{8.19}$$

where $i_{d,LS}$ is the drain current of the low-side (active) switch and $v_{ds,HS}$ is the drain voltage of the high-side (synchronous) switch.

When using the recommended high-bandwidth passive probes, these four measurements cannot be made simultaneously due to grounding conflicts. However, (8.19) can conveniently be split into two separate integrals with the same bounds:

$$E_{sw} = \int_{t_a}^{t_b} V_{dc} i_{d,LS} dt - \int_{t_a}^{t_b} I_L v_{ds,HS} dt. \tag{8.20}$$

Now, the required measurements can be made without conflicting ground connections. First, the measurements V_{dc} and $i_{d,LS}$ can be made using the setup shown in Figure 8.26(a). This configuration is similar to the conventional setup discussed previously, with the primary difference that the DC voltage is measured here with a passive probe grounded at the $-V_{dc}$ terminal, requiring the polarity of the current shunt measurement to be reversed. After a DPT is performed using this configuration, a second DPT can be performed with the modified measurement

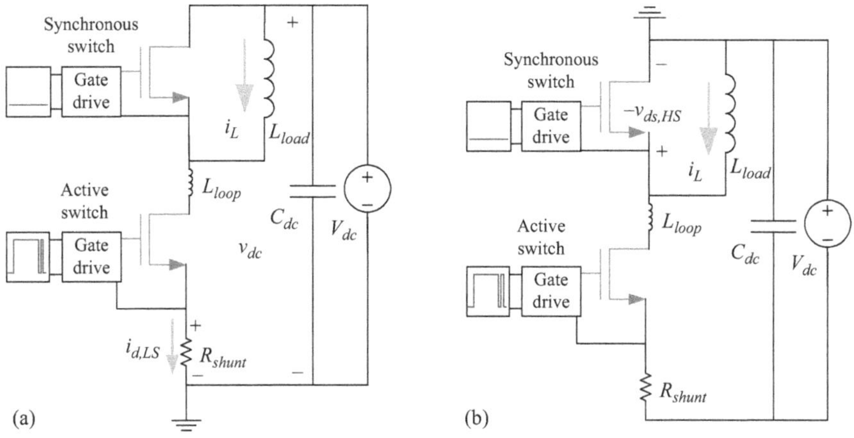

Figure 8.26 Modified DPT configuration for alternative switching loss measurement method, with measured variables indicated: (a) test 1, to capture low-side drain current; (b) test 2, to capture high-side drain voltage

connections shown in Figure 8.26(b). In this configuration, the only measurements taken are the high-side drain voltage and load current. It is important to capture $v_{ds,HS}$ with a high-bandwidth passive probe, which requires connecting the oscilloscope's ground to the $+V_{dc}$ terminal and measuring the voltage with reversed polarity. The load current can be measured in both (a) and (b) configurations since it is an isolated probe with no ground.

Once all measurements have been obtained, (8.20) can be calculated over the desired integral bounds (t_a to t_b) to obtain the total switching loss in both devices. The bounds of integration must be selected to include two complete switching transients (turn-on and turn-off) as well as the time between them. If a typical double pulse is applied, the integration bounds could extend from the end of the first pulse to the beginning of the second pulse or from the beginning to the end of the second pulse.

The I_L measurement, captured during both tests, can be used to time-align the two waveforms approximately. However, the calculation is immune to a reasonable degree of time misalignment as long as both terms in (8.20) are integrated over the same time duration t_b–t_a. At any instant when the circuit is operating in steady state (i.e., when no switching transient is occurring), the two terms sum to zero. Therefore, the immunity to time misalignment simply requires that the same total duration of steady-state data be included in both integrals. However, compared with the conventional technique, this technique may be more susceptible to error at light load caused by CM bias and measurement ringing because of the longer time scale of integration. Figure 8.27 demonstrates this alternative switching loss measurement and calculation for a SiC MOSFET.

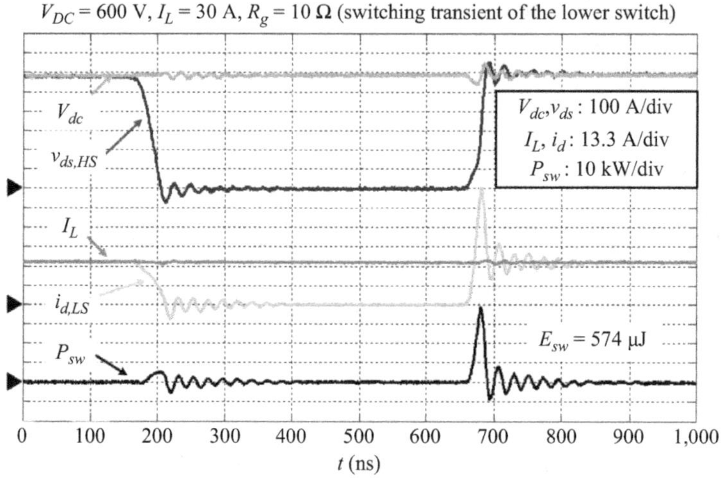

Figure 8.27 Implementation example of alternative switching loss measurement using results from a SiC MOSFET DPT

8.13 Calculation of overshoot voltages and slew rates

The shape of the drain voltage waveform is another important deliverable from dynamic characterization, sometimes even more critical than switching loss. With suitably high-bandwidth drain voltage measurements, v_{ds} can be processed to determine the peak overshoot voltage as well as the slew rate dv_{ds}/dt. In the turn-off transient, both the peak overshoot and the slew rate are strongly related to load current, in addition to relationships with gate drive speed, bus voltage, and temperature. Figures 8.28 and 8.29 show examples of some of these trends for a SiC MOSFET and a GaN FET over voltage, current, and temperature. The peak voltage can be measured directly, and the magnitude of overshoot can be calculated by subtracting the nominal DC bus voltage. The slew rate can be calculated instantaneously throughout the transient according to

$$\frac{dv_{ds}}{dt}(t) = \frac{v_{ds}(t+\Delta t) - v_{ds}(t-\Delta t)}{2\Delta t} \tag{8.21}$$

where Δt is an integer multiple of the sampling period. The number of samples used for this calculation determines the degree of noise filtering. Once the instantaneous slew rate is calculated, the peak can be found from each transient. The average slew rate is typically reported as the slope between the 10% and 90% points of the v_{ds} waveform, according to

$$\frac{dv_{ds}}{dt}(average) = \frac{0.8\,V_{dc}}{t_{(v_{ds}=0.9V_{dc})} - t_{(v_{ds}=0.1V_{dc})}}. \tag{8.22}$$

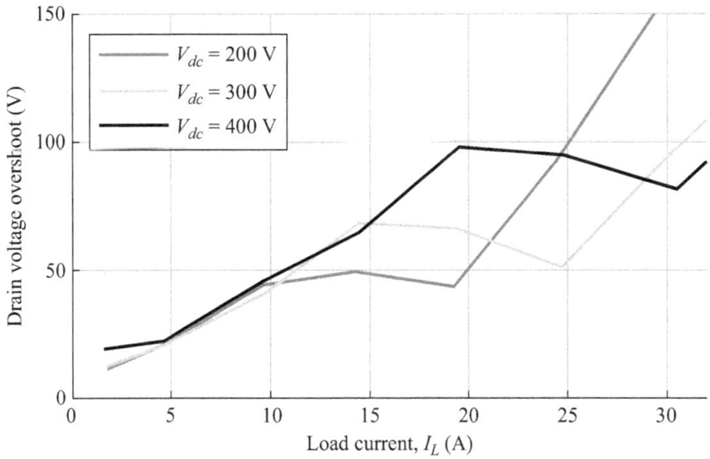

Figure 8.28 Example of turn-off drain voltage overshoot trend versus current and voltage using DPT data for a GaN FET

Figures 8.30 and 8.31 show the peak and average v_{ds} slew rates during turn-off and turn-on transients for the SiC and GaN examples, as a function of voltage, current, and temperature.

Although the typical DPT configuration can give both turn-on and turn-off slew rates, only one of the two overshoot voltages can be acquired with this measurement setup. During the turn-on transient, the synchronous switch also experiences a voltage overshoot, but synchronous switch drain voltage is not measured in the typical configuration. However, this measurement can easily be made using the separate test given in Figure 8.26(b). In order to use a passive probe for this measurement, all other passive measurements must be removed from the circuit, and the oscilloscope ground/common must be connected to the stable voltage of $+V_{dc}$ (the drain of the high-side switch).

An alternative scheme for measuring synchronous switch overvoltage is given in Figure 8.32. Rather than moving the voltage probe to the high-side switch, this option makes the low-side switch behave as the synchronous device. With this modification, the drain voltage, gate voltage, and drain current of the synchronous device can all be measured simultaneously for both turn-on and turn-off transients. The waveforms from this special test configuration can be time-aligned with their corresponding active switch measurements by using the i_d waveform, which should be identical for the active and synchronous switches except for a DC offset equal to the load current. Figure 8.33 shows an example. This test gives a wealth of additional information about the overall switching behavior of the phase leg, and it also provides accurate measurement of the v_{ds} overshoot during a hard turn-on transient. Figure 8.34 shows an example of the overall trend of this overshoot using DPT results from a GaN FET.

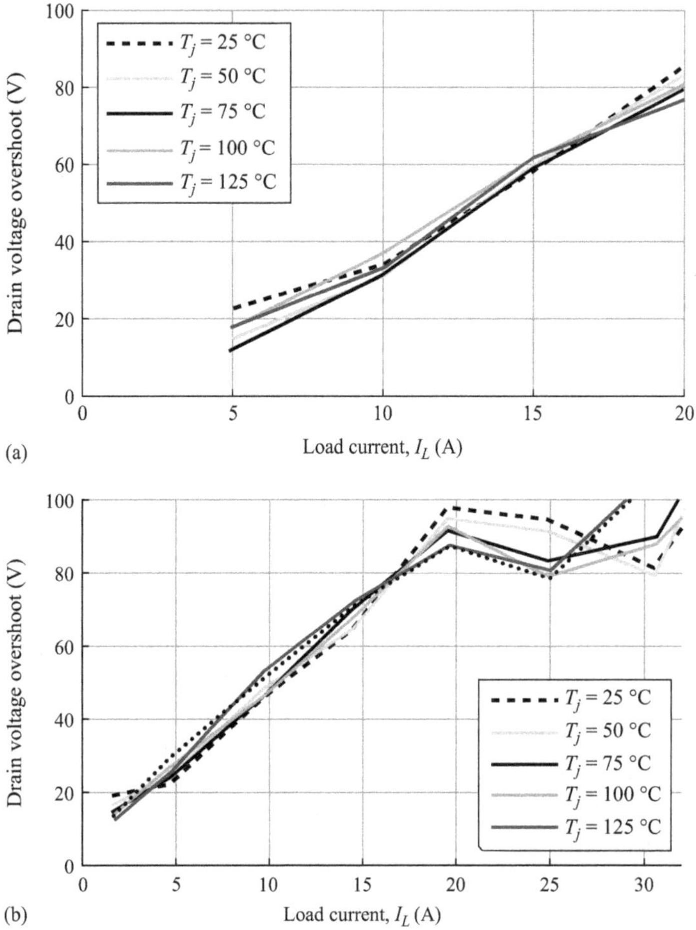

Figure 8.29 Example of turn-off drain voltage overshoot trends versus temperature using DPT data for (a) a SiC MOSFET and (b) a GaN FET

Other waveform features that may be of interest include peak gate voltage, peak drain current overshoot, or drain current slew rate. These parameters are not as commonly reported in WBG device characterizations, but may be important for certain applications and devices.

8.14 Internal waveform estimation using static and dynamic results

Combining the results of static and dynamic characterization can enable further post-processing analysis. As previously explained, the slew rates of drain voltage

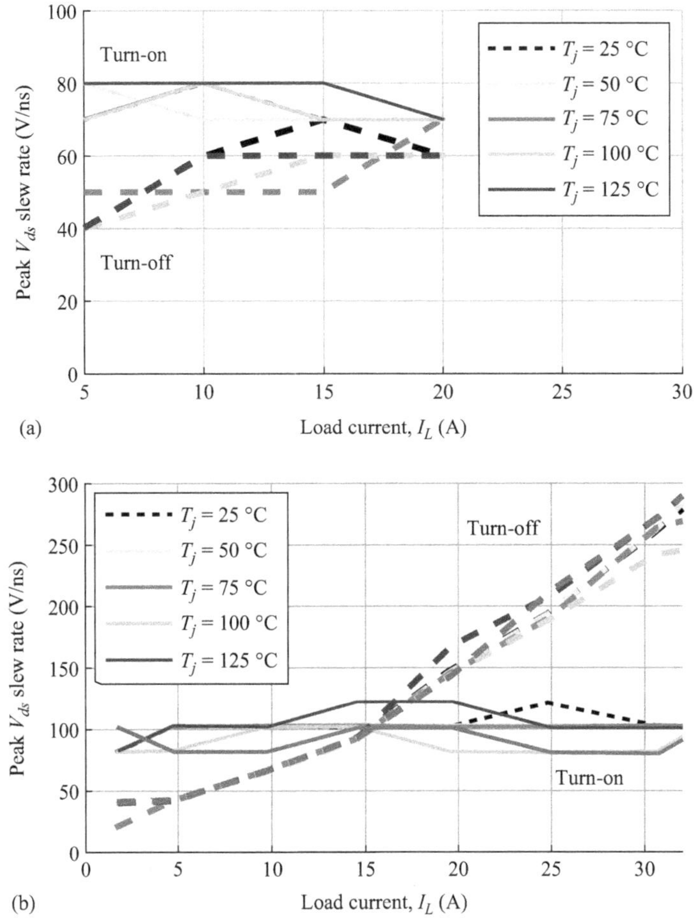

Figure 8.30 Example of drain voltage peak slew rate trends versus current and voltage using DPT data for (a) a SiC MOSFET and (b) a GaN FET

and current can be calculated using (8.12) and (8.21) with a suitable degree of noise filtering. This technique can be applied to calculate dv_{ds}/dt, dv_{gs}/dt, and di_d/dt. The current through the output capacitance of the measured switch can then be calculated using these slew rates and the voltage-dependent C_{oss} curve collected from static characterization, as given by

$$i_{Coss} = C_{oss}\frac{dv_{ds}}{dt}. \tag{8.23}$$

Additional stray capacitances should also be considered for improved accuracy. The values of stray high-side and low-side capacitances can be determined from experimental fit of the E_{on} and E_{off} no-load convergence points, according to (8.17) and (8.18).

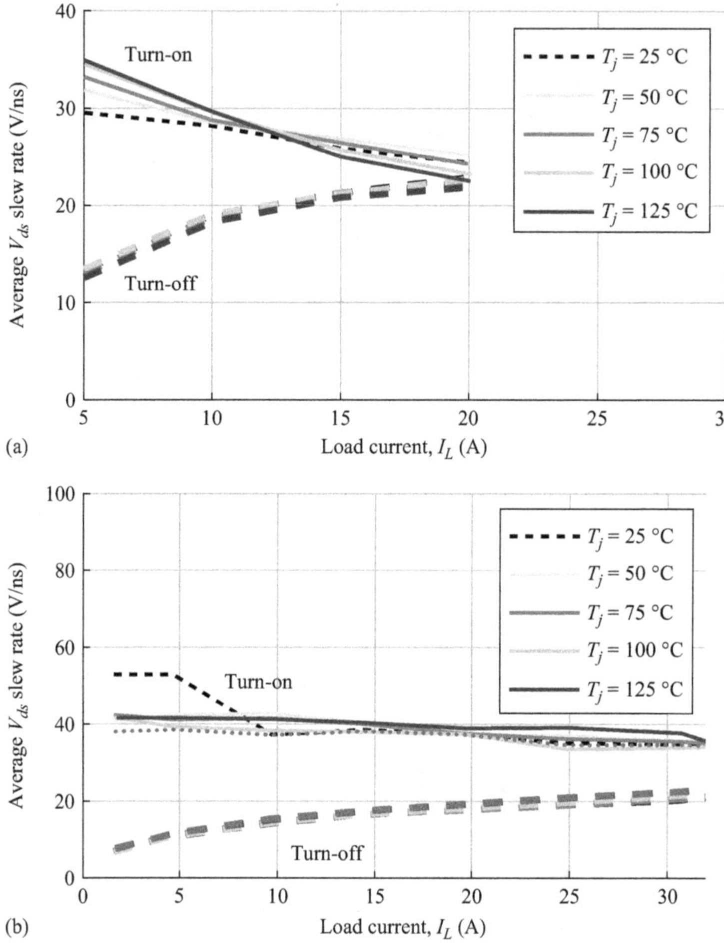

Figure 8.31 Example of drain voltage average slew rate trends versus current and voltage using DPT data for (a) a SiC MOSFET and (b) a GaN FET

Using the C_{oss} current, the internal channel current can further be derived by subtracting it from the drain current, as given by

$$i_{chan} = i_d - i_{Coss}. \tag{8.24}$$

Similarly, the gate current can be calculated from the gate and drain voltage slew rates, the v_{gs}-dependent gate capacitance, and the v_{ds}-dependent Miller capacitance, according to

$$i_g = C_{gs}\frac{dv_{gs}}{dt} + C_{gd}\left(\frac{dv_{gs}}{dt} - \frac{dv_{ds}}{dt}\right). \tag{8.25}$$

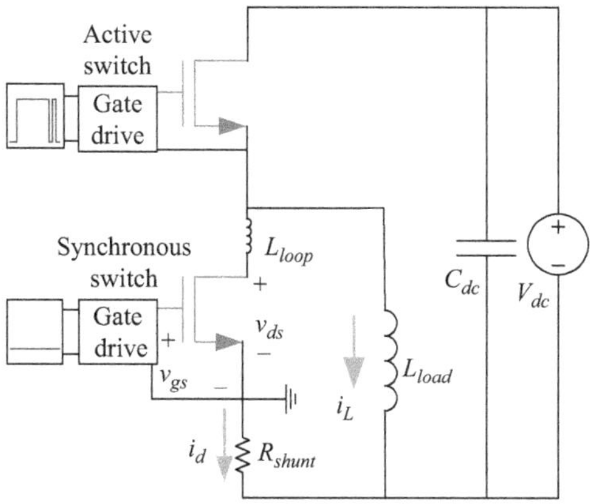

Figure 8.32 Modified DPT configuration for measurement of synchronous switch voltage and current

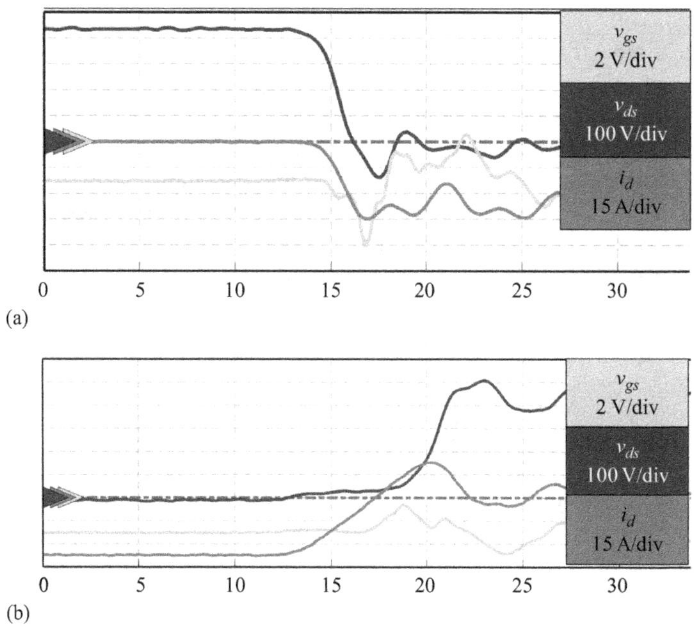

Figure 8.33 Example synchronous switch waveforms for a GaN FET during (a) a hard turn-off of its complementary switch and (b) a hard turn-on of its complementary switch

Now, the internal gate voltage calculated by compensating for the internal gate resistance of the device package according to

$$v_{gs,internal} = v_{gs,measured} - R_{g,internal} i_g. \tag{8.26}$$

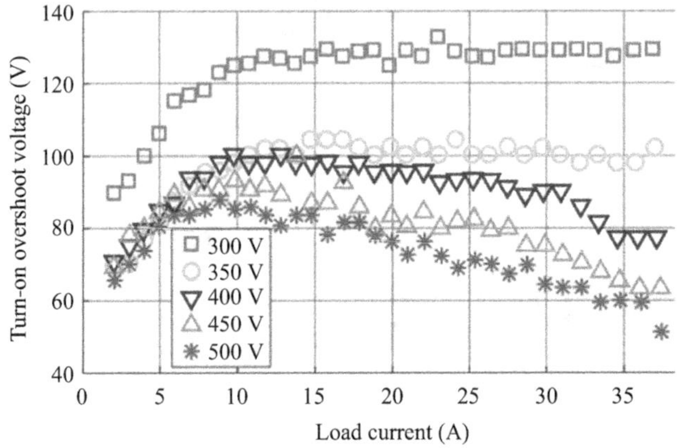

Figure 8.34 Example of synchronous switch drain voltage overshoot trend versus current and voltage using DPT data for a GaN FET

At each step in these calculations, it is crucial that the nonlinear capacitances are determined at each instant during the transient, according to simultaneous measurements of v_{ds} and v_{gs} and the nonlinear curves collected from static characterization.

Next, the drain current slew rate can be used to find the voltage drop across any parasitic inductances in the power loop. The internal drain and source packaging inductance is often listed in device datasheets, and can also be measured using a precision impedance analyzer during static characterization. These parameters can be used to estimate the actual voltage at the drain and source terminals of the die, according to

$$v_{ds,internal} = v_{ds,measured} - (L_d + L_s)\frac{di_d}{dt}. \tag{8.27}$$

The voltage drop across the total power loop inductance can also be calculated using the L_{loop} estimated during the deskew procedure in Section 8.9. This value can be combined with the measured drain voltage and the DC bus voltage (considered as a constant) to find the synchronous switch v_{ds}, according to

$$v_{ds,sync} = V_{dc} - v_{ds,active} - L_{loop}\frac{di_d}{dt}. \tag{8.28}$$

It is important to realize that all of these calculations in (8.23)–(8.28) are only estimations, because of the assumptions required and because of likely noise-induced error. Perhaps the greatest limitation is the second-order (or higher) relationships between these different variables. In fact, (8.27) indicates a source of error in the calculations of (8.23) and (8.25), since the v_{ds} being differentiated includes some voltage drop across internal package inductance rather than just the junction capacitance. Similarly, (8.26) should also include some parasitic inductance, which would require a second-order derivative of v_{gs}. However, it is not recommended to use second-order derivatives in these estimations.

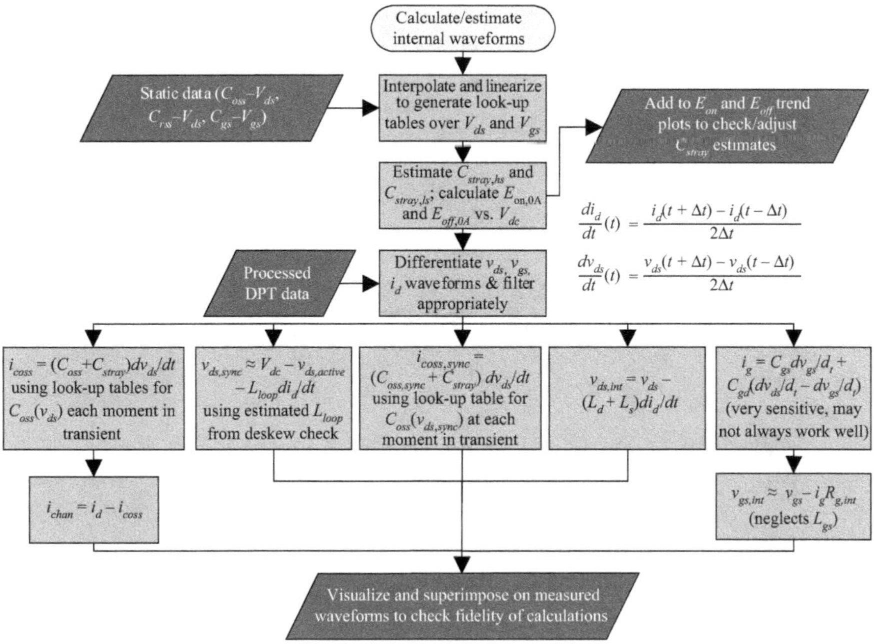

Figure 8.35 Flow chart for computational calculation of unmeasured or internal device variables using DPT data and static characterization results. Different shapes are used to indicate processes, decisions, and data input/output

In most cases, this technique can enhance the post-processing analysis and shed light on interesting waveform phenomena. For example, (8.24) indicates how high the channel current is able to spike during the turn-on transient, and how quickly the channel current drops to zero in the turn-off transient. Comparing these values with the compensated v_{gs} from (8.26) allows estimation of the dynamic transconductance of the device. The gate current from (8.25) can point to potential problems with the gate drive circuit, particularly if the peak value is much lower than the specification of the gate drive IC. And of course, the drain voltages given in (8.27) and (8.28) help to interpret the turn-on and turn-off overshoot voltage trends described in Section 8.13.

Figure 8.35 describes the overall procedure for estimation of internal or unmeasured waveforms, and Figure 8.36 shows an example of some turn-on and turn-off waveforms for a GaN FET with estimated waveforms superimposed on top of measured waveforms. The limitations of this estimation are evident here, as the channel current shows significant ringing after the device has turned off. This is not likely to be real, since the channel is typically clamped off at this time and cannot support negative current while the voltage is positive. This particular error is caused by ringing in the i_d and v_{ds} waveform from parasitic inductances not considered in the calculation. But despite this limitation, the channel turn-off time is clearly visible.

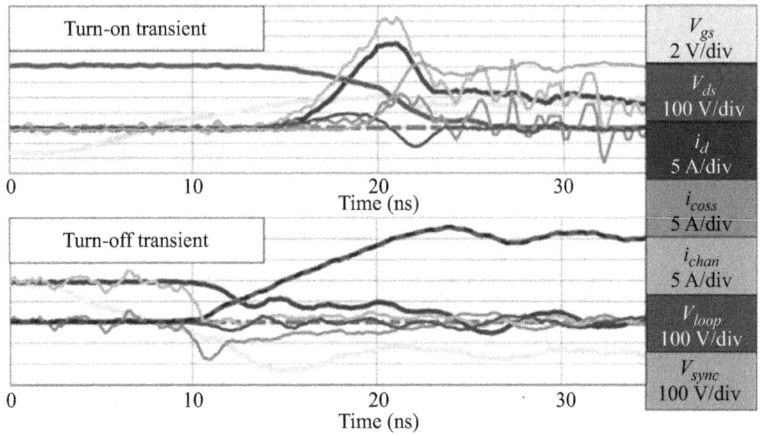

Figure 8.36 Example turn-on and turn-off waveforms including internal/unmeasured variables calculated using DPT data and static characterization results for a GaN FET

8.15 Summary

Post-processing and careful analysis of the DPT measurement data is a critical step in the overall characterization. Even a well-designed DPT board and test setup can produce erroneous results if the *I–V* alignment, ringing, and DC bias are not given proper attention. Effective post-processing can yield accurate results, while also furthering the experimenter's understanding of the device under test. The switching energy losses and subinterval durations can be calculated, and their trends over operating conditions often highlight meaningful characteristics of the device. Further analysis of the waveforms is enabled by combining it with static characterization results, producing estimations of additional waveforms that are not directly measurable. Appendix B gives an example of this procedure implemented in MATLAB code, which can be used as a reference to get started in post-processing DPT data or to incorporate new features into an existing computational tool.

References

[1] Tektronix, "ABCs of probes—primer, Application Note," 2005. [Online] Available: http://www.tektronix.com. Accessed in December 2017.

[2] Tektronix, "Power supply measurement and analysis, Application Note," 2011. [Online] Available: http://www.tektronix.com. Accessed in December 2017.

[3] Tektronix, "Improving vertical resolution in Tektronix digital phosphor oscilloscopes, Application Note," 2011. [Online] Available: http://www.tektronix.com. Accessed in December 2017.

[4] Tektronix, "Improve measurement accuracy and reduce cost with passive probes from Tektronix, Application Note," 2011, [Online] Available: http://www.tektronix.com, Accessed in December 2017.

[5] C. Zhen, "Characterization and modeling of high-switching-speed behavior of SiC active devices," M.S thesis, Department of Electrical and Computer Engineering, Virginia Polytechnic Institute and State University, Blacksburg, USA, 2009.

[6] J. B. Witcher, "Methodology for switching characterization of power devices and modules," M.S thesis, Department of Electrical and Computer Engineering, Virginia Polytechnic Institute and State University, Blacksburg, USA, 2003.

[7] Tektronix, "High-voltage differential probes," [Online] Available: http://www.tek.com. Accessed in December 2017.

[8] Tektronix, "Passive high voltage probes," [Online] Available: http://www.tek.com. Accessed in December 2017.

[9] PEARSON Current monitor 2877 Datasheet, 2007. [Online] Available: http://www.pearsonelectronics.com/pdf/2877.pdf. Accessed in December 2017.

[10] PEM CWT mini specification, November. 2010. [Online] Available: http://www.pemuk.com/Userfiles/CWTmini/cwt_mini_1110.pdf. Accessed in December 2017.

[11] T & M research products, SSDN series coaxial shunt. [Online] Available: http://www.tandmresearch.com. Accessed in December 2017.

[12] Tektronix, TCP0030A datasheet, April, 2013. [Online] Available: http://www.tek.com. Accessed in December 2017.

Chapter 9
Cross-talk consideration

When introducing fundamentals of dynamic characterization in Chapter 4, we started from a three-phase voltage source converter, which is a widely applied circuit for many applications, and derived a simplified configuration to equivalently evaluate the switching behavior of WBG devices, as illustrated in Figure 9.1. With this assumption, details of the DPT have been comprehensively discussed including DPT basics, gate drive, layout design, protection, measurement, and data processing in previous chapters.

However, the observed switching performance of WBG devices in power converters is almost always worse than test results by DPT with the common observation of slower switching speed and higher switching losses [1–3]. Therefore, starting from this chapter, we will revisit the assumptions made for the simplification of the dynamic characterization in DPT, and better understand the impacting factors and limitations of the switching performance in a practical system. As a result, we will either modify the conventional DPT to characterize the switching behavior more accurately for an actual converter or will develop solution(s) to address the adverse effects of the identified limitations on the switching performance such that the conventional DPT results can still be valid and used.

Figure 9.2 highlights key elements that can potentially affect the switching characteristics in a three-phase PWM voltage source inverter, which are specified in Table 9.1. Within the switch/switch-based phase-leg configuration of a single phase, gate drive, parasitics, and interference between the lower and upper switches (i.e., cross-talk) are the critical elements for high-speed switching performance of WBG devices; while in the entire three-phase converter system, other phase-legs, heatsink, and motor/cable load are potential limitations for fast switching speed behavior.

Among these six potential impacting factors, gate drive and parasitics have been covered in the conventional DPT and discussed specifically in Chapters 5 and 6. This chapter focuses on the cross-talk. Later in Chapter 10, limiting factors in a three-phase system will be presented.

Within this chapter, first, the mechanism causing the cross-talk is discussed and unique challenges for WBG devices are analyzed. Then, practical solutions to mitigate the cross-talk are summarized and their effectiveness is compared. Finally, the methodology to evaluate the impact of cross-talk on the dynamic characterization is described. Two case studies based on (1) GaN transistor and (2) SiC MOSFET are presented to quantify the influence of cross-talk for the WBG dynamic characteristics.

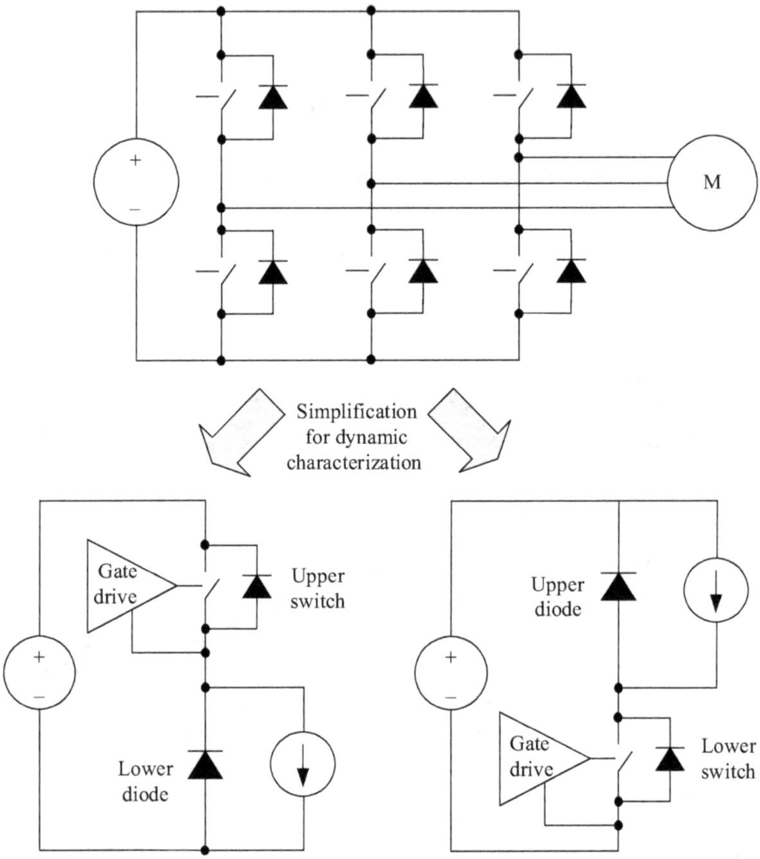

Figure 9.1 Review of derivation of DPT configuration for dynamic characterization

9.1 Mechanism causing cross-talk

In the conventional DPT based on either switch/diode configuration or switch/switch configuration described in Chapter 4, the upper device is always assumed to behave as the freewheeling diode. However, practically, in the switch/switch-based phase-leg configuration, the switching process of the lower switch will affect the status of the upper one, which in turn might worsen the switching performance of both the lower and upper switches. This interaction between two switches is called cross-talk and illustrated in Figure 9.3 [4].

Specifically, during the turn-on transient of the lower switch, the drain-source voltage rise on the upper switch induces a current that is coupled through its Miller capacitance C_{gd_H}. This induced current generates a voltage drop across the gate

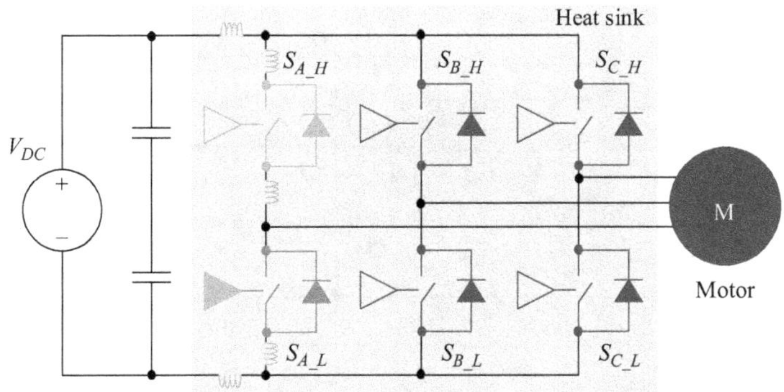

Figure 9.2 Illustration of key components potentially affecting the switching performance in PWM voltage source inverter fed motor drives (assume the phase A lower switch S_{A_L} is the DUT)

Table 9.1 Summary of potential impacting factors and limitations of high-speed switching of WBG devices

Configuration	Impacting factors and limitations		
Single-phase	Gate drive	Parasitics	Interference between switches (cross-talk)
Three-phase	Other phase-legs	Heatsink	Motor/cable load

resistance R_{g_H}. If this positive spurious gate voltage exceeds the threshold voltage of the upper switch, the upper switch will partially turn on. As a result, a shoot-through current will flow from the upper switch to the lower one, leading to excessive switching losses in both switches.

During the turn-off transient of the lower switch, similarly, a negative spurious gate voltage will be induced. This negative spurious gate voltage will not generate a shoot-through problem, but it may degrade the upper switch's reliability if its magnitude exceeds the maximum allowable negative biased gate voltage acceptable to the semiconductor device itself.

Consequently, concerning the cross-talk, the fast switching speed capability of power devices in the phase-leg is affected. Based on the aforementioned analysis, the core element of cross-talk is the spurious gate voltage of the upper switch. The peak value of this spurious gate voltage is shown as [5–7],

$$v_{gs\_H(\max)} = \frac{dv}{dt} \times R_{g\_H} C_{gd\_H} \times \left(1 - e^{-\frac{V_{DC}}{\frac{dv}{dt} \times C_{iss\_H} R_{g\_H}}}\right) \tag{9.1}$$

where *dv/dt* is the slew rate of drain-to-source voltage of the lower switch, partially representing the switching speed. R_{g_H}, C_{gd_H} and C_{iss_H} refer to the gate resistance,

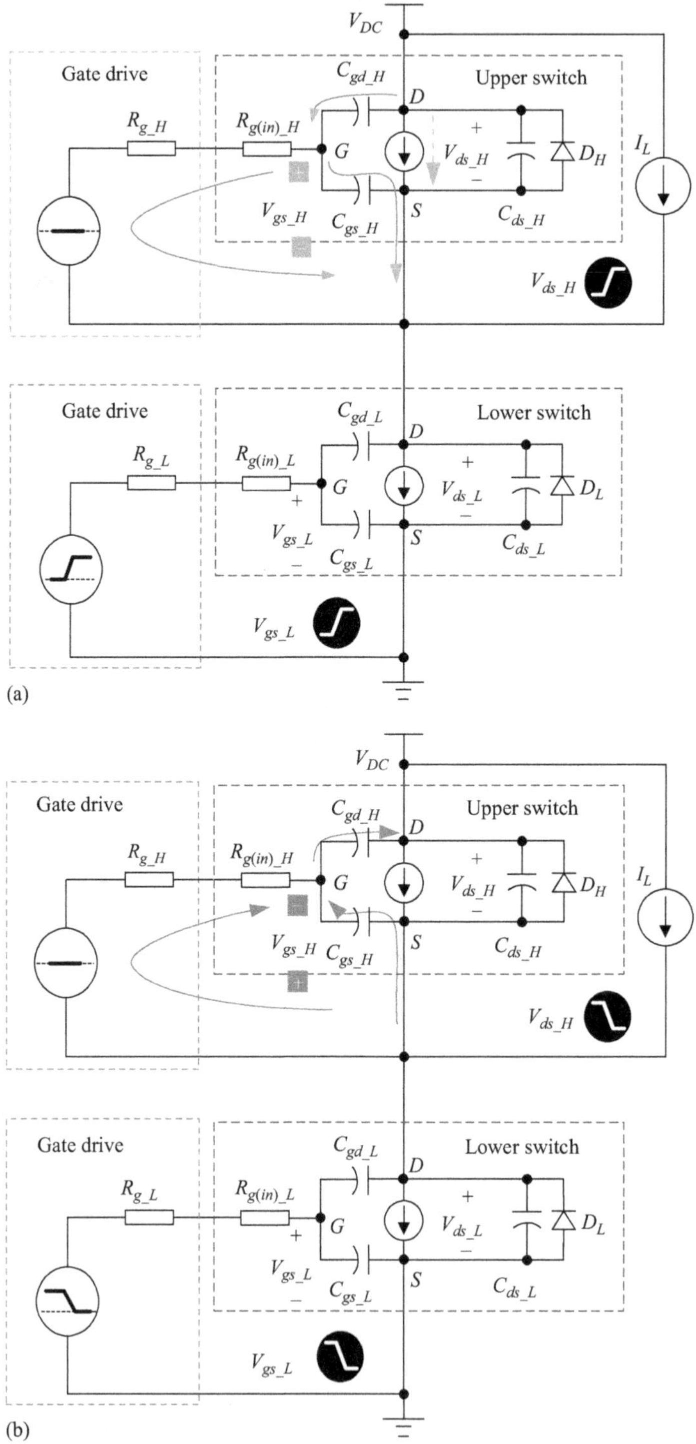

Figure 9.3 Mechanism causing cross-talk interference: (a) turn-on transient of lower switch; (b) turn-off transient of lower switch

Table 9.2 Characteristics of several comparable Si/WBG power devices [8–13]

Type	Manufacturer	Model	V_{DS}/I_D (25 °C)	$V_{gs(th)}$ (25 °C) (V)	$V_{gs\_max(-)}$	$R_{g(in)}$ (Ω)
Si IGBT	Infineon	IKW15N120T2	1,200 V/30 A	5.8	−20 V	N/A
Si MOSFET	Microsemi	APT34M120J	1,200 V/35 A	4.0	−30 V	N/A
SiC MOSFET	Wolfspeed	C3M0075120K	1,200 V/30 A	2.5	−4 V/−8 V (dynamic)	10.5
SiC MOSFET	ROHM	SCT3080KL	1,200 V/31 A	2.7	−4 V	12
GaN HEMT	EPC	EPC2047	200 V/32 A	1.6	−4 V	0.5
GaN HEMT	GaN Systems	GS66508B	650 V/30 A	1.7	−10 V/−20 V (dynamic)	1.5

Miller capacitance and input capacitance of the upper switch. Thus, dv/dt, R_{g_H}, C_{gd_H}, C_{iss_H} and V_{DC} determine the peak spurious gate voltage of the upper switch.

According to their inherent properties, WBG devices are more susceptible to cross-talk as compared to their Si counterparts due to the following reasons: (1) fast switching-speed, which results in high dv/dt; (2) relatively large internal gate resistance $R_{g(in)}$ for SiC MOSFETs, which increases resistance associated in the gate loop; (3) lower threshold voltage $V_{gs(th)}$, causing the positive spurious gate voltage to more easily incur shoot-though failure during turn-on; (4) lower magnitude of maximum allowable negative gate voltage $V_{gs\_max(-)}$, causing the negative spurious gate voltage to more easily overstress the gate-source terminals of power devices during turn-off. Table 9.2 lists the comparison of $R_{g(in)}$, $V_{gs(th)}$, and $V_{gs\_max(-)}$ between commercial WBG devices and their Si counterparts.

Therefore, cross-talk plays a significant role on the switching characteristics of WBG devices, particularly when the gate drive circuit is designed for the maximum achievable switching speed. However, the conventional DPT, which merely focuses on the operating switch, cannot be directly utilized to characterize the influence of cross-talk, including the spurious gate voltage and potential cross-talk losses, since cross-talk affects both switches of the phase-leg. Therefore, a modified methodology for characterization of cross-talk-related switching performance is necessary. To quantify the adverse effect of cross-talk on the switching characteristics, it is essential to establish a benchmark of the switching behavior with the cross-talk being suppressed. Thus, the approach for cross-talk mitigation will be the first step to fully evaluate the switching performance with the consideration of cross-talk. Also, it is practically crucial to implement the suppression technique to avoid cross-talk and optimize the switching of WBG devices.

9.2 Solutions for cross-talk suppression

The core element of cross-talk is the spurious gate voltage. To suppress the cross-talk from the device user's perspective, gate drive design is critical. The preferred gate drive design criteria are summarized as follows.

Table 9.3 Ideal gate voltages and resistances for cross-talk suppression

Switch states	Gate voltage of the interfered switch (V)	Gate resistance of the interfered switch (Ω)
Turn-on transient	$V_{gs\_max(-)}$	0
Turn-off transient	0	0

As discussed earlier, the amplitude of the spurious gate voltage is determined by the portion of displacement current from the Miller capacitance that is used to charge/discharge the gate-source capacitance of the upper switch. Once its gate impedance becomes small during the switching transient, most displacement current will be bypassed by the gate loop, resulting in less current that would induce spurious gate voltage, and thus cross-talk is mitigated.

Additionally, the gate voltage of the upper switch before cross-talk is important to the amplitude of the spurious gate voltage. Based on the mechanisms causing cross-talk, before the turn-on transient of the lower switch, the gate voltage of the upper switch is preferred to be the maximum allowable negative gate voltage $V_{gs\_max(-)}$ to reduce positive spurious gate voltage. On the other hand, before the turn-off transient of the lower switch, 0 V gate voltage of the upper switch is preferred to minimize the peak value of negative spurious gate voltage within the required range.

In total, the ideal gate voltages and resistances to best mitigate cross-talk in the phase-leg configuration are summarized in Table 9.3.

Based on the design criteria highlighted before, multiple implementation circuits have been developed, which generally can be divided into two categories: passive solutions and active solutions. Detailed descriptions of these two groups of solutions are summarized as follows.

9.2.1 Passive solutions

Three commonly used approaches for cross-talk suppression are listed next.

1. Add additional capacitor between gate-source terminals to shunt the Miller current [14–17], as shown in Figure 9.4. This external gate-source capacitor, which equivalently reduces the gate loop impedance, is to take up additional charge originated from the Miller capacitance, resulting in spurious gate voltage reduction and cross-talk mitigation for the affected power device. While the device is under operation, this capacitor will worsen the switching performance with slower switching speed and higher switching losses.
2. Apply a negative biased turn-off gate voltage [14–17], as shown in Figure 9.5. This method introduces more safety margin for the positive spurious gate voltage below the device threshold and suppresses the cross-talk-related loss during the turn-on transient. But the overstress of power devices' gate terminal becomes worse due to the higher negative spurious gate voltage induced during the turn-off transient, especially for the WBG device when considering its lower magnitude of the maximum allowable negative gate voltage compared with the Si devices (see Table 9.2).

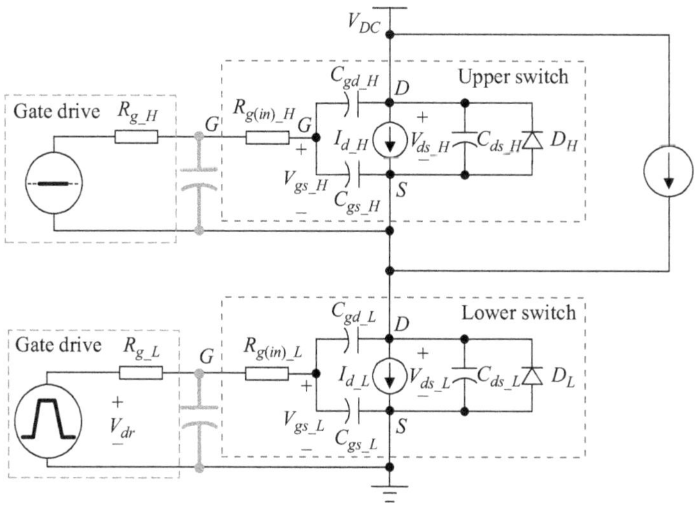

Figure 9.4 Solution for cross-talk suppression: external capacitor across gate-source terminals

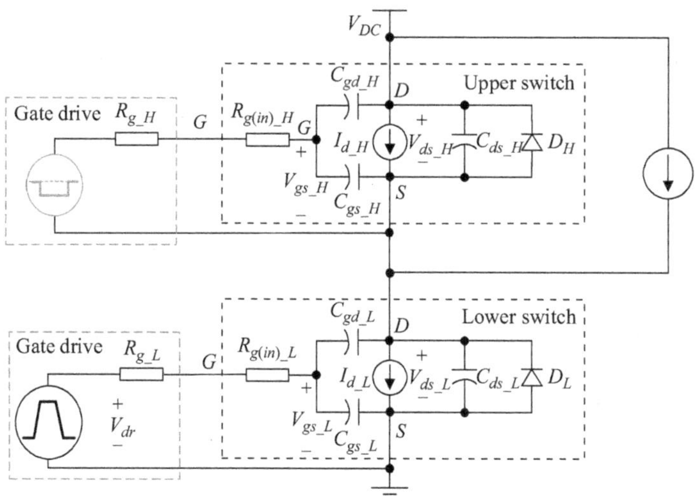

Figure 9.5 Solution for cross-talk suppression: negative turn-off gate voltage

3. Use an asymmetric gate drive, with a discrete gate resistor for turn-on that is bypassed by a low-impedance Schottky diode in the turn-off path, as shown in Figure 9.6. The cross-talk-related gate loop during both turn-on and turn-off transients is determined by the turn-off path. Therefore, similar to adding external capacitor across the gate-source terminals, this approach depends on the reduction of gate loop impedance for cross-talk suppression. However, it

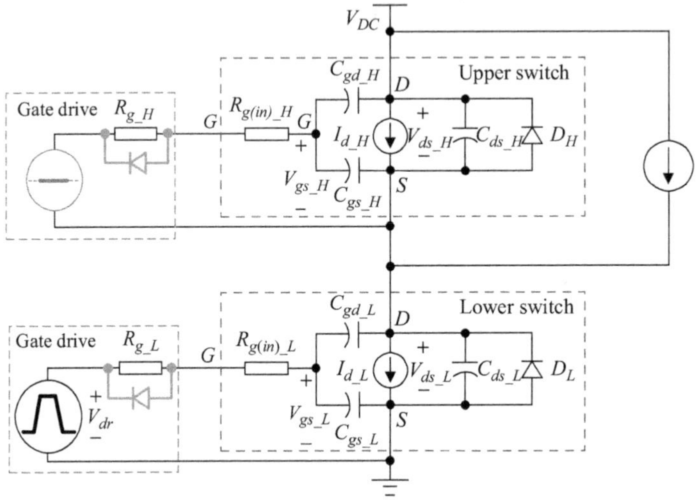

Figure 9.6 Solution for cross-talk suppression: asymmetric gate drive with low impedance turn-off gate loop

limits the selection of the turn-off gate resistance to be low, making the turn-off switching speed not tunable. Also, this approach only minimizes the external gate loop impedance. Considering the relatively large internal gate resistance for SiC MOSFETs, the effectiveness of this approach is limited.

In summary, passive solutions are simple and easy to implement but always with the penalty of reduced switching performance. The underlying reason is that the preferred gate drive for cross-talk suppression during turn-on and turn-off transients is different. The gate drive cannot at the same time best serve the switches under operation. Passive solutions cannot actively tune the gate drive behavior for different switches under different switching transitions. In the end, the global optimization of the switching performance for both switches during turn-on and turn-off transients is difficult to achieve.

9.2.2 Active solutions

To cope with the limitations of the passive solutions, active schemes with auxiliary small rating transistors embedded within the conventional gate drive have been developed to adaptively tune gate loop impedance and/or voltage [18].

One well-known active solution is active Miller clamping [17,19]. The operation principle of this method is to activate an additional transistor between the gate-source terminals to short the gate-source region (i.e., providing low impedance gate loop) after detecting a gate voltage of the affected power device above the clamp threshold [17]. Next, two examples are highlighted based on the concepts of active gate loop impedance regulation and active gate voltage control (GVC), respectively. Other active approaches with similar concepts but different

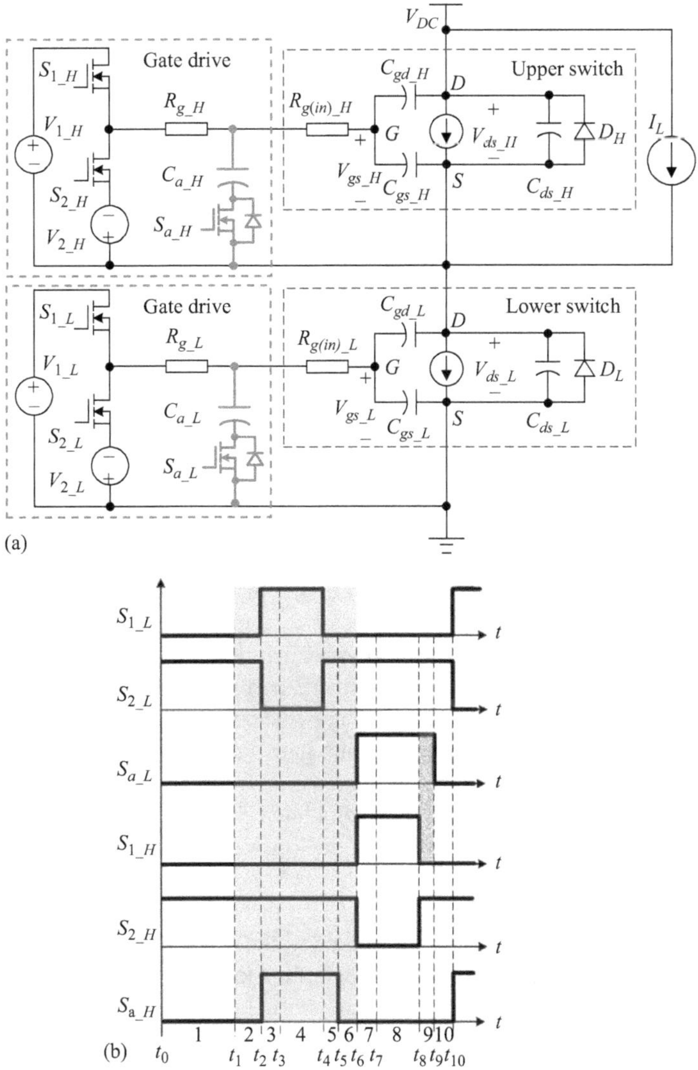

Figure 9.7 GIR-based active gate drive: (a) GIR-based gate assist circuit (S_{a_H}, C_{a_H} or S_{a_L}, C_{a_L}); (b) corresponding logic signals of main device (S_{1_H}, S_{2_H} or S_{1_L}, S_{2_L}) and auxiliary transistor (S_{a_H} or S_{a_L})

implementation circuits have been developed [20–24] and will not be described in detail here.

9.2.2.1 Gate impedance regulation (GIR)-based active gate drive

The active gate drive based on GIR along with the logic signals of its corresponding transistors are shown in Figure 9.7. Compared with the conventional gate drive circuit (S_{1_H}, S_{2_H} or S_{1_L}, S_{2_L}), only one auxiliary transistor (S_{a_H} or S_{a_L}) in series with one capacitor (C_{a_H} or C_{a_L}) is added between the gate-source terminals of each device. The operation principle of the GIR-based gate drive is described next, as shown in Figure 9.8.

Subinterval 1 [$t_0 - t_1$]: Auxiliary capacitor pre-charge. Before the main power devices start to operate, the auxiliary capacitors C_{a_L} and C_{a_H} need to be pre-charged. Via the body diode of auxiliary transistors S_{a_L}, S_{a_H} and gate resistor R_{g_L}, R_{g_H}, the voltage across C_{a_L} and C_{a_H} is established at the negative gate voltage V_2 at the end of t_1. The duration of this subinterval is determined by the time constant R_gC.

Subinterval 2 [$t_1 - t_2$]: Initialization completion. Auxiliary transistors S_{a_H}, S_{a_L} remain off to wait for the operation of the main power devices. At the end of t_2, the lower switch starts to turn on.

Subinterval 3 [$t_2 - t_3$]: Turn-on transient of the lower switch. S_{1_L} turns on, and both S_{2_L} and auxiliary transistor S_{a_L} turn off. At this moment, the gate-source terminals of the lower switch are connected with the output capacitance of S_{a_L} in series with C_{a_L}. Since the output capacitance of the auxiliary transistor is usually an order of magnitude smaller than the input capacitance of the main power device, this auxiliary circuit has negligible impact on the turn-on switching behavior of the lower switch. Meanwhile, auxiliary transistor S_{a_H} turns on. C_{a_H} directly connects with the gate-source terminals of the upper switch. This large external capacitance offers a low impedance loop to *Cdv/dt*-induced current during the turn-on transient of the lower switch, that is, the gate loop impedance of the upper switch is largely decreased. Therefore, the positive spurious gate voltage will be minimized, and then the cross-talk during the turn-on transient is suppressed. At the end of t_3, the turn-on transient of the lower switch is completed.

Subinterval 4 [$t_3 - t_4$]: Full turn-on of the lower switch. The switching status of all transistors remains the same as that during subinterval 3. The negative power supply V_{2_H} starts to discharge C_{a_H} and C_{gs_H} through R_{g_H} until voltages across C_{a_H} and C_{gs_H} return to V_{2_H}. At the end of t_4, the lower switch starts to turn off.

Subinterval 5 [$t_4 - t_5$]: Turn-off transient of the lower switch. S_{1_L} turns off, S_{2_L} turns on, and auxiliary transistor S_{a_L} remains off. Just as in subinterval 3, the auxiliary circuit does not affect the turn-off switching performance of the lower switch. Meanwhile, auxiliary transistor S_{a_H} remains on. The gate loop impedance of the lower switch is decreased due to the low impedance C_{a_H}. Also, the negative spurious gate voltage is minimized. At the end of t_5, the turn-off transient of the lower switch is completed.

Subinterval 6 [$t_5 - t_6$]: Full turn-off of the lower switch. Auxiliary transistor S_{a_H} turns off. C_{a_H} is disconnected from the gate-source terminals of the upper switch for the upcoming switching transient of the upper switch. The negative power supply V_{2_H} starts to charge C_{gs_H} through R_{g_H} until V_{gs_H} returns to V_{2_H}. At the end of t_6, the upper switch starts to turn on.

Based on the circuit symmetry, the operation principle during a switching transient of the upper switch, that is, subintervals 7–10, as shown in Figure 9.7(b), are similar to that during subintervals 3–6. Specifically, subintervals 7–10 represent

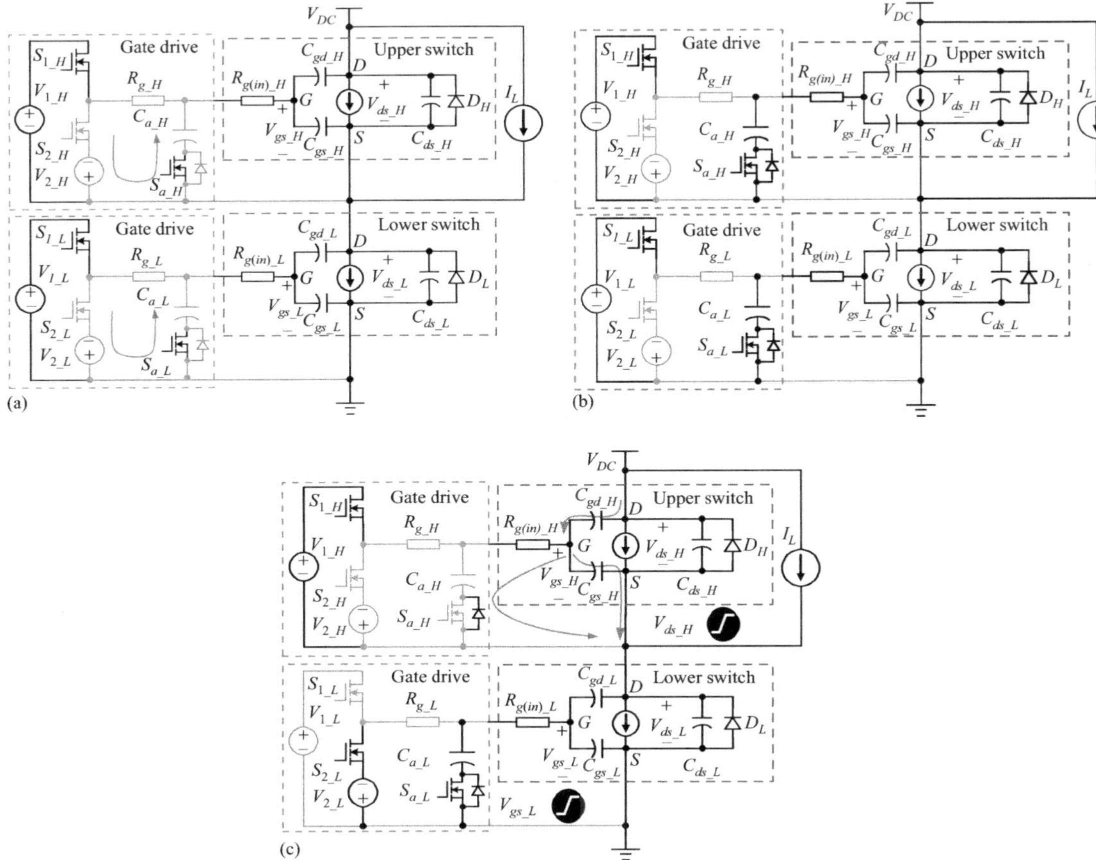

Figure 9.8 Operation principle of the GIR circuit: (a) subinterval 1 $[t_0 - t_1]$; (b) subinterval 2 $[t_1 - t_2]$; (c) subinterval 3 $[t_2 - t_3]$; (d) subinterval 4 $[t_3 - t_4]$; (e) subinterval 5 $[t_4 - t_5]$; (f) subinterval 6 $[t_5 - t_6]$

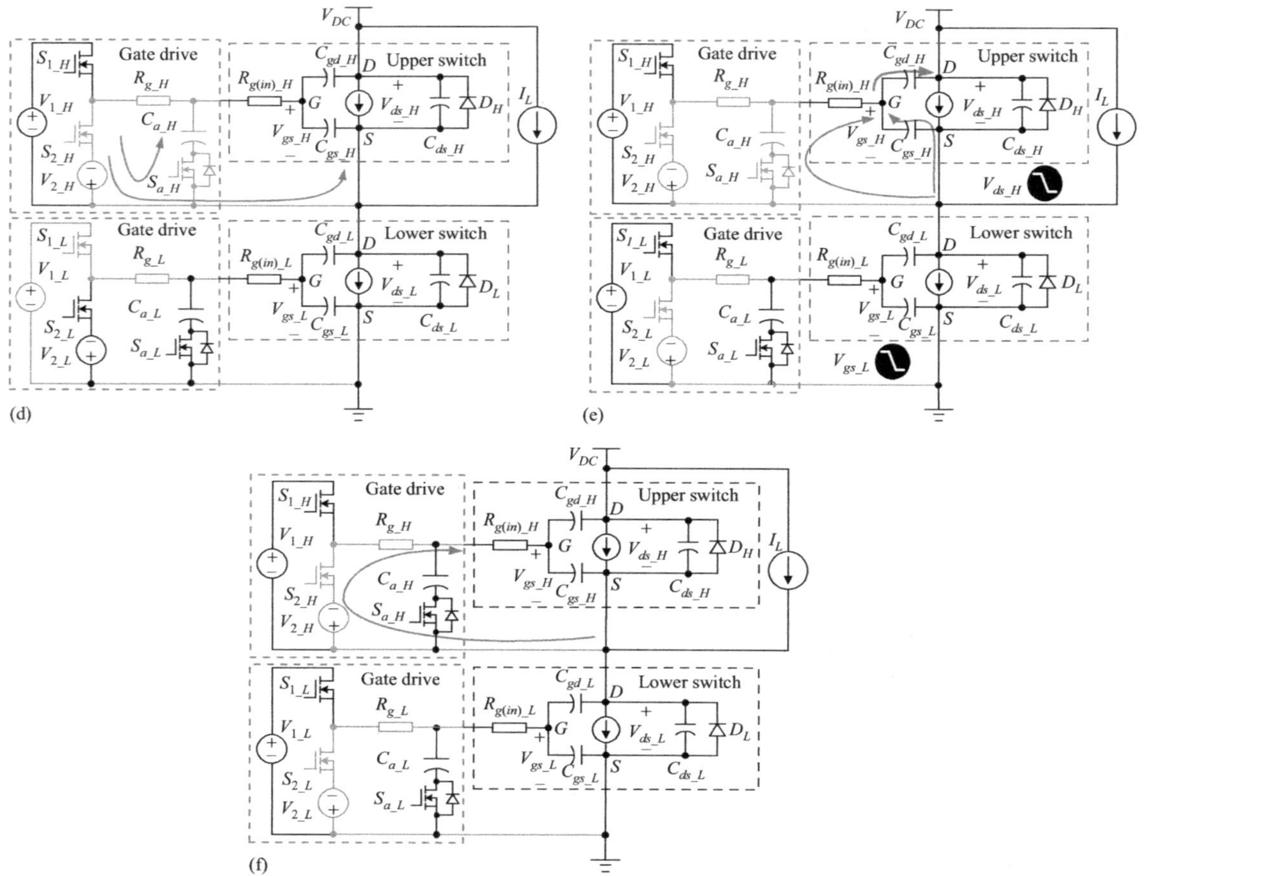

Figure 9.8 (Continued)

turn-on transient of the upper switch, full turn-on of the upper switch, turn-off transient of the upper switch, and full turn-off of the upper switch, respectively.

According to the operation principle, the logic signals of auxiliary transistors are almost the same as those of their complementary main power devices, except that the auxiliary transistors will remain on until the end of the turn-off transient of the complementary main power devices, as shown in Figure 9.7(b) (shaded area). Consequently, logic signals of auxiliary transistors can be easily synthesized based on the logic signals of the main power devices, and this feedforward control scheme is suitable for the WBG power devices with fast switching capability.

9.2.2.2 GVC-based active gate drive

In addition to regulating gate loop impedance for cross-talk mitigation, a more direct solution to avoid cross-talk is to actively control the gate voltage of the affected device during the switching transient of the opposite switch in a phase-leg. Specifically, we can negatively pre-charge the gate-source capacitance of the upper switch before the turn-on transient of the lower switch so that *Cdv/dt*-induced positive spurious gate charge will be cancelled by this pre-stored negative gate charge. Once the pre-stored gate-source charge of the upper switch is sufficient, cross-talk during the turn-on transient of the lower switch can be eliminated. Also, during the turn-off transient of the lower switch, considering the potential hazard of shoot-through, instead of pre-charging gate-source capacitance of the upper switch with positive charge, the gate loop impedance is reduced to minimize the negative spurious gate voltage.

Based on the aforementioned concept, an active gate drive based on GVC along with its corresponding logic signals for control of the transistors have been developed, as shown in Figure 9.9. Compared with the conventional gate drive circuit (S_{1_H}, S_{2_H} or S_{1_L}, S_{2_L}), two auxiliary transistors (S_{a1_H} and S_{a2_H} or S_{a1_L} and S_{a2_L}) together with one diode (D_{a_H} or D_{a_L}) are added between the gate-source terminal of each device, while the commonly used negative isolated power supply (V_{2_H} or V_{2_L}) is not needed (see Figure 9.3). Figure 9.10 displays the operating procedure of the GVC-based active gate drive, described as follows.

Subinterval 1 [$t_1 - t_2$]: Gate-source capacitance pre-charge of the upper switch. Before the lower switch starts to turn on, S_{1_H} and auxiliary transistor S_{a2_H} turn off; S_{2_H} and auxiliary transistor S_{a1_H} turn on. Via the loop formed by S_{2_H}, S_{a1_H} and R_{g_H}, the positive turn-on gate voltage V_{1_H} is used to pre-charge the gate-source capacitance of the upper switch C_{gs_H} to the required negative voltage for the cross-talk elimination at the end of t_2. The duration of this subinterval is determined by the required pre-charge negative voltage, which will be discussed in detail in the next section. At the end of t_2, the lower switch starts to turn on.

Subinterval 2 [$t_2 - t_3$]: Turn-on transient of the lower switch. S_{1_L} turns on, S_{2_L} turns off, auxiliary transistor S_{a2_L} remains on, and S_{a1_L} remains off. Since the junction capacitance of the auxiliary diode D_{a_L} is usually an order of magnitude smaller than the input capacitance of the main power device, this auxiliary circuit has negligible impact on the turn-on switching behavior of the lower switch. Meanwhile, auxiliary transistor S_{a1_H} turns off and S_{a2_H} remains off. Due to the

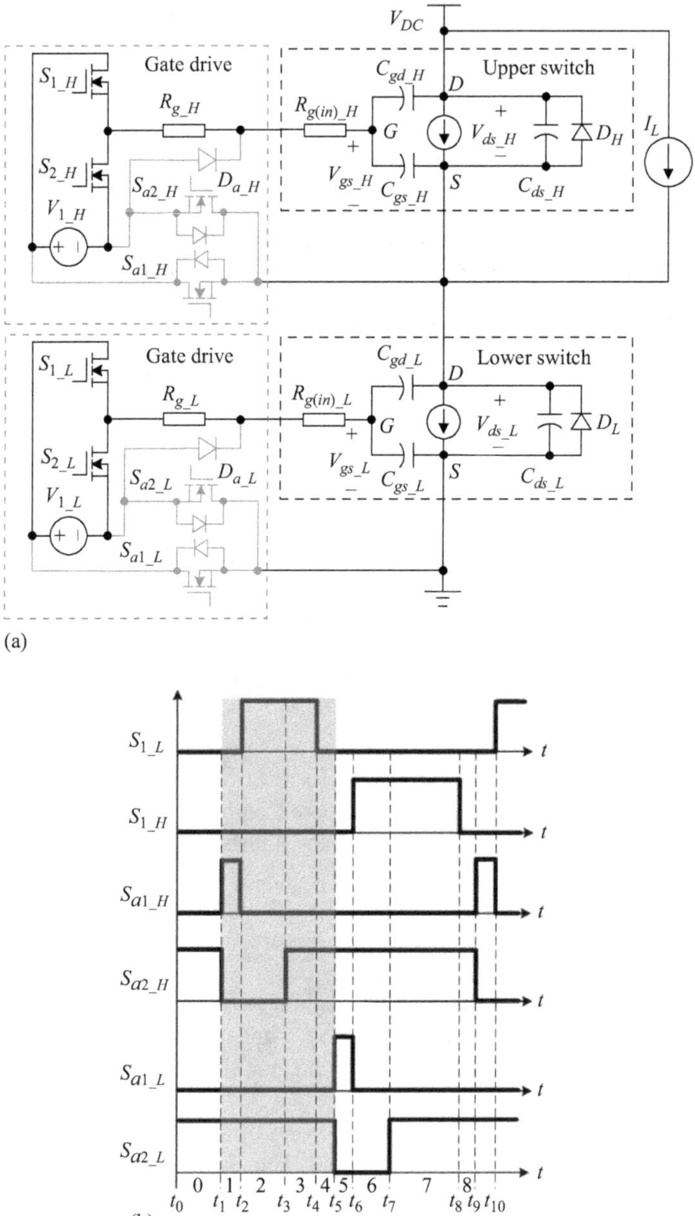

(b)

Figure 9.9 GVC-based active gate drive: (a) GVC-based gate drive (S_{a1_H}, S_{a2_H}, D_{a_H} or S_{a1_L}, S_{a2_L}, D_{a_L}); (b) corresponding logic signals of main device (S_{1_H} or S_{1_L}) and auxiliary transistors (S_{a1_H}, S_{a2_H} or S_{a1_L}, S_{a2_L})

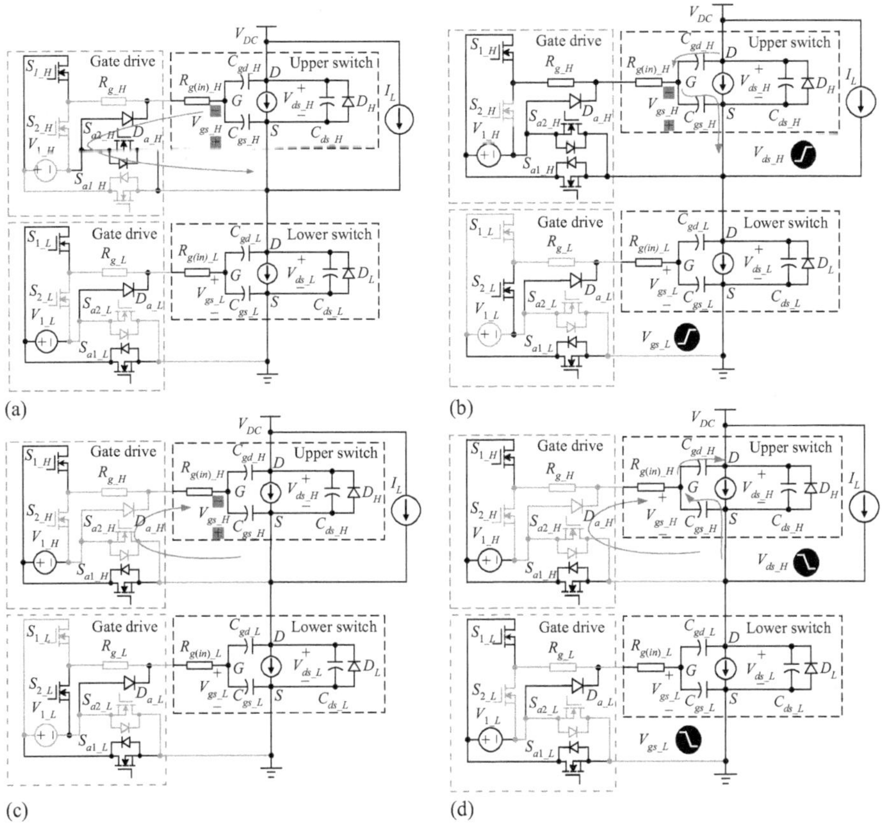

Figure 9.10 Operation principle of the GVC circuit: (a) Subinterval 1 [$t_1 - t_2$]; (b) subinterval 2 [$t_2 - t_3$]; (c) subinterval 3 [$t_3 - t_4$]; (d) subinterval 4 [$t_4 - t_5$]

body diode of S_{a2_H} and auxiliary diode D_{a_H}, instead of bypassing through the external gate loop, the negative charge pre-stored in C_{gs_H} is fully utilized to cancel the positive charge transferred from the Miller capacitance of the upper switch C_{gd_H}. Once this pre-stored negative gate-source charge (i.e., negative gate voltage) is sufficient, cross-talk during the turn-on transient of the lower switch can be eliminated. At the end of t_3, the turn-on transient of the lower switch is completed.

Subinterval 3 [t_3 – t_4]: Full turn-on of the lower switch. Auxiliary transistor S_{a2_H} turns on. The residual negative voltage across C_{gs_H} at subinterval 2 returns to 0 for the preparation of cross-talk mitigation during the upcoming turn-off transient of the lower switch. At the end of t_4, the lower switch starts to turn off.

Subinterval 4 [$t_4 - t_5$]: Turn-off transient of the lower switch. S_{1_L} turns off and S_{2_L} turns on. Switching status of all auxiliary transistors is the same as that during subinterval 3. This auxiliary circuit does not affect the turn-off switching performance of the lower switch, just as in subinterval 2. Meanwhile, auxiliary transistor

Table 9.4 Summary of key features of the cross-talk suppression solutions

Preferred features	Passive solutions			Active solutions	
	External gate-source capacitor	**Negative turn-off gate voltage**	**Asymmetric gate resistance**	**GIR**	**GVC**
Turn-on cross-talk suppression	Yes	Yes	Yes	Yes	Yes
Turn-off cross-talk suppression	Yes	No	Yes	Yes	Yes
No limitation on switching	No	Yes	No	Yes	Yes
Less implementation complexity	Yes	Yes	Yes	No	No

S_{a2_H} and diode D_{a_H} offer a low impedance loop to *Cdv/dt*-induced current during the turn-off transient of the lower switch, that is, the gate loop impedance of the upper switch is largely decreased. Therefore, the negative spurious gate voltage is minimized. At the end of t_5, the turn-off transient of the lower switch is finished.

Based on the circuit symmetry, the operation principle during switching transient of the upper switch, that is, subintervals 5–8, as shown in Figure 9.9(b), are similar to that during subintervals 1–4. Specifically, subintervals 5–8 correspond to gate-source capacitance pre-charge of the lower switch, turn-on transient of the upper switch, full turn-on of the upper switch, and turn-off transient of the upper switch, respectively.

According to the operation principle, the logic signals for the control of auxiliary transistors can be synthesized merely based on the logic signals of the main power devices using several logic gates. Thus, through feedforward control, the GVC assist circuit is well suited for WBG power devices with fast switching capability.

9.2.3 Summary of cross-talk suppression solutions

Table 9.4 compares the key features of the solutions for cross-talk suppression with respect to the effectiveness of cross-talk suppression, influence on switching performance, and implementation complexity. Practically, if the auxiliary circuit of the active solution (e.g., small rating transistor and/or diode) can be embedded within the conventional gate drive IC, it can achieve the best tradeoff between the performance and complexity.

9.3 Methodology for characterization of cross-talk-related switching performance

Cross-talk affects the switching performance of both switches in the phase-leg during the turn-on and turn-off transients differently. During the turn-on, the operating switch may dissipate excessive loss induced by cross-talk. In the meantime, the

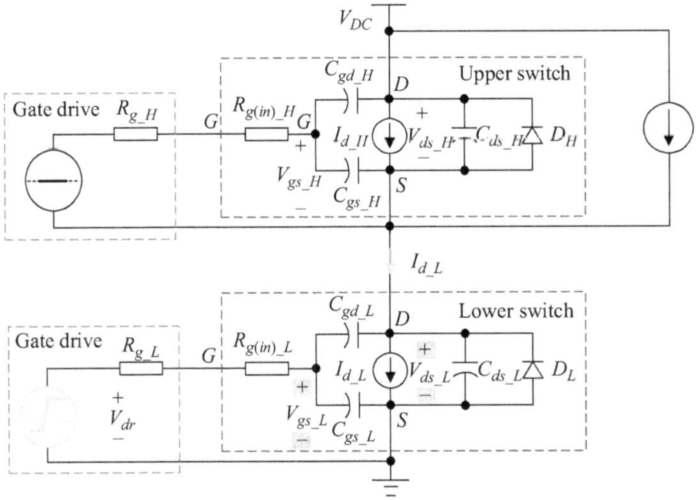

Figure 9.11 Characterization of cross-talk-related turn-on switching loss for the lower switch

synchronous switch, as the interfered device, could also have unexpected loss due to the partial shoot-through current. During the turn-off, the switching performance of the operating switch should remain similar while the spurious gate voltage of the synchronous switch can be observed. Accordingly, three testing schemes based on the conventional DPT circuit are designed to characterize the impact of the cross-talk on the switching performance.

Figure 9.11 shows the scheme to evaluate the influence of cross-talk on the switching performance of the lower switch (i.e., operating device) during the turn-on transient. Similar to the conventional DPT, the drain-source voltage, drain current, and the gate-source voltage of the lower switch are measured.

Figure 9.12 shows the scheme to evaluate the influence of cross-talk on the upper switch performance during the turn-on transient. Unlike the aforementioned scheme, based on the same DPT test circuit, the drain-source voltage and drain current of the upper switch are measured. It is suggested to use the recommended voltage and current measurement techniques (passive or single-ended voltage probe plus the coaxial shunt) discussed in Chapter 8 to measure the drain-source voltage and drain current of the upper switch.

Regarding the grounding of the voltage probe, instead of the source terminal of the upper switch, which is the switching node with high dv/dt during the switching transient, the drain terminal of the upper switch, that is, the positive DC bus, with the stable potential should be selected. To safely implement this measurement scheme, an isolated DC power supply has to be employed; otherwise, the DC bus will be shorted through the grounding path provided by the scope.

For the current measurement, two approaches can be adopted: (1) a dedicated coaxial shunt placed between the positive DC bus and drain terminal of the upper switch to directly measure the upper switch drain current; or (2) an indirect

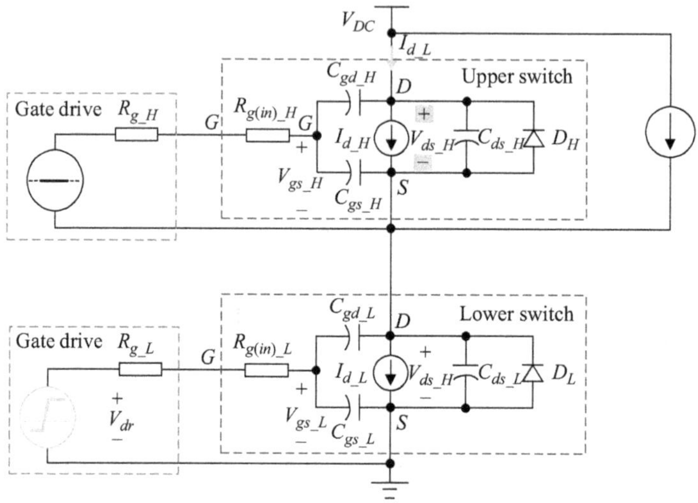

Figure 9.12 Characterization of cross-talk-related turn-on switching loss for the upper switch

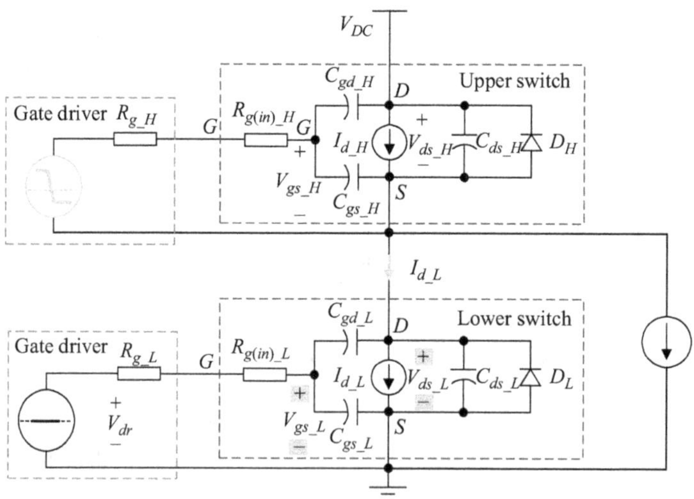

Figure 9.13 Characterization of cross-talk-related turn-off spurious gate voltage

estimation based on the measured drain current of the lower switch with the assumption that during the fast switching transient, the load current almost stays constant. Thus, the upper switch drain current should be equal to the drain current of the lower switch minus the load current.

Figure 9.13 shows the scheme to evaluate the spurious gate-source voltage of the interfered switch during the turn-off transient. Considering the challenge of accurately measuring the gate-source voltage of the upper switch due to the unstable potential of the source terminal of the upper switch, the alternative scheme is to shift the operating switch from the lower switch to the upper one and measure the gate-source voltage of the lower switch. Accordingly, the upper switch will be controlled by the double-pulse with the lower switch constantly off. Also, the load inductor has to be connected in parallel with the lower switch.

Furthermore, the effect of cross-talk on the switching performance is highly dependent on the gate drive. It is necessary to conduct the comparison experiment to quantify the cross-talk-related switching performance with different gate drives. Two case studies are presented next to illustrate the aforementioned methodology for the characterization of cross-talk-related switching performance. In comparison with the conventional gate drive, the first case study bases on GaN HEMT with passive solutions for cross-talk suppression; while the other example depends on SiC MOSFET with active gate drive circuit.

9.3.1 Case study 1: GaN HEMT with passive solution

9.3.1.1 Test setup

A DPT with GaN Systems GS66508P 650 V/30 A enhancement-mode GaN HEMT is established and a comprehensive study of the cross-talk mechanism is performed. This study focuses on experimental results at 400 V and 25 °C, because these conditions produce the fastest switching and the worst case for cross-talk with the consideration of the inherent properties of GaN HEMTs [25]. The driving voltage was chosen as 7 V, the maximum rated steady-state gate voltage from the device manufacturer.

Three separate gate drive cases were studied. The circuit shown in Figure 9.14(a) was used to measure the overlap and C_{oss}-based switching losses, with limited impact from cross-talk. For this case, the synchronous device gate and source sense terminals were shorted together directly at the device pads, providing a low-impedance gate loop path for the C_{gd} displacement current to flow. Because of the internal device resistance and inductance, there may still be enough ringing on $v_{gs,int}$ to turn the synchronous device on, but this could be considered the "best case" for limiting cross-talk. However, it is not practical to use this design for a synchronous switching buck converter or phase-leg topology.

The circuit shown in Figure 9.14(b) is based on the conventional gate drive, and could be considered the "worst case" for cross-talk, because the same gate resistance was used for both turn-on and turn-off. This is generally not recommended for WBG devices, but it is still sometimes used for the sake of simplicity.

Finally, the circuit shown in Figure 9.14(c) is the one using an asymmetric gate drive, with a discrete gate resistor for turn-on that is bypassed by a low-impedance Schottky diode in the turn-off path. This could be considered the "practical case" because it allows the designer to mitigate cross-talk while still controlling the switching of both devices.

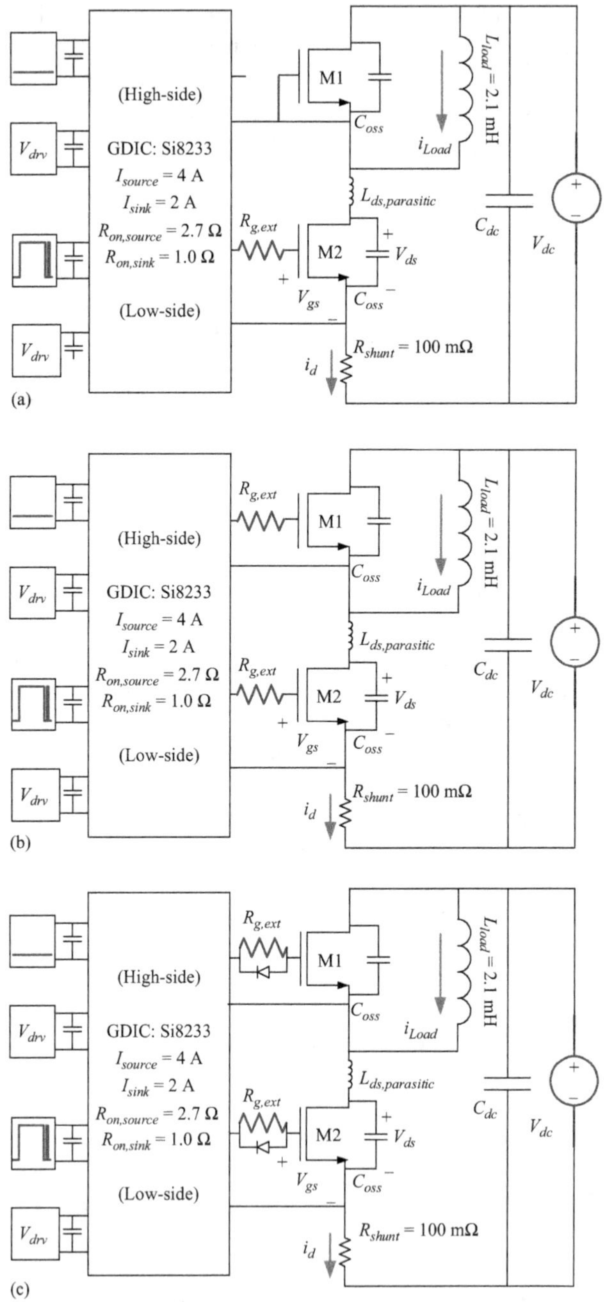

Figure 9.14 DPT circuit schematic for GaN HEMT-based characterization of cross-talk-related switching characteristics: (a) with no cross-talk; (b) with symmetric gate resistance; (c) with asymmetric gate resistance

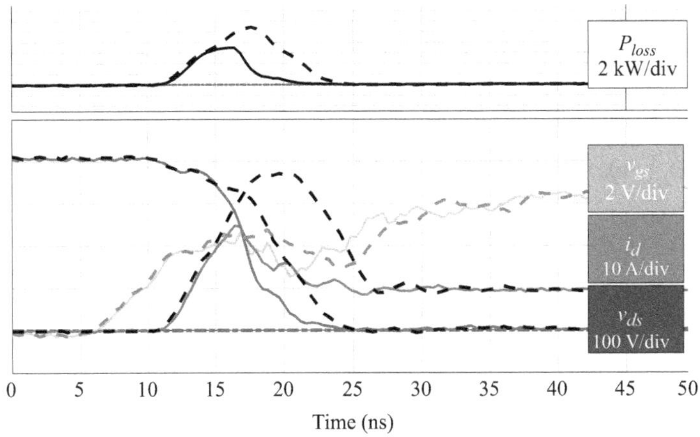

Figure 9.15 *Turn-on transient waveforms of GS66508P at 400 V and 10 A with $R_{g,ext} = 10\ \Omega$, showing the best case as a solid line and the worst case as a dashed line*

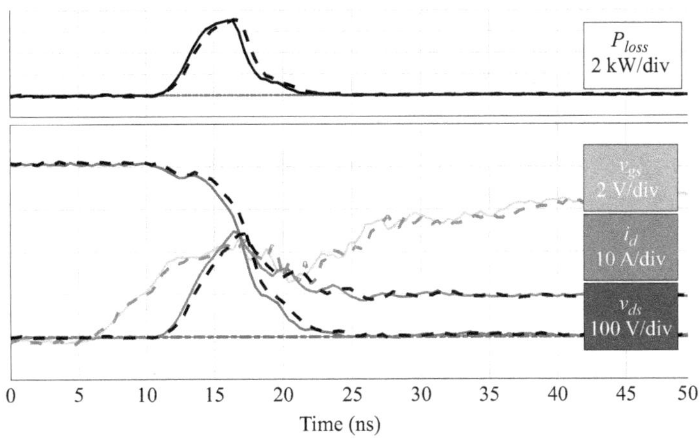

Figure 9.16 *Turn-on transient waveforms of GS66508P at 400 V and 10 A with $R_{g,ext} = 10\ \Omega$, showing the cross-talk best case as a solid line and the asymmetric drive case as a dashed line*

9.3.1.2 Effect of cross-talk on switching performance

The effect of cross-talk on the turn-on transient can be seen clearly in Figures 9.15 and 9.16. Figure 9.15 compares the best case and the worst case with a gate resistance $R_{g,ext}$ of 10 Ω, showing a longer turn-on time and a shoot-through current of about 20 A in the worst case. Figure 9.16 compares the best case and practical case, showing very little effect from cross-talk with an asymmetric gate drive. However, the overall turn-on loss can still be greatly reduced by dropping the gate

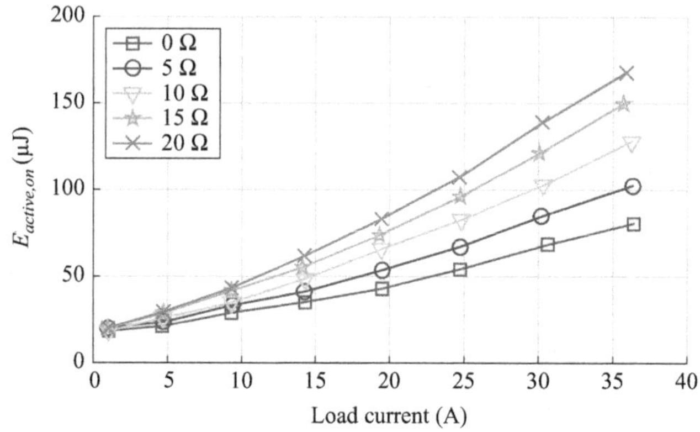

Figure 9.17 Turn-on loss of GS66508P for an active device, with the cross-talk best case gate drive

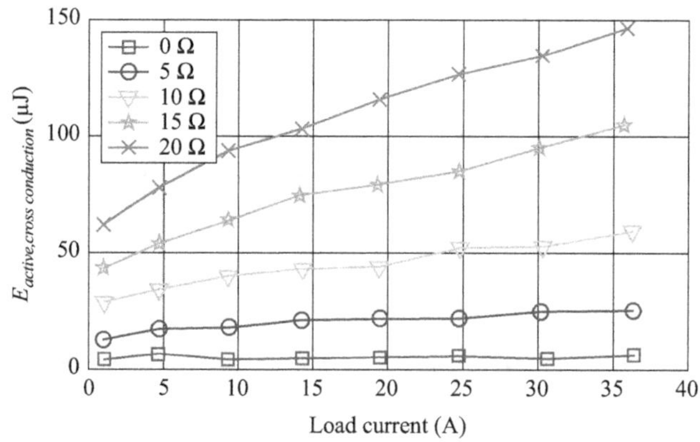

Figure 9.18 Additional turn-on loss of GS66508P due to cross-talk, with the cross-talk worst case gate drive

resistance to 0 Ω. In this case, using an asymmetric gate drive is not necessary, since the 0 Ω resistance is the primary path for turn-on or turn-off and bypasses the Schottky diode completely.

As expected, the turn-on loss of the active device is correlated to both load current and gate resistance. This is true for all three of the cases described. Figure 9.17 shows the turn-on loss for the best cross-talk case, with the synchronous device gate-source input shorted. Figure 9.18 shows the additional turn-on losses caused by cross-talk with the worst case, estimated by subtracting the best case loss from the worst case loss. These cross-talk losses become prohibitive with even as small as 5 Ω a gate resistance. Part of this additional energy is lost as the shoot-through current flows through the saturated active device channel, and some extra overlap energy is lost, when the shoot-through current extends the turn-on transition.

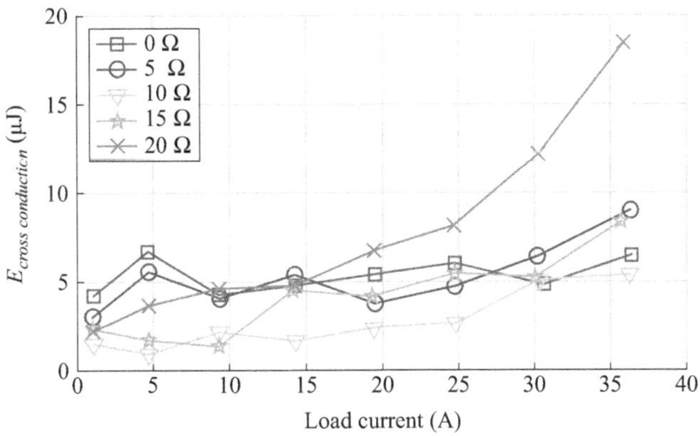

Figure 9.19 Additional turn-on loss of GS66508P due to cross-talk, with the practical case using an asymmetric gate drive

Figure 9.19 shows the additional turn-on losses caused by cross-talk in the practical case, estimated by subtracting the best case loss from the calculated practical case loss. Unfortunately, Figure 9.19 does not show a clear trend of loss versus gate resistance or load current. This calculation assumes that no spurious switching occurs in the best case, which may not be accurate due to ringing on the internal gate loop impedance. However, some general conclusions can be drawn. The results show that cross-talk losses with an asymmetric gate drive are approximately 5 μJ in the active device. An inverse correlation is expected between this loss and gate resistance, because the overall dv_{ds}/dt of both active and synchronous devices is strongly determined by the turn-on gate resistance of the active device.

The losses occurring in the synchronous switch due to cross-talk can be estimated by comparing the measured energy lost to the synchronous device during the turn-on of the active device to the energy recovered from the synchronous device during the active device turn-off transient. Because the synchronous device switching transitions are ZVS, there should be no overlap losses and the speed of the turn-on and turn-off should not impact the switching energy in either case. Therefore, any loss occurring in the synchronous switch will be due to cross-talk. This loss is shown for a sweep of operating conditions in Figure 9.20. It is interesting to note that in the case of cross-talk, the load current does not have a significant effect. Since the voltage slew rate (dv_{ds}/dt) during the hard turn-on transient is primarily a function of the gate drive speed and the DC bus voltage, these are the parameters that show the strongest relationship with the cross-talk energy loss.

Two example turn-on transients with 0 Ω and asymmetric 20 Ω drives are compared in Figure 9.21. The asymmetric 20 Ω drive produces a much longer turn-on time and loss, but the loss is reduced by several μJ due to reduced cross-talk.

After the turn-off dv_{ds}/dt transient is over, some additional energy is lost due to cross-talk in the 5–20 ns following the transient. The C_{gd} displacement current in

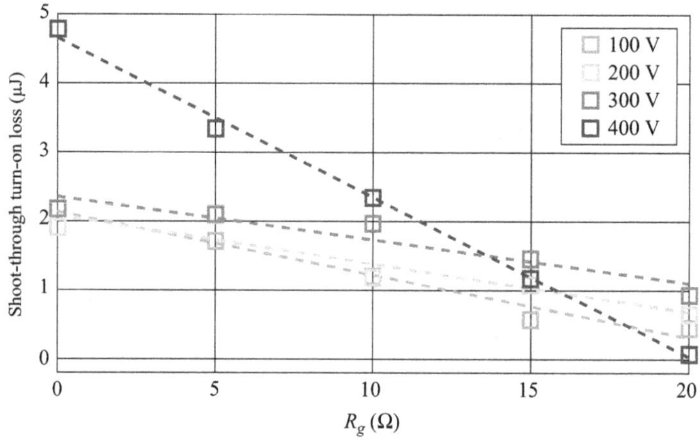

Figure 9.20 Cross-talk loss in the synchronous switch with GS66508P, as a function of R_g and V_{DC}

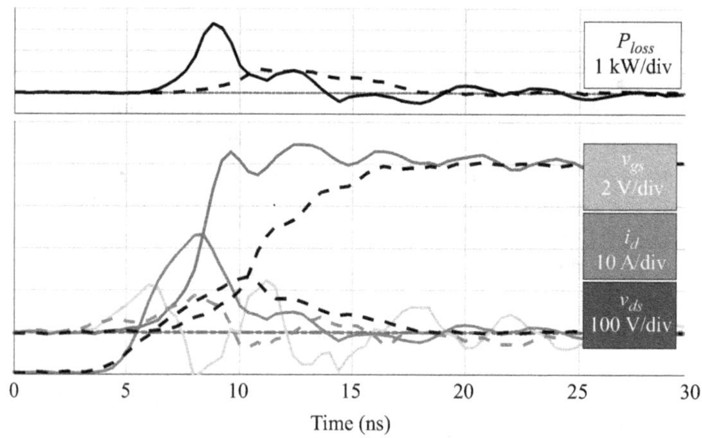

Figure 9.21 Synchronous device transient waveforms for GS66508P, during active device turn-on at 400 V and 10 A with an asymmetric gate drive circuit, showing $R_{g,ext} = 0\ \Omega$ with a solid line and $R_{g,ext} = 20\ \Omega$ with a dashed line

the synchronous device induces a negative gate voltage for a short time after the turn-off, and the magnitude of this negative v_{gs} increases with higher gate resistance and higher load current. It is interesting to note that this negative spurious voltage occurs after the dv_{ds}/dt transient, not during the transient as is typical for cross-talk. Figure 9.22 compares the turn-off transient waveforms with 0 and 20 Ω, showing a negative gate voltage of about −10 V in the 20 Ω case with 10 A load current.

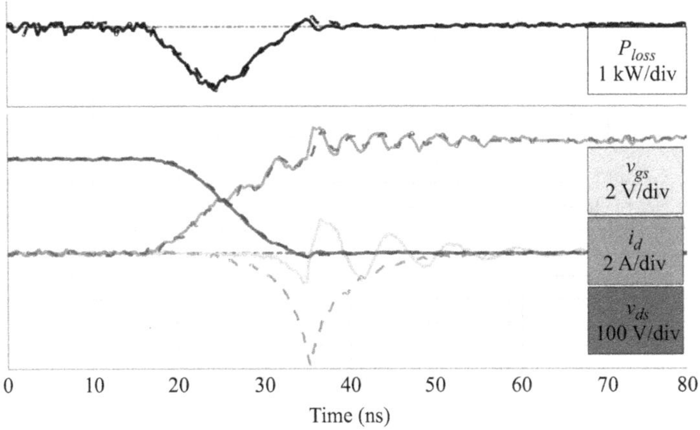

Figure 9.22 Synchronous device transient waveforms for GS66508P, during active device turn-off at 400 V and 10 A with an asymmetric gate drive circuit, showing $R_{g,ext} = 0\ \Omega$ with a solid line and $R_{g,ext} = 20\ \Omega$ with a dashed line

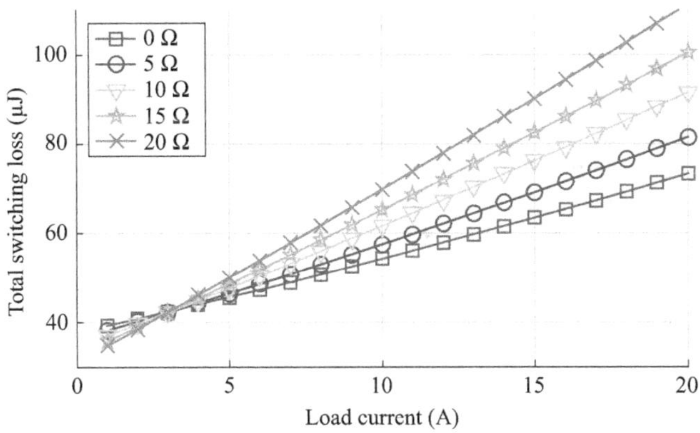

Figure 9.23 Overall switching energy loss as a function of R_g and I_L, demonstrating tradeoff between cross-talk and overlap losses

9.3.1.3 Discussion

From the results presented here, it is clear that a conventional gate drive circuit with nonzero gate resistance is not an effective option for this device, due to heavy overlap and cross-talk losses during both turn-on and turn-off transients. An asymmetric gate drive circuit presents an alternative, with lower cross-talk losses at turn-on and overlap losses at turn-off. However, the turn-on overlap loss is still increased significantly with higher gate resistance.

To choose the optimal asymmetric turn-on gate resistance, it is important to consider the tradeoff between overlap switching loss and cross-talk-induced shoot-through loss. Figure 9.23 shows the total switching energy loss expected from a

Table 9.5 Four comparison groups for characterization of cross-talk-related switching performance of SiC MOSFET

First group	Conventional gate drive with V_{gs} of 20/0 V
Second group	Conventional gate drive with V_{gs} of 20/−2 V
Third group	GIR-based gate drive with V_{gs} of 20/0 V
Fourth group	GVR-based gate drive with V_{gs} of 20/0 V

Table 9.6 Auxiliary components of GIR and GVC-based active gate drives

GIR circuit	Auxiliary transistor	Si7308DN N-channel Si MOSFET
	Auxiliary capacitor	100 nF/35 V ceramic capacitor
GVC circuit	Auxiliary transistors	Si5902BDC dual N-channel Si MOSFETs
	Auxiliary diode	MSS1P3 Schottky barrier diode

sweep of R_g and I_L conditions at 400 V DC bus, considering the cross-talk loss of both switches, as well as their C_{oss}-related loss and overlap loss.

An asymmetric gate drive with 20 Ω of turn-on gate resistance may reduce overall cross-talk loss by 5–10 μJ when compared to a conventional gate drive circuit with 0 Ω. However, this adds as much as 80 μJ at the full rated load. Based on these experimental results, an asymmetric gate drive circuit with 5–10 Ω may be appropriate for light load applications below 10 A with the device studied. However, this only results in approximately the same overall losses, and not a significant reduction. This design will also lower the peak dv_{ds}/dt during turn-on, which will be helpful in preventing spurious switching events due to interference in the signal path.

9.3.2 Case study 2: SiC MOSFET with active gate drive circuit

9.3.2.1 Test setup

A DPT with Wolfspeed CMF20120D 1,200 V/24 A SiC MOSFETs in a phase-leg configuration is established to characterize the cross-talk-related switching performance. Comparison experiments under four different groups are conducted according to different gate drives: two conventional gate drives with different turn-off gate voltages along with GIR and GVC-based active gate drives for cross-talk suppression, as listed in Table 9.5. For each group, 12 cases with gate resistances of 10/4.7 Ω, operating voltages of 800/600 V, and operating currents of 5/10/20 A are tested. The auxiliary components of active gate drive circuits are listed in Table 9.6.

Figure 9.24 illustrates the test setup. Sharing one motherboard, two gate drive daughter boards integrating different active schemes are designed to guarantee that the switching performance comparison is merely contributed by the gate drive and cross-talk rather than different layouts and their induced parasitics in the power loop.

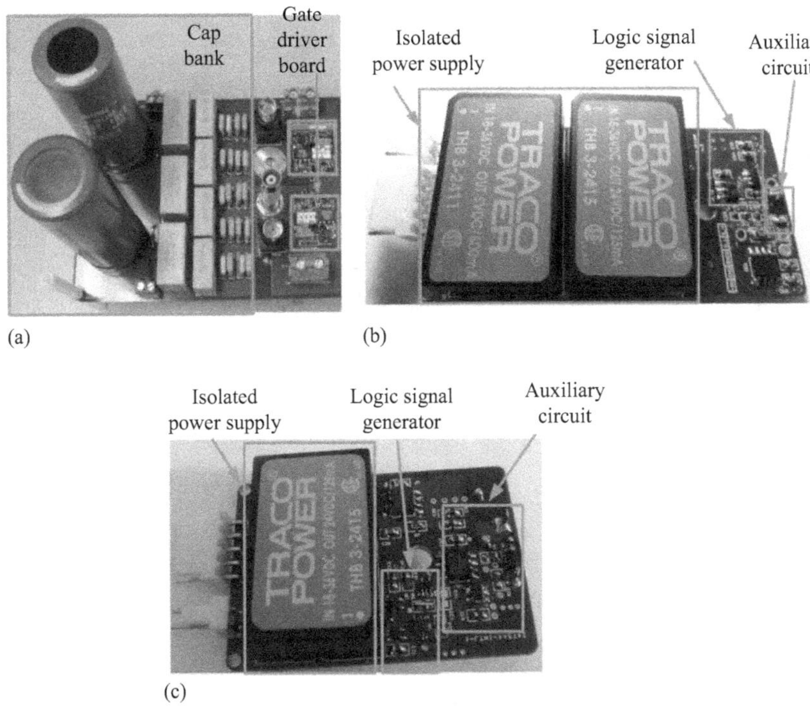

Figure 9.24 DPT circuit for SiC MOSFET-based characterization of cross-talk-related switching characteristics: (a) DPT board with two gate drive daughter boards; (b) GIR-based gate drive circuit; (c) GVC-based gate drive circuit

9.3.2.2 Effect of cross-talk on switching performance

Figure 9.25 displays the comparison waveforms under four different gate drives with the operating condition of 800 V/10 A with 10-Ω gate resistance. During turn-on transient of the lower switch, as shown in Figure 9.25(a), the turn-on energy losses of the lower switch with either GIR or GVC are less than those with the conventional gate drive, even if −2 V turn off gate voltage is applied. Meanwhile, both active gate drives improve the slew rate of the drain-source voltage thanks to the cross-talk suppression. Also, as can be observed in Figure 9.25(b), the switching losses of the upper switch generated by *Cdv/dt*-induced shoot-through current are decreased by using active gate drives. Moreover, during the turn-off transient of the upper switch, both active gate drives enable the spurious negative gate voltages of the lower switch to be largely suppressed, as shown in Figure 9.25(c).

For example, under V_{gs} of 20/0 V, with GIR-based active gate drive, the turn-on energy loss of the lower switch and shoot-through energy loss of the upper switch are reduced by 42.8 and 39.3 μJ, respectively, enabling total turn-on energy loss savings of approximately 17.1%. In addition, *dv/dt* increases from 22.7 to 24.4 V/ns.

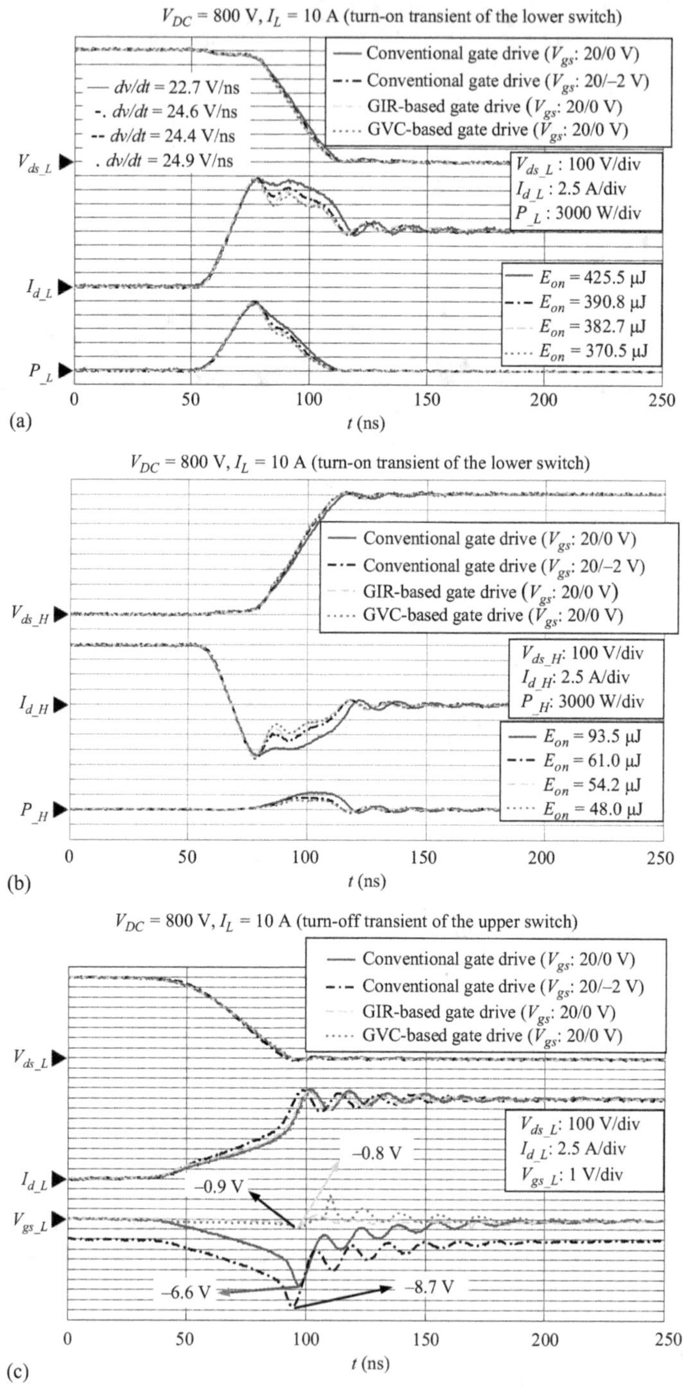

Figure 9.25 Cross-talk-related switching waveforms comparison: conventional gate drive versus active gate drive with cross-talk suppression circuit: (a) lower switch waveforms during the turn-on transient of the lower switch; (b) upper switch waveforms during the turn-on transient of the lower switch; (c) lower switch waveforms during the turn-off transient of the upper switch

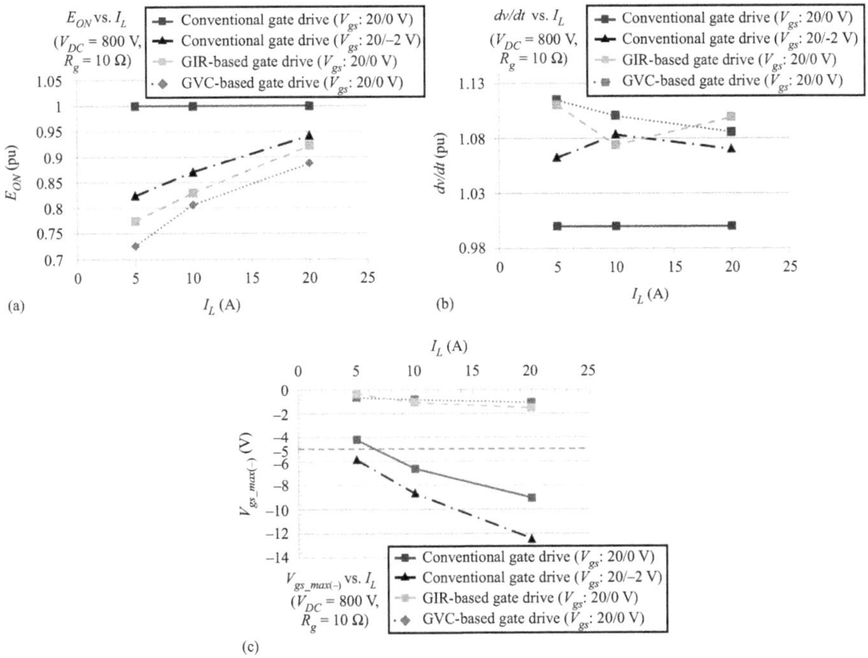

Figure 9.26 Comparison test results dependence on load current: (a) normalized turn-on energy losses during the turn-on transient of the lower switch; (b) normalized dv/dt during the turn-on transient of the lower switch; (c) negative peak value of spurious gate voltage during the turn-off transient of the upper switch

Also, the negative peak value of the spurious gate voltage is minimized from −6.6 V, which exceeds −5 V, the maximum rating of negative gate voltage of the CMF20120D, to −0.8 V.

Furthermore, the GVC-based active gate drive is more effective for the switching performance enhancement: total turn-on energy loss reduction up to 19.4%, *dv/dt* improvement from 22.7 to 24.9 V/ns, and spurious negative gate voltage suppression below −1.0 V. The reason why GVC-based active gate drive has better performance than the GIR circuit for cross-talk elimination during the turn-on transient is due to the large internal gate resistance of SiC devices, which limits the total gate impedance minimization; obviously the GIR-based circuit can only eliminate the impact of the external gate impedance.

Figure 9.26 presents the comparison test results dependence on the load current under the DC bus voltage of 800 V with 10-Ω gate resistance. In order to clearly demonstrate the influence of cross-talk on the switching performance dependence on different gate drives, the turn-on energy loss and *dv/dt* under the second, third, and fourth groups are normalized based on that under the first group, which are listed in Table 9.7. It shows that under different load currents, the turn-on energy losses with active gate drives of 0 V turn-off gate voltage are always less than that with a conventional gate drive, and *dv/dt* is higher, even for the case of when −2 V

Table 9.7 E_{on} and dv/dt versus I_L under first group with V_{DC} of 800 V and R_g of 10 Ω

I_L (A)	5	10	20
E_{ON} (μJ)	381.3	519.0	852.6
dv/dt (V/ns)	23.5	22.7	21.1

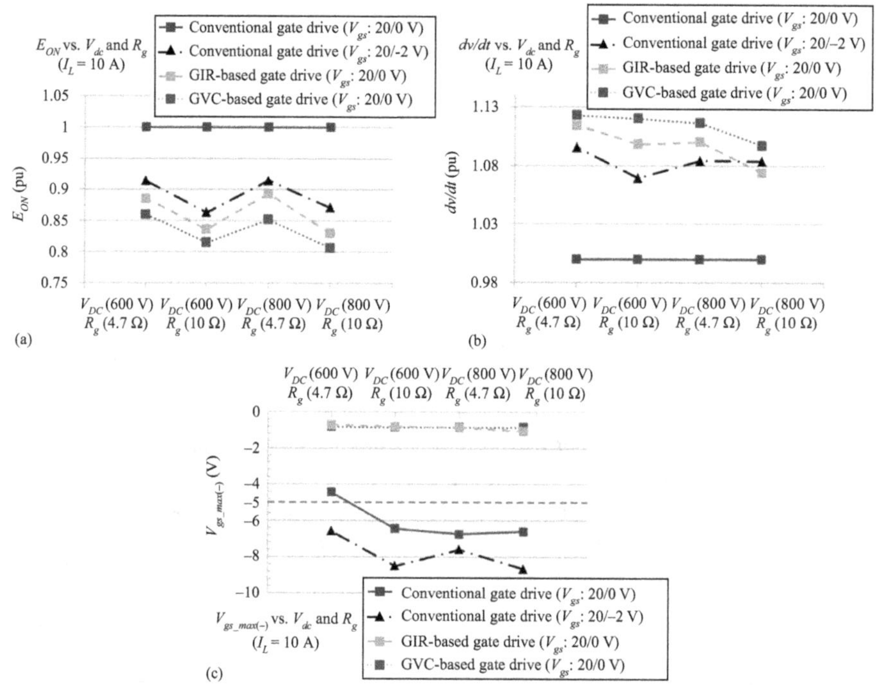

Figure 9.27 Comparison test results dependence on DC bus voltage and gate resistance: (a) Normalized turn-on energy losses during the turn-on transient of the lower switch; (b) normalized dv/dt during the turn-on transient of the lower switch; (c) negative peak value of spurious gate voltage during the turn-off transient of the upper switch

turn-off gate voltage is applied in the conventional gate drive. Note also that both active gate drives have the capability of limiting the peak value of negative spurious gate voltages within its required range. Furthermore, better switching behavior during the turn-on transient by using the GVC-based gate drive verifies again that the large internal gate resistance of SiC devices limits the performance of GIR circuit for cross-talk suppression.

Figure 9.27 illustrates the comparison test results dependence on DC bus voltage and gate resistance under the load current of 10 A. According to the

Table 9.8 E_{on} and dv/dt versus V_{DC} and R_g under first group with I_L of 10 A

V_{DC}/R_g (V/Ω)	600/4.7	600/10	800/4.7	800/10
E_{ON} (μJ)	263.8	332.0	416.3	519.0
dv/dt (V/ns)	29.7	19.7	33.1	22.7

conditions under the first group listed in Table 9.8, turn-on energy loss and *dv/dt* under the second, third, and fourth groups are managed the same as that in Figure 9.26. The experimental results show that under different DC bus voltage with varying gate resistances, the active gate drives have the capability to improve the switching behavior of power devices in a phase-leg configuration in comparison to the conventional gate circuit, even with −2 V turn-off gate voltage. Moreover, compared with GIR circuit, the GVR circuit is still more effective for cross-talk suppression.

9.3.2.3 Discussion

In summary, switching performance of SiC MOSFETs can be adversely affected by cross-talk. It is critical to characterize and understand the cross-talk-related switching behavior. In practice, it is recommended to employ a gate drive with cross-talk suppression functionality to maximize the switching capability of SiC MOSFETs. Active gate drives, as illustrated earlier, are able to effectively mitigate the cross-talk but with the penalty of additional implementation complexity. Specifically, considering the GIR and GVC circuits in the case study, each gate drive circuit has its own application. Thanks to its fewer auxiliary components and easier logic signal synthesis, the GIR circuit is convenient to integrate with a conventional gate drive on a printed circuit board. But the implementation issue of embedding a large capacitor on chip impedes its chip-level integration when considering gate drive integrated circuit design. On the contrary, the all-transistors-based GVC circuit is a suitable option for gate drive chip-level integration, which largely simplifies the implementation of relatively complicated logic signal synthesis.

9.4 Summary

1. Cross-talk induces a spurious gate voltage on its complementary device, which may generate shoot-through current and degrade the device, and in turn may worsen the switching performance of both the lower and upper switches in the phase-leg configuration. According to the inherent properties, WBG devices are easily affected by cross-talk due to (1) fast switching-speed and high *dv/dt*; (2) relatively large internal gate resistance for SiC MOSFETs; (3) lower threshold voltage; and (4) lower magnitude of maximum allowable negative gate voltage.

2. GIR and GVC are two main schemes for cross-talk suppression. The implementation circuits generally can be divided into two categories: passive solutions and active solutions. Passive solutions are simple and easy to implement but always with the penalty of the reduced switching performance. In the end, the global optimization of the switching performance for both switches during turn-on and turn-off transients is difficult to achieve. To cope with the limitations of the passive solutions, the active schemes with auxiliary small rating transistors embedded with the conventional gate drive have been developed to adaptively tune gate loop impedance and/or voltage. If the auxiliary circuit of the active solution (e.g., small rating transistor and/or diode) can be embedded with the conventional gate drive IC, then these active solutions, which can achieve the best tradeoff between the performance and complexity, are preferred.
3. Conventional DPT, which merely focuses on the operating switch, cannot be directly utilized to characterize the influence of cross-talk, including the spurious gate voltage and potential cross-talk losses, since cross-talk affects both switches in the phase-leg. Therefore, the modified methodology for characterization of cross-talk-related switching performance is necessary.
4. Generally, three aspects should be focused on for the characterization of cross-talk-related switching performance: (1) excessive switching loss of the operating switch due to cross-talk-induced shoot-through current during the turn-on, (2) extra switching loss of the synchronous switch during the turn-on, and (3) spurious gate voltage of the synchronous switch during the turn-off. Three testing schemes based on the conventional DPT circuit are designed and illustrated in Figures 9.11–9.13 to characterize the impact of the cross-talk on the switching performance.
5. Two case studies based on GaN HEMTs and SiC MOSFETs are presented to illustrate the methodology for characterization of cross-talk-related switching performance.

References

[1] C. J. Cass, W. Yi, R. Burgos, T. P. Chow, F. Wang, and D. Boroyevich, “Evaluation of SiC JFETs for a three-phase current-source rectifier with high switching frequency,” in *Applied Power Electronics Conference, APEC 2007 – Twenty Second Annual IEEE*, Anaheim, CA, USA, 2007, pp. 345–351.

[2] I. Josifovic, J. Popovic-Gerber, and J. A. Ferreira, “Improving SiC JFET switching behavior under influence of circuit parasitics,” *IEEE Transactions on Power Electronics*, vol. 27, pp. 3843–3854, 2012.

[3] Z. Zhang, F. Wang, L. M. Tolbert, B. J. Blalock, and D. Costinett, “Evaluation of switching performance of SiC devices in PWM inverter fed induction motor drives,” *IEEE Transactions on Power Electronics*, vol. 30, no. 10, pp. 5701–5711, 2015.

[4] Z. Zhang, W. Zhang, F. Wang, L. M. Tolbert, and B. J. Blalock, "Analysis of the switching speed limitation of wide band-gap devices in a phase-leg configuration," *Energy Conversion Congress and Exposition (ECCE)*, Raleigh, NC, USA, 2012 IEEE, 2012, pp. 3950–3955.

[5] T. Wu. International rectifier, Cdv/dt induced turn-on in synchronous buck regulations [Online]. Available: http://www.irf.com/.

[6] A. Elbanhawy, "MOSFET susceptibility to cross conduction," in *Power Electronics and Technology*, 2005, pp. 26–33.

[7] Fairchild Semiconductor AN-7019, Limiting cross-conduction current in synchronous buck converter designs [Online]. Available: http://www.fairchildsemi.com/an/AN/AN-7019.pdf#page=1. Accessed in September 2015.

[8] Infineon IKW15N120T2 Datasheet, Rev 2.2, Dec. 2013 [Online]. Available: https://www.infineon.com/dgdl/Infineon-IKW15N120T2-DS-v02_02-EN.pdf?fileId=db3a304412b407950112b426d2d43acd. Accessed in December 2013.

[9] Microsemi APT34M120J Datasheet, 050-8088 Rev C, 7-2011 [Online]. Available: http://www.microsemi.com/en/sites/default/files/datasheets/APT34M120J_C.pdf. Accessed in July 2013.

[10] Wolfspeed/ Cree C3M0075120K Datasheet, Rev A, Feb. 2017 [Online]. Available: https://www.wolfspeed.com/media/downloads/946/C3M0075120K.pdf. Accessed in February 2017.

[11] ROHM SCT3080KL Datasheet, Rev C, Aug. 2017 [Online]. Available: http://www.rohm.com/web/global/datasheet/SCT3080KL. Accessed in August 2017.

[12] EPC EPC2047 Datasheet, Preliminary 2017 [Online]. Available: http://epc-co.com/epc/Portals/0/epc/documents/datasheets/epc2047_preliminary.pdf.

[13] GaN Systems GS66508B Datasheet, Rev 171101, 2017 [Online]. Available: http://www.gansystems.com/datasheets/GS66508B%20DS%20Rev%20171101.pdf.

[14] B. Callanan. CREE CPWR AN-08. Application considerations for silicon carbide MOSFETs [Online]. Available: http://www.cree.com. Accessed in January 2011.

[15] Semisouth AN-SS1. Silicon carbide enhancement-mode junction field effect transistor and recommendations for use [Online]. Available: http://www.semisouth.com.

[16] R. Kelley, F. Rees, and D. Schwob, "Optimized gate driver for enhancement-mode SiC JFET," in *PCIM*, 2009.

[17] Avago Technologies AN-5314. Active Miller clamp [Online]. Available: http://www.avagotech.com. Accessed in December 2014.

[18] Z. Zhang, F. Wang, L. M. Tolbert, and B. J. Blalock, "Active gate driver for cross talk suppression of SiC power devices in a phase-leg configuration," *IEEE Transactions on Power Electronics*, vol. 29, no. 4, April 2014, pp. 1986–1997.

[19] C. Zheng, M. Danilovic, D. Boroyevich, and Z. Shen, "Modularized design consideration of a general-purpose, high-speed phase-leg PEBB based on SiC MOSFETs," in *Power Electronics and Applications (EPE 2011),*

Proceedings of the 2011-14th European Conference on, Birmingham, UK, 2011, pp. 1–10.

[20] B. Wrzecionko, D. Bortis, J. Biela, and J. W. Kolar, "Novel AC-coupled gate driver for ultrafast switching of normally off SiC JFETs," *IEEE Transactions on Power Electronics*, vol. 27, no. 7, pp. 3452–3463, 2012.

[21] Y. Zushi, S. Sato, K. Matsui, Y. Murakami, and S. Tanimoto, "A novel gate assist circuit for quick and stable driving of SiC-JFETs in a 3-phase inverter," in *Proceedings of the IEEE Applied Power Electronics Conference and Exposition (APEC)*, Orlando, FL, USA, February 2012. pp. 1734–1739.

[22] J. Wang and H. S. Chung, "A novel RCD level shifter for elimination of spurious turn-on in the bridge-leg configuration," *IEEE Transactions on Power Electronics*, vol. 30, pp. 976–984, 2015.

[23] H. Li, X. Zhang, Z. Zhang, *et al.*, "Design of a 10 kW GaN-based high power density three-phase inverter," *Energy Conversion Congress and Exposition (ECCE) 2016 IEEE*, Milwaukee, WI, USA, 2016, pp. 1–8.

[24] F. Gao, Q. Zhou, P. Wang, and C. Zhang, "A gate driver of SiC MOSFET for suppressing the negative voltage spikes in a bridge circuit," *IEEE Transactions on Power Electronics*, vol. 33, pp. 2339–2353, 2018.

[25] E. A. Jones, F. Wang, D. Costinett, Z. Zhang, and B. Guo, "Cross conduction analysis for enhancement-mode 650-V GaN HFETs in a phase-leg topology," *2015 IEEE 3rd Workshop on Wide Bandgap Power Devices and Applications (WiPDA)*, Blacksburg, VA, USA, 2015, pp. 98–103.

Chapter 10
Impact of three-phase system*

This chapter focuses on the limitations and impacting factors of switching performance of WBG devices in a three-phase power converter, and their influences on the accuracy of the dynamic characterization results based on the conventional DPT.

In prior chapters, including the cross-talk consideration, dynamic characterization using DPT is limited to the two-level half-bridge phase-leg configuration. It seems that the switching performance of power devices can be accurately evaluated via DPT in such a phase-leg. Maybe this is true to the relatively slow Si devices, but not anymore for fast switching WBG devices. Due to intrinsic characteristics of WBG devices, such as small junction capacitance, small specific on-resistance, and high *di/dt* and *dv/dt* during fast switching transients, their switching behavior becomes more susceptible to parasitics and noise of the application circuit, compared with slower Si devices. Thus, the simple and layout-optimized DPT with an optimally designed load inductor in the phase-leg configuration may not sufficiently represent a more complex configuration of actual converters and loads.

Therefore, the switching performance of the WBG devices in actual power converters needs to be systematically evaluated as compared to the switching behavior in DPT. The most critical limitations and impacting factors should be identified and understood. Then either the conventional DPT has to be modified with the additional practical application considerations or the solution needs to be developed to suppress the adverse effects of the identified limitations, allowing the conventional DPT results to represent the true switching behavior of WBG devices in practice. In this chapter, a SiC MOSFET-based three-phase voltage source inverter-fed motor drive is selected as the representative under investigation since it is a widely applied topology in a popular application. Also, the three-phase voltage source inverter in motor drives covers almost all the components in an actual phase-leg-based power converter. Thus, the generality of the following analysis is not lost.

*Part of the chapter content is from a previously published paper by the authors. from Z. Zhang, F. Wang, L.M. Tolbert, B.J. Blalock, D. Costinett, “Evaluation of switching performance of SiC devices in PWM inverter fed induction motor drives,” *IEEE Transactions on Power Electronics*, vol. 30, no. 10, Oct. 2015, pp. 5701–5711.

10.1 Impacting factors and limitations in the actual three-phase system

As shown in Figure 10.1, there are three primary differences between the configurations of a three-phase inverter in motor drives and a DPT: different inductive loads, two more phase-legs for the inverter, and cooling systems (e.g., heat sink). Accordingly, first, the high-frequency impedance of a motor (e.g., an induction motor) and the optimally designed inductor employed in DPT is compared and analyzed here to illustrate the resulting switching behavior differences. Second, the impact of the other two phase-legs on the switching performance of the phase-leg under test is described for the three-phase inverter. Third, the parasitic capacitive coupling effect between power devices and heat sink, and its influence on the switching performance, is discussed for motor drives. Finally, for experimental verification, the switching performance of Wolfspeed 1,200 V/24 A SiC MOSFETs is evaluated under pulse test and continuous operating condition in a three-phase inverter connected to an induction motor.

10.1.1 Parasitics of inductive load

As can be observed in Figure 10.1(a), the load inductor of the DPT circuit is paralleled with the upper switch, assuming that the lower switch is selected as the DUT. During the switching transient, the upper switch stays off, and can be modeled as a conducting diode when the switching current is commutating (i.e., current rise subinterval t_{cr} during the turn-on transient and current fall subinterval t_{cf} during the turn-off transient in Figure 10.2), or as an output capacitance while the switching voltage is changing (i.e., voltage fall subinterval t_{vf} during the turn-on transient and voltage rise subinterval t_{vr} during the turn-off transient in Figure 10.2). Thus, the DPT circuit can be redrawn as Figure 10.3, where Z_L represents the

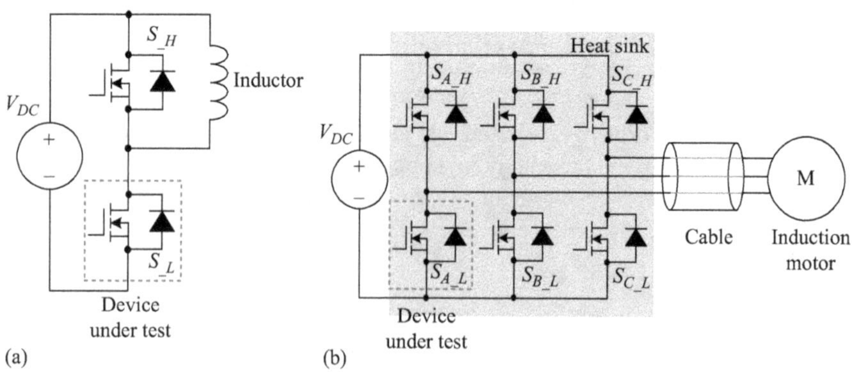

Figure 10.1 Configuration comparison between double pulse tester and three-phase voltage source converter: (a) DPT circuit; (b) three-phase voltage source inverter-fed motor drives

inductive load impedance. If Z_L is much greater than the equivalent impedance of the upper switch during the switching transient, then the impact of inductive load on the switching performance can be neglected; otherwise, Z_L must be taken into consideration. For the specific analysis of the comparisons between DPT load

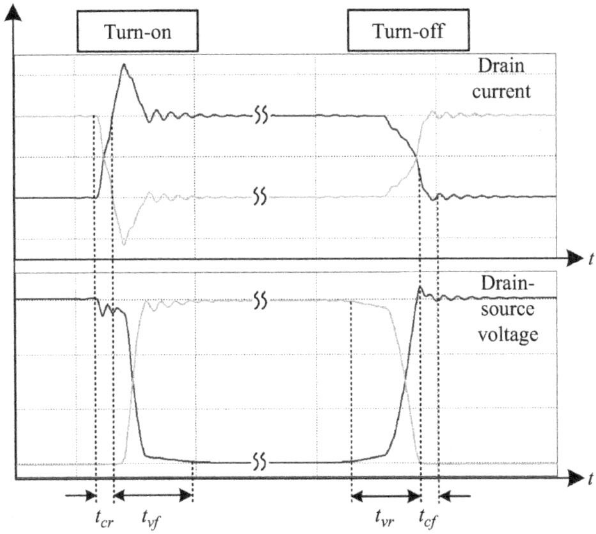

Figure 10.2 Typical switching current/voltage waveforms

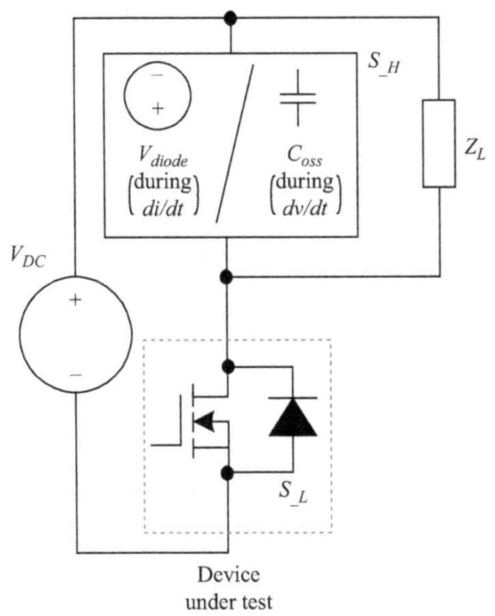

Figure 10.3 Simplified DPT circuit

inductor and induction motor load, four 115 μH inductors in series are employed to form a DPT load inductor, which is a typical way of designing DPT load inductor with small parasitic capacitance; a 7.5 kW induction motor and 2-m power cable are selected as the representative of the motor load, since it is a proper load in terms of power rating for the SiC devices under evaluation (Wolfspeed CMF20120D 1,200 V/24 A SiC MOSFETs). In addition, for the higher power rating induction motor with longer power cable in many practical applications, the associated impedance Z_L at high-frequency becomes lower [1]. Thus, the motor load selected in this case study indicates a conservative impact of induction motor on the switching performance. Then the generality of the following analysis is not lost.

10.1.1.1 Impact of induction motor on switching performance

As illustrated in Figure 10.3, during the switching current commutation subinterval, the impedance of the upper switch is extremely small, since it can be considered as a voltage source. During the switching voltage change subinterval, the impedance of the upper switch depends on the output capacitance of the device. Normally, the equivalent impedance of the upper switch is also small in the switching-related frequency range that is determined by the switching speed [2]. Typically, the frequency range is several MHz to tens of MHz for SiC devices considering switching intervals of tens of nanoseconds.

In the typical DPT, the load inductor during a switching transient can be approximated as a current source. Ideally, due to the infinite impedance of the current source during this dynamic process, the load inductor has no impact on the switching behavior of the DUT. Practically, the load inductor of the DPT is optimally designed so that its parasitics, such as equivalent parallel capacitance (EPC), are very small [3], making the load inductor impedance much larger than the upper switch equivalent impedance at frequencies of interest with respect to the switching transition. Therefore, the load inductor has little effect on the switching performance in the DPT.

However, in adjustable speed drive systems, the parasitics of the induction motor cause its high-frequency impedance to be significantly decreased, as shown in Figure 10.4(a). Measured using the Agilent 4294A impedance analyzer, the impedance of the DPT load inductor is greater than the induction motor equivalent impedance above 200 kHz, although the DPT load inductance is smaller than the induction motor inductance. In addition, the measured induction motor impedance in our example system is even smaller than the equivalent impedance of the SiC device used in this case (Wolfspeed CMF20120D SiC MOSFET) below 11.5 MHz. Note the device equivalent impedance is mainly determined by its output capacitance and the Wolfspeed/Cree device has an output capacitance of 120 pF [4]. On the contrary, above 700 kHz, the DPT load impedance is greater than the device's impedance due to its output capacitance. Consequently, the induction motor will play a significant role in the switching behavior of the DUT due to the small motor impedance at high frequencies. Also, it is important to note that SiC has smaller output capacitance than its Si counterpart, resulting in higher

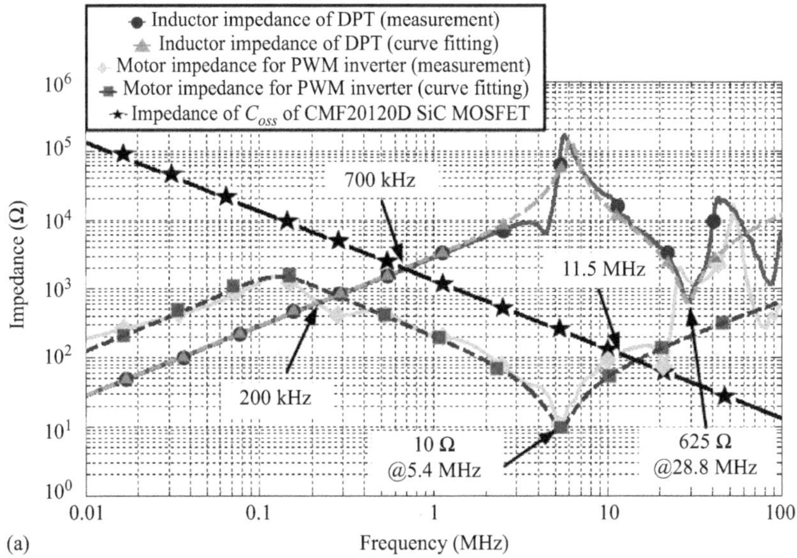

(a)

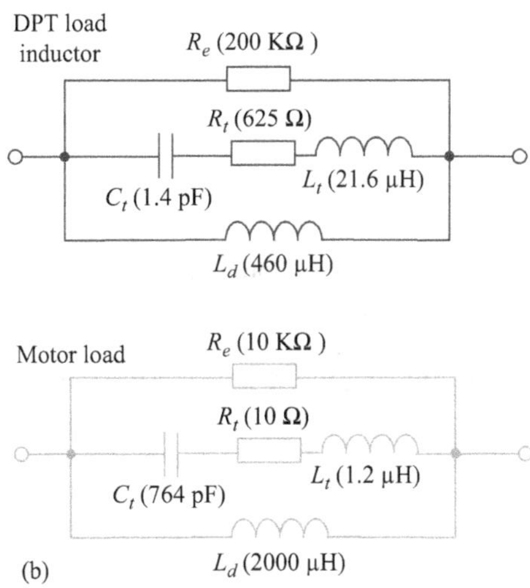

(b)

Figure 10.4 Impedance and circuit model of inductive loads: (a) impedance of inductive loads versus C_{oss} of SiC device; (b) circuit model of inductive loads

equivalent impedance and making the SiC switching behavior more susceptible to the parasitics of the induction motor.

Based on the measured impedance, circuit models of the DPT load inductor and motor load can be built, as shown in Figure 10.4(b) following the methods

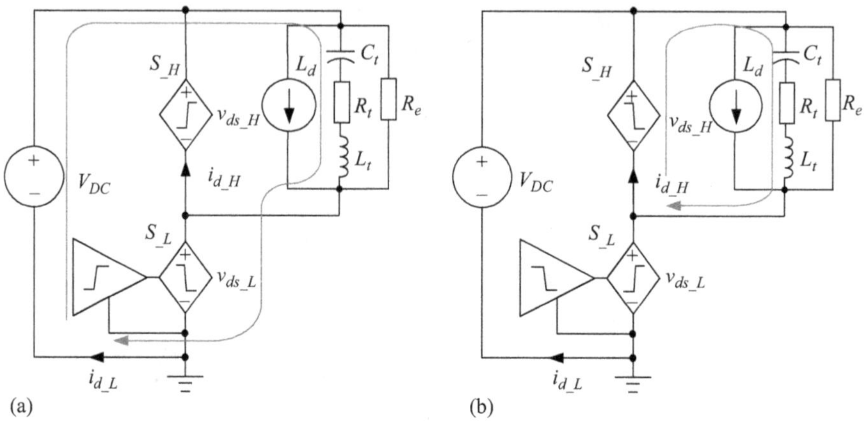

Figure 10.5 Impact of parasitics of induction motor on switching current: (a) turn-on transient; (b) turn-off transient

in [1,5–7]. Each parameter in the circuit model has physical meaning and significance. L_d refers to the stator winding leakage inductance; R_e represents the high-frequency iron loss of the stator winding; L_t, C_t, and R_t are introduced to capture the second resonance in the motor impedance characteristics, according to [7], that may be caused by the skin effect and turn-to-turn capacitance of the stator windings. As can be observed in Figure 10.4(a), the fitted curves of the inductive load impedance based on the circuit models can represent the critical characteristics of the motor and DPT impedances. During the switching transient, the large inductance L_d can be modeled as a current source, as shown in Figure 10.5. It is the series resonant network formed by L_t, C_t, and R_t that affects the switching performance. Compared with the DPT load inductor with extremely small parasitic capacitance C_t (1.2% C_{oss} of SiC MOSFET under evaluation) and large damping resistance R_t (625 Ω), the parasitic capacitance C_t of the induction motor is 6 times that of C_{oss} of the SiC MOSFETs, and its corresponding damping resistance R_t (10 Ω) is relatively small. Therefore, in the time domain, considering the high *dv/dt* drain-source voltage of the upper switch V_{ds_H} across the *LRC* series resonant network, the induction motor with large C_t and small R_t causes more serious parasitic ringing with longer duration. This parasitic ringing causes additional resonant current to flow in the switching commutation loop. Thus, the switching performance of the lower switch is affected. Unlike the typical high-frequency parasitic ringing (typically tens of MHz) due to the resonance between the power loop inductance and C_{oss} of power devices, the ringing caused by parasitics of the induction motor has relatively low frequency with 5.4 MHz in this case study (see Figure 10.4(a)). Also, this lower frequency ringing will be observed in the drain current of the lower switch during the turn-on transient and drain current of the upper switch during the turn-off transient, as illustrated in Figure 10.5.

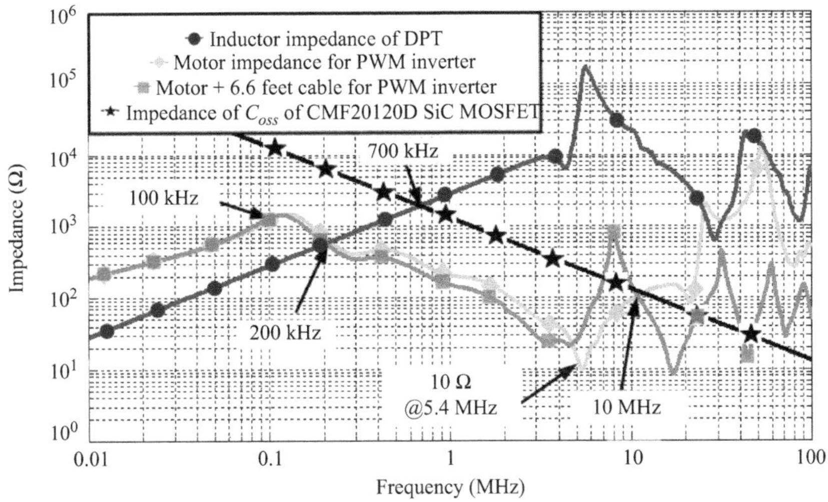

Figure 10.6 Impedance of induction motor with 2 m cable versus induction motor

10.1.1.2 Impact of induction motor with power cable on switching performance

Induction motor drive systems fed with a power cable are widely used in many industrial applications. It is therefore necessary to evaluate the impact of an induction motor with a power cable on the switching performance. Figure 10.6 displays the impedance comparison between the induction motor with and without a 2-m power cable. The impedances of the induction motor and the induction motor plus power cable are identical below 100 kHz, but significant differences occur at high frequency. Especially above 10 MHz, the impedance of the induction motor with power cable is smaller than that of the induction motor, and is even comparable to the impedance of the SiC MOSFET's C_{oss}. Based on the aforementioned analysis in the frequency domain, more serious impact on switching performance can be expected of the induction motor with power cable. Also, since the cable length in practical applications is generally much longer than the 2-m length used in this case study, more attention must be paid to the power cable's impact. In the time domain, due to the complicated model of power cable with distributed *LC* networks, it is difficult to illustrate the precise relationship between the power cable and the switching behavior based on a high-order circuit model. Generally speaking, the capacitive coupling among the conductors of the power cable extends the charging/discharging subinterval during switching transients, leading to lower *dv/dt*, longer switching time, and larger switching losses.

10.1.2 Interaction of phase-legs

As can be observed in Figure 10.1, in comparison with DPT consisting of only one phase-leg, a three-phase inverter has two additional phase-legs. Thus, additional

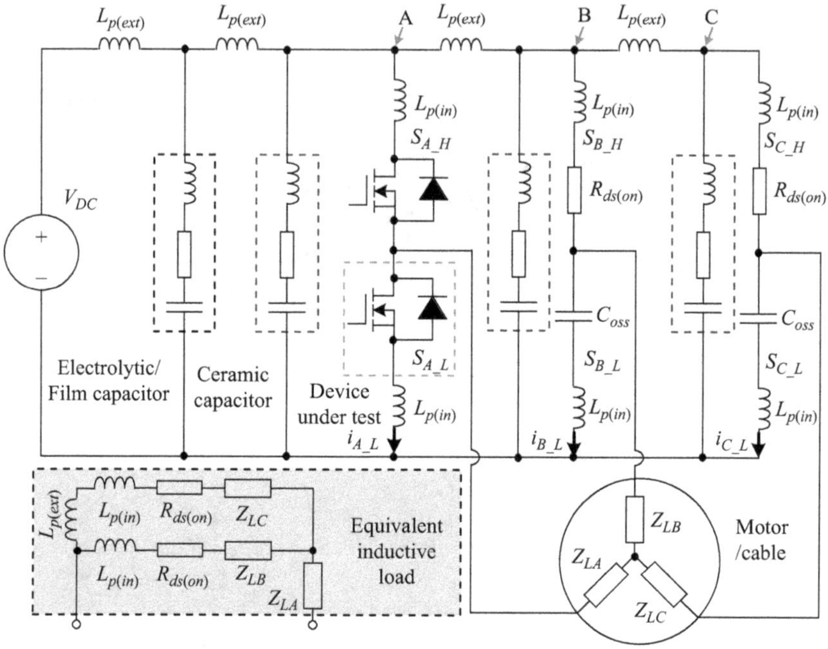

Figure 10.7 Equivalent circuit of three-phase PWM inverter considering the parasitics

device and interconnection parasitics will be involved in the circuit. Assuming the upper switches of phases B and C are in on-state while their lower switches in off-state during the switching transients of the phase A devices, Figure 10.7 displays the equivalent circuit of a three-phase inverter considering the parasitics, where the upper switches of phases B and C are represented by their on-state resistance $R_{ds(on)}$, and the junction capacitance C_{oss} represents the lower switches of phases B and C. Also, $L_{p(in)}$ is the power loop parasitic inductance within each phase-leg, while $L_{p(ext)}$ is the parasitic inductance distributed in the DC bus. Z_{LA}, Z_{LB}, and Z_{LC} are the impedances of the inductive load of each phase.

As can be observed in Figure 10.7, the equivalent inductive load consists of parasitic inductance, on-state resistance, and three-phase inductive loads. Considering the relatively small parasitic inductance (i.e., $L_{p(in)}$, $L_{p(ext)}$) and the small $R_{ds(on)}$ of SiC devices in comparison with the impedance of inductive loads (Z_{LB}, Z_{LC}), the impact of parasitics induced by the upper switches of phases B and C on the inductive loads can be negligible. In addition, the potentials at nodes B and C will almost stay constant during the switching transient of the phase-A DUT, due to the existence of ceramic capacitors with small equivalent series inductance (ESL) utilized as decoupling capacitors mounted close to each phase-leg in practical applications. Therefore, for the lower switches of phases B and C, the voltages across C_{oss} have little variation so that the switching performance of the DUT is hardly affected by the lower switches of phases B and C.

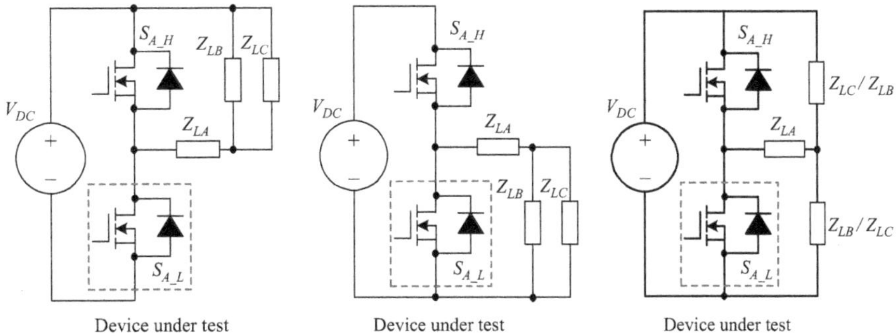

Figure 10.8 Connection structures of inductive loads according to different sections of space vector modulation

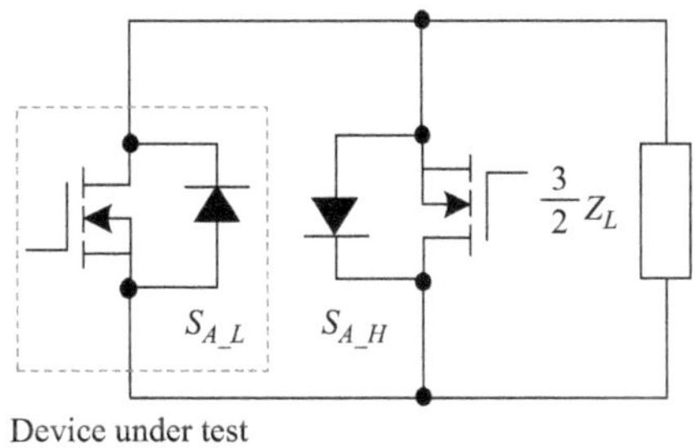

Figure 10.9 Uniform equivalent circuit considering constant DC bus voltage thanks to decoupling capacitors

In addition to the parasitics induced by phases B and C, the on/off states of switches in the other phase-leg alter the connection structure of inductive loads with respect to the phase-leg under test. Therefore, different from a fixed inductive load position within the DPT, the tested device in a three-phase voltage source inverter during the commutation interval will see different combinations of the inductive loads. According to the on/off states of the phases B and C switches, based on the space vector modulation scheme, there are three connection structures of inductive loads with respect to phase A during a switching transient, as shown in Figure 10.8. Fortunately, as previously mentioned, the DC bus voltage during the switching transient stays nearly constant thanks to the decoupling capacitors close to each phase-leg. Hence, according to circuit theory, the DC bus voltage source is shorted during the dynamic process analysis. Assuming the impedances of inductive loads for all phases are identical and denoted by Z_L, three different circuits in Figure 10.8 can be simplified as a uniform equivalent circuit, as shown in Figure 10.9. Consequently, different connection structures due to the switching states of phases B and C have little effect on the switching performance of phase A devices in practical applications.

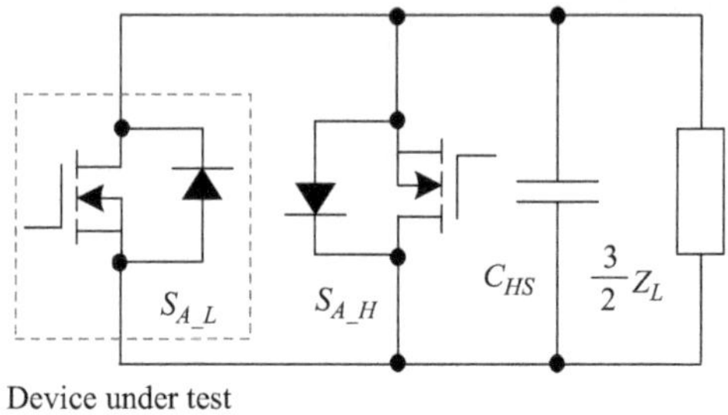

Figure 10.10 Uniform equivalent circuit considering capacitive coupling effect induced by heat sink

10.1.3 Coupling effect by heat sink

In a three-phase inverter, SiC devices are attached to a heat sink for thermal management. Usually, a thin layer of insulating material is used to separate the SiC devices from the electrically conductive heat sink. Thus, a parasitic capacitance is formed between the drain base plate of the SiC devices and the common heat sink plate [8]. In the end, this capacitance C_{HS} is paralleled with devices, which equivalently increases their effective C_{oss}. However, in a motor drive in this study, in addition to the coupling capacitance between devices and heat sink, the parasitics of the induction motor is paralleled with devices as well, as shown in Figure 10.10. Compared with the motor parasitic capacitance C_t (764 pF in Figure 10.4(b)), the coupling capacitance (39.75 pF based on our test setup) is only 5.2% of C_t. Therefore, the capacitive coupling effect is insignificant when evaluating the switching performance of SiC MOSFETs in a motor drive system with relatively large capacitive parasitics.

10.1.4 Experimental verification

10.1.4.1 Hardware setup

To evaluate the influence of the aforementioned impacting factors, a three-phase inverter test circuit was designed to satisfy the following requirements: (1) sufficient flexibility such that the three-phase inverter can be evolved from a layout-optimized DPT; (2) during this design evolution, the layout associated with the DUT must remain consistent; and (3) a common heat sink is used for all power devices. Figure 10.11 displays a three-phase voltage source inverter with Wolfspeed 1,200 V CMF20120D SiC MOSFETs, including one uniform mother board, six gate drive daughter boards integrated with devices, and one common heat sink. Hence, a DPT configuration can be achieved by means of merely connecting the gate drive boards of phase A with the mother board. In addition, a three-phase PWM inverter can be assembled with all six gate drive boards of three phase-legs mounted to the mother board. Also, using the lower switch of phase A as the DUT, it can be observed that the layouts of both the power and gate loop for the DUT under either DPT configuration or inverter circuit are nearly the same.

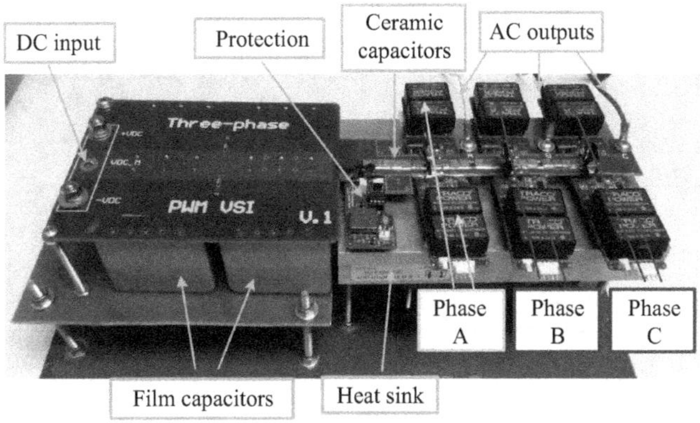

Figure 10.11 SiC-based three-phase voltage source inverter

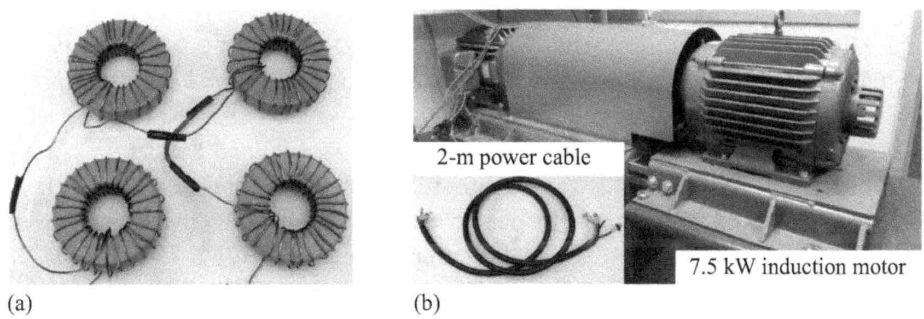

Figure 10.12 Different inductive loads: (a) DPT load inductor; (b) 7.5 kW induction motor with 2-m power cable

Figure 10.12 displays two types of inductive loads: DPT load inductor and a 7.5 kW induction motor with a 2-m power cable for evaluation of the difference in switching performance for DPT and motor drive conditions. Note that the DPT load inductor consists of four small inductors in series to minimize parasitics. For the fast transient measurement, the following equipment is used to detect the switching waveforms: a digital oscilloscope (Tektronix DPO 401) with frequency bandwidth of 1 GHz and maximum sampling rate of 5 GS/s, two THDP0200 high-voltage differential probes with frequency bandwidth of 200 MHz for drain-source voltage measurement of the upper and lower switches in a phase-leg, and two TCP0030A current probes with frequency bandwidth of 120 MHz for drain current measurement of the upper and lower switches in a phase-leg.

10.1.4.2 Methodology for experimental verification

A four-step verification procedure has been devised, that starts with the DPT circuit and ends with the three-phase PWM inverter-fed induction motor drive. Each step will now be described.

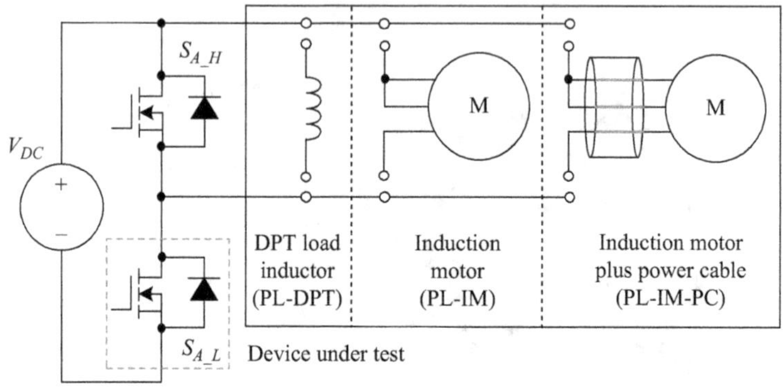

Figure 10.13 Test circuits for evaluation of switching performance with different inductive loads

1. Evaluation of the impact of different inductive loads on the switching performance. Here the gate drive boards of phase A are connected with the mother board to form a DPT circuit, and the switching performance of the phase A lower switch is tested with the optimally designed load inductor, 7.5 kW induction motor, and 7.5 kW induction motor plus 2-m power cable. As shown in Figure 10.13, the switching performance differences in this step are only attributed to the inductive loads. For simplicity, PL-DPT, PL-IM, and PL-IM-PC refer to the test circuits with DPT load inductor, induction motor, and induction motor plus power cable, respectively.
2. Evaluation of the interaction of phase-legs on switching performance. The gate drive boards of phases B and C are added to the mother board to establish a three-phase inverter, and the switching performance of the DUT is tested by turning on the upper switches of phases B and C, as shown in Figure 10.14(a). Hence, compared with the switching behavior of the DUT with the induction motor in step 1 (i.e., PL-IM), the impact of parasitics induced by phases B and C can be identified. Then, according to the modulation scheme, the switches of phases B and C are controlled in different on/off states during the switching transient of the DUT, as shown in Figure 10.14(b) and (c), to evaluate the impact of different inductive load connection structures on switching performance. For simplicity, 3Φ-IM-1, 3Φ-IM-2, and 3Φ-IM-3 indicate the test circuits as shown in Figure 10.14(a), (b), and (c), respectively.
3. Evaluation of the impact of the heat sink on switching performance. A common heat sink is attached with all the power devices in the three-phase inverter, and the switching behavior of the DUT is tested by controlling the upper switches of phases B and C to be on, as shown in Figure 10.14(d). Thus, in comparison with the switching performance of the DUT with the same on/off states of the switches of phases B and C in step 2 (i.e., 3Φ-IM-1), the capacitive coupling effect induced by the heat sink can be identified. For simplicity, 3Φ-IM-HS-1 represents the test circuits in Figure 10.14(d).

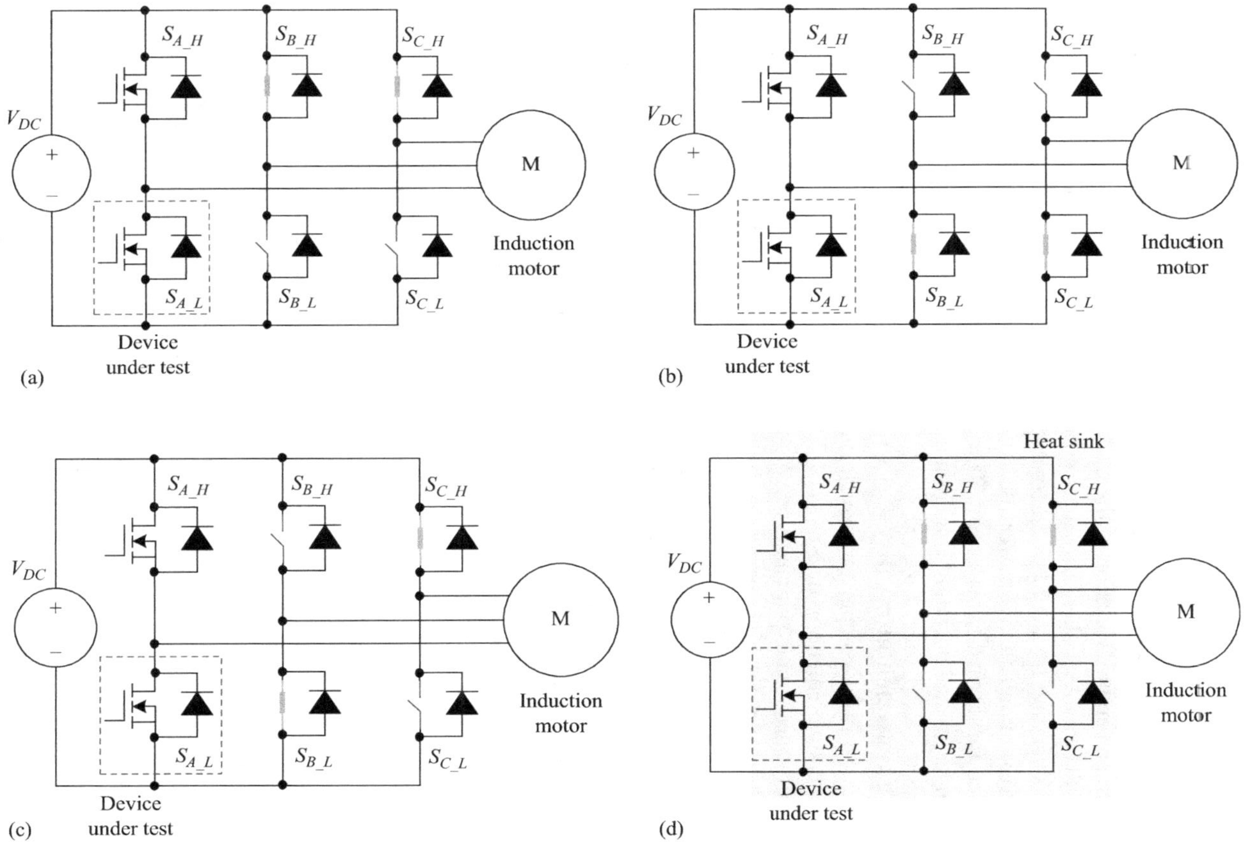

Figure 10.14 Test circuits for evaluation of the impact of phase-legs and heat sink on the switching performance: (a) 3Φ-IM-1; (b) 3Φ-IM-2; (c) 3Φ-IM-3; (d) 3Φ-IM-HS-1

4. Evaluation of the impact of the induction motor rotation on the switching performance. The switching behaviors in the previous three steps are evaluated based on pulse test, which means that the motor is not rotating, although certain pulse currents flow through the stator windings. In step 4, the switching performance of the DUT is tested when the induction motor is rotating. Based on the same hardware setup as in step 3, the difference of the switching performance under the motor running test and pulse test can be evaluated. Also note that the junction temperature variation of the DUT under the continuous operating condition will become another factor that impacts the switching behavior. In this test, the low switching frequency of 5 kHz and small AC RMS current of 11 A are selected to minimize active power loss. Also, the heat sink with forced air cooling is utilized for thermal management. Thus, the junction temperature rise of the DUT under no load operating condition for 1 to 2 minutes is insignificant.

10.1.4.3 Experimental results

Figure 10.15 shows the comparison waveforms with three different inductive loads under the operating condition of 600 V/10 A with 5 Ω gate resistance. As can be observed from the overall switching waveforms during both turn-on and turn-off transients, there is a 5.4 MHz frequency ringing in the switching current with the induction motor, which corresponds to the resonance frequency created by the parasitics of the induction motor (see Figure 10.4). In addition, as predicted, during a turn-on transient, this low frequency ringing current is found from the drain current of the lower switch; while during a turn-off transient, it flows through the upper one. Also note that the ringing duration is up to 1.5 μs, which is much longer than the tens-of-nanoseconds' switching time, as shown in the zoomed-in switching waveforms. Furthermore, as compared to the switching behavior with the DPT load inductor, the parasitics of the induction motor cause the turn-on and turn-off switching time to increase from 26 to 29 ns and 32 to 38 ns, respectively. The total energy loss slightly increases as well. After inserting a 2-m power cable between the inverter and motor, the switching performance becomes worse: switching time increases up to 42% during turn-on, and doubles during turn-off; an additional 32% of energy loss is dissipated during the switching transient.

More test results with different load currents and gate resistances are shown in Figure 10.16. In order to clearly demonstrate the impact of different inductive loads on the switching performance, the switching time t_{sw} and energy loss E_{sw} with induction motor and motor plus power cable are normalized based on results with DPT load inductor, which are listed in Tables 10.1 and 10.2. These results show that under different load currents with varying gate resistances, t_{sw} and E_{sw} with either induction motor or induction motor plus power cable are increased compared to the test data with DPT load inductor. In addition, the impact of the induction motor plus power cable on the switching performance is more serious. All of the experimental data agree with the aforementioned theoretical analysis.

Switching waveforms in Figures 10.17 and 10.18 illustrate the impact of phases B and C on the switching performance of the lower switch in phase A. Neither the parasitics induced by phases B and C nor the different inductive load

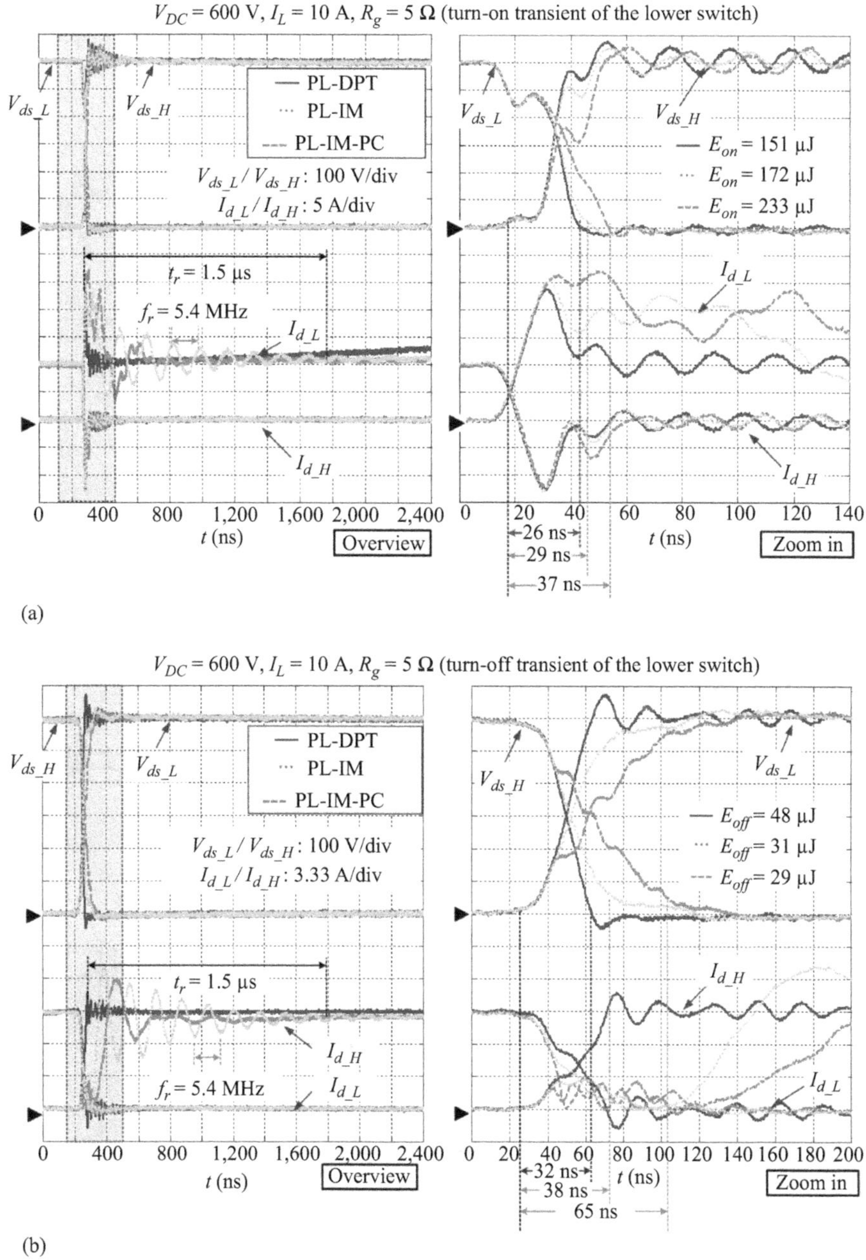

Figure 10.15 Impact of different inductive loads on switching performance: (a) turn-on transient; (b) turn-off transient

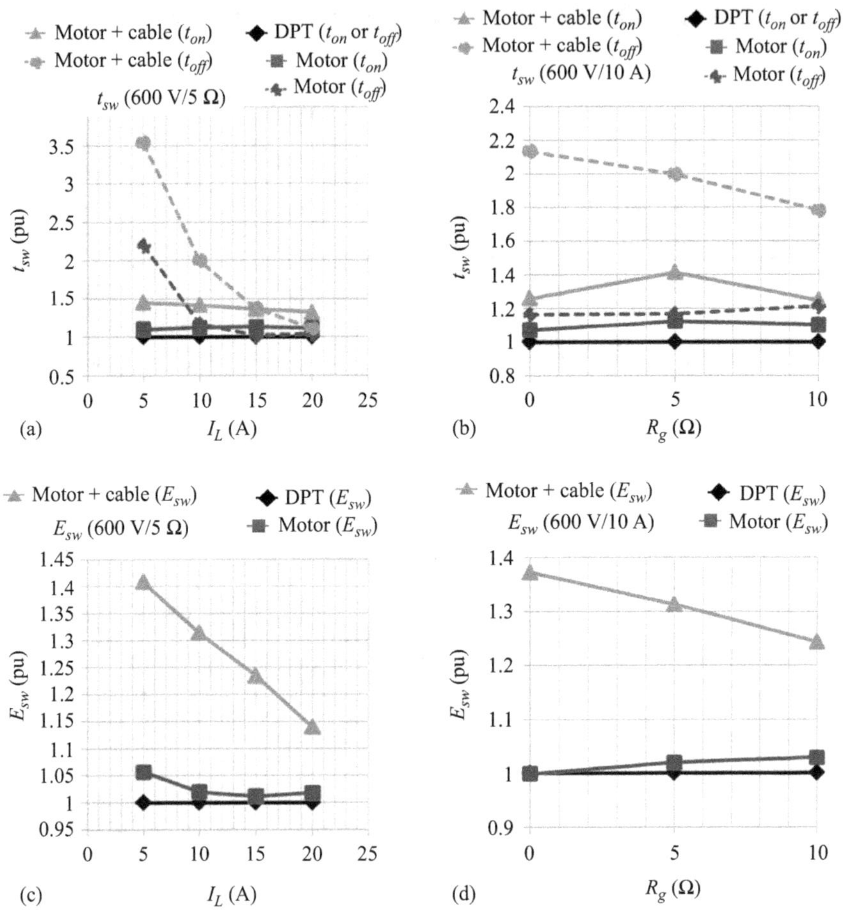

Figure 10.16 Comparison of t_{sw} and E_{sw} among different inductive loads under different I_L and R_g: (a) t_{sw} dependence on I_L; (b) t_{sw} dependence on R_g; (c) E_{sw} dependence on I_L; (d) E_{sw} dependence on R_g

Table 10.1 t_{sw} and E_{sw} versus I_L under DPT condition with V_{DC} of 600 V and R_g of 5 Ω

I_L (A)	5	10	15	20
t_{ON} (ns)	16	17	18	20
t_{OFF} (ns)	51	29	24	23
E_{sw} (μJ)	104	141	195	265

Table 10.2 t_{sw} and E_{sw} versus R_G under DPT condition with V_{DC} of 600 V and I_L of 10 A

R_G (Ω)	0	5	10
t_{ON} (ns)	17	26	41
t_{OFF} (ns)	29	32	38
E_{sw} (μJ)	132	176	240

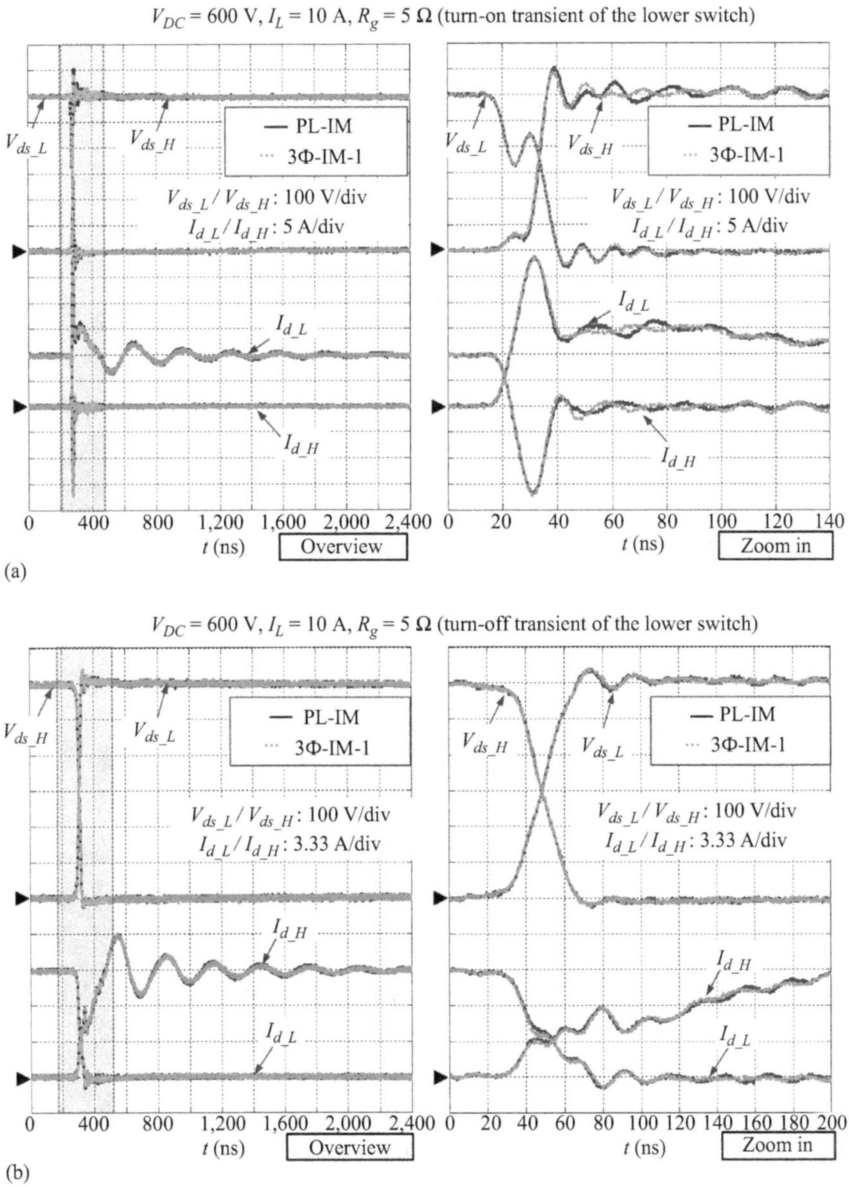

Figure 10.17 Impact of parasitics induced by phases B and C on switching performance: (a) turn-on transient; (b) turn-off transient

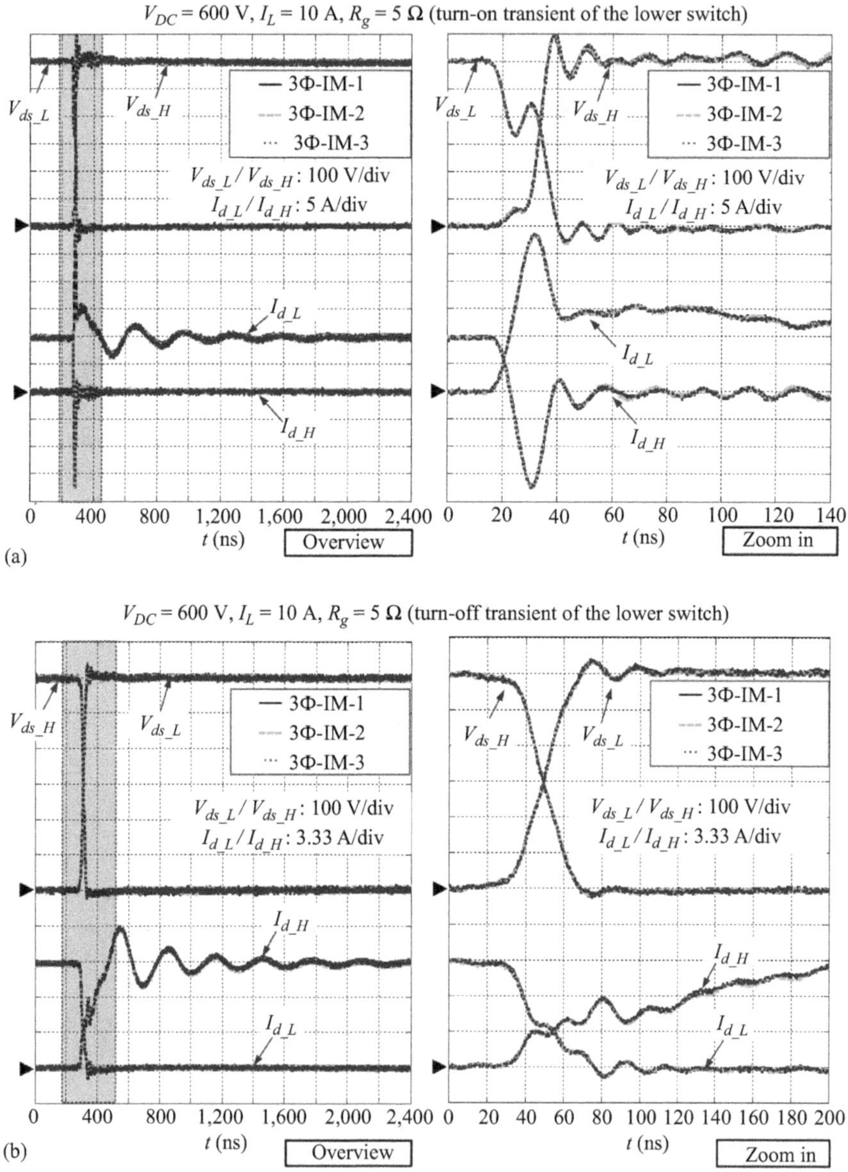

Figure 10.18 Impact of different connection structures of inductive loads on switching performance: (a) turn-on transient; (b) turn-off transient

connection structures due to the switches' on–off states have significant effect on the switching performance since the switching current and voltage under different test configurations are almost identical. Similarly, as can be observed in Figure 10.19, the impact of the heat sink on the switching performance is insignificant.

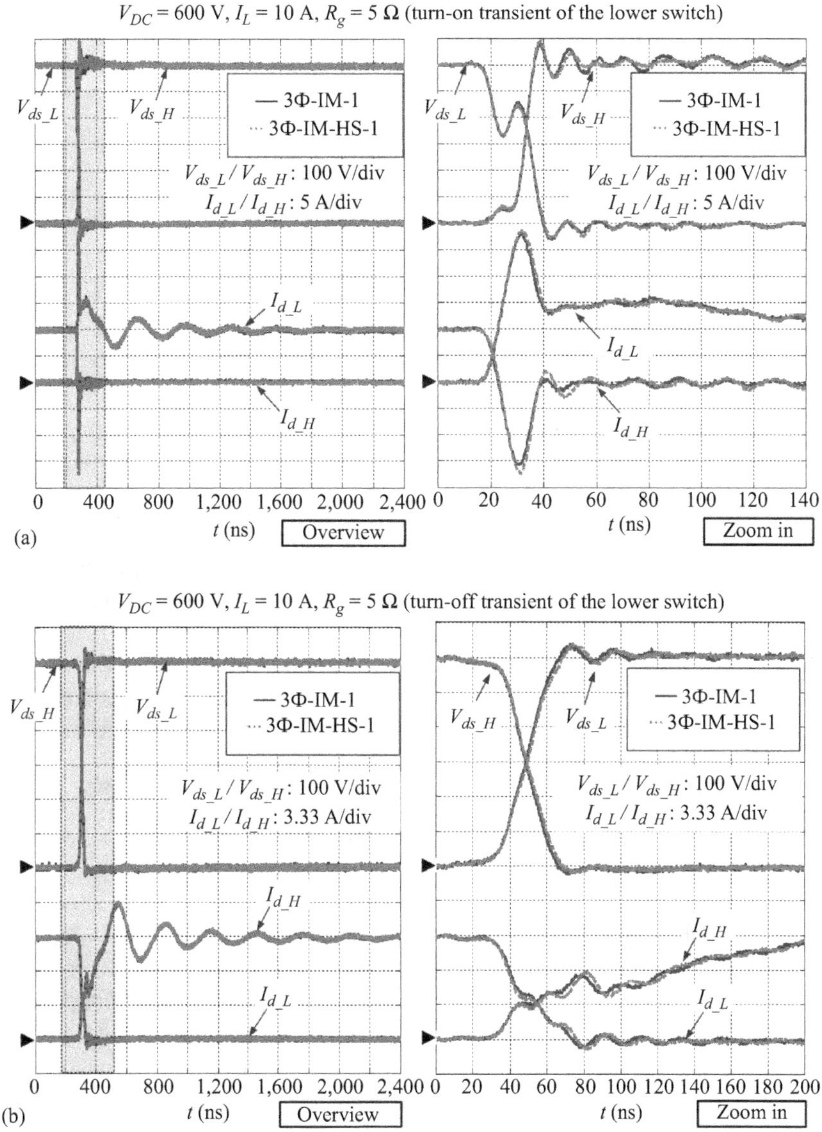

Figure 10.19 Impact of capacitive coupling effect on switching performance: (a) turn-on transient; (b) turn-off transient

Since the aforementioned test results are based on the pulse test, we cannot necessarily draw a conclusion for a rotating induction motor until additional tests are done. Figure 10.20 shows the gate signal, line-to-line voltage and phase current of the three-phase inverter when driving the induction motor with no load. According to the modulation scheme, we picked one switching cycle when the

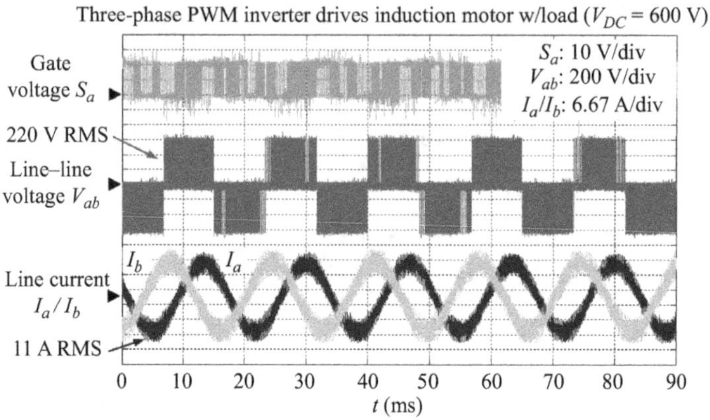

Figure 10.20 Typical waveforms of three-phase inverter under continuous operating condition

circuit configuration during a switching transient is the same as that in 3Φ-IM-HS-1 (see Figure 10.14(d)), and compared the switching waveforms between the running test and previous pulse test. The switching waveforms in Figure 10.21 verify that the induction motor rotation has no appreciable influence on the switching performance. The reason is that the flux penetrating into the rotor magnetic circuit at high frequency is very small such that the high-frequency impedance of induction motor is almost independent of the rotor speed [7].

10.2 Dynamic characterization with practical application considerations

According to the analysis discussed earlier, non-ideal high-frequency characteristics of inductive load is identified as the most critical impacting factor of the switching performance. Therefore, two approaches can be employed to improve the dynamic characterization considering the impact of practical loads.

The most straightforward solution is to conduct DPT with the practical loads connected instead of the dedicated DPT inductor, as shown in Figure 10.22. However, this makes the dynamic characterization of WBG devices using DPT application dependent and less general. In the end, to obtain the valid switching characteristics results, DPT has to be conducted separately for different applications.

The other approach is to develop solution to suppress the adverse effects of the impact of practical loads, allowing the conventional DPT results with DPT load inductor to represent the true switching behavior of WBG devices in practice. The following discussion will follow this path and focus on the decoupling of interaction between WBG devices and inductive load.

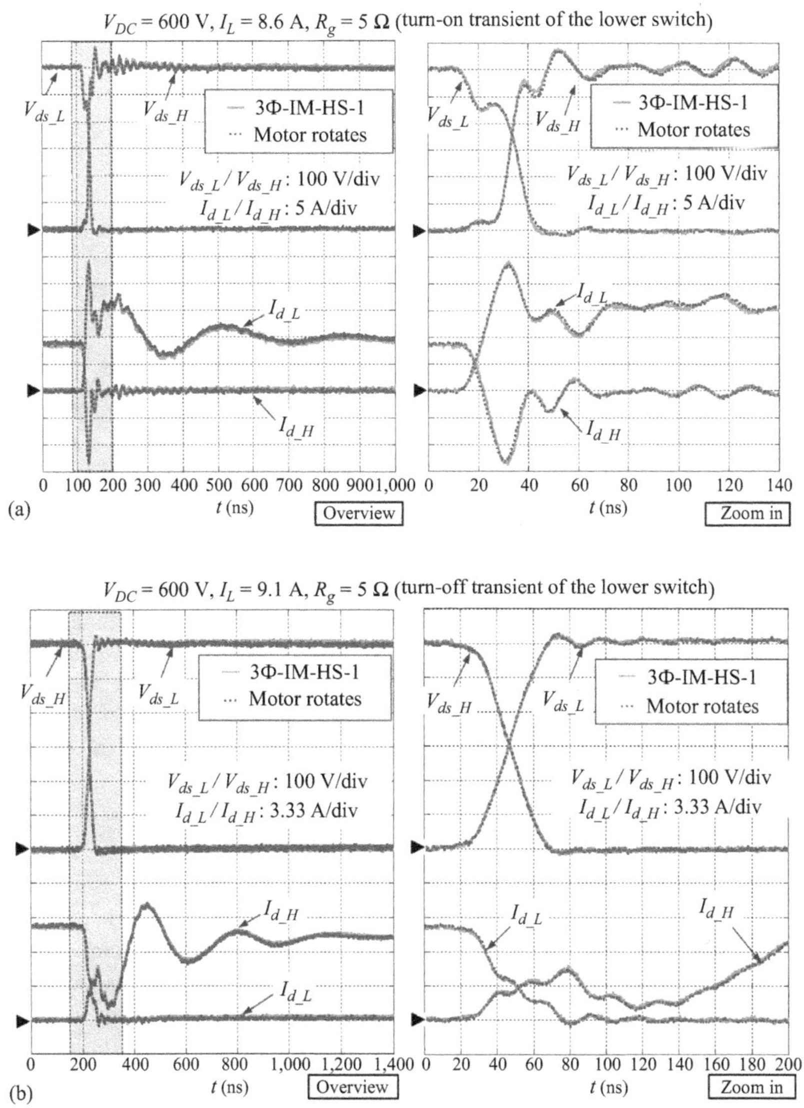

Figure 10.21 Impact of mechanical state of induction motor on switching performance: (a) turn-on transient; (b) turn-off transient

As discussed earlier, the impact of inductive load on the switching performance mainly comes from the relatively low impedance at the switching transient-related frequency range. The analysis and experimental results show that the inductive load consisting of the motor and power cable significantly worsen the switching speed and losses. Consequently, this section focuses on the motor plus

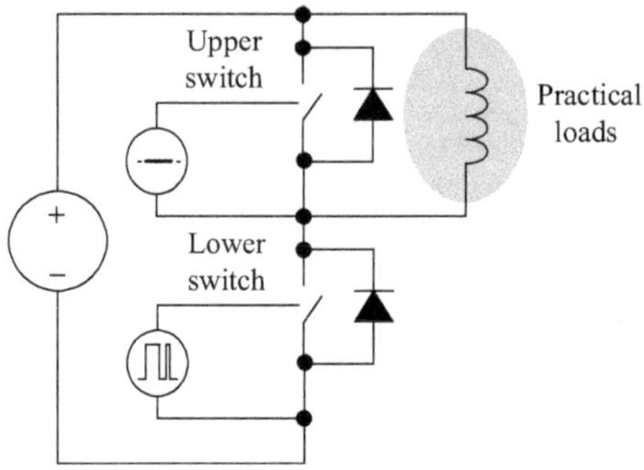

Figure 10.22 Modified DPT with the consideration of practical loads to improve the dynamic characterization results of WBG devices in actual system

power cable-based inductive load, and aims at mitigating its negative influence during the switching transient. First, the high-frequency inductive load modeling is presented, as the fundamental knowledge for the following solution development. Second, a basic idea of decoupling the interaction between the inductive load and power device in the voltage source converter (VSC) during the switching transient is discussed. Based on this idea, an auxiliary filter is introduced between the converter and inductive load so that the parasitics of the load will not be "seen" from the converter side during the switching transient. Finally, a DPT with Wolfspeed 1,200 V/20 A SiC MOSFETs is utilized to demonstrate the validity and effectiveness of this auxiliary filter.

10.2.1 High-frequency modeling of inductive load

According to the high-frequency impedance of the inductive load consisting of motor plus cable measured by impedance analyzer, a circuit model is derived, as shown in Figure 10.23(a). Several *LRC* series resonant networks are employed in a paralleled structure to simulate the high-frequency behavior of the inductive load under investigation. As can be observed in Figure 10.23(b), the fitted curves of the inductive load impedance based on the circuit model can represent the critical characteristics of the motor plus cable impedances in frequency domain.

Based on the behavior model derived earlier, a simulation circuit is built by MATLAB®/Simulink® to verify the accuracy of this high-frequency model in time domain. In Figure 10.24(a), the voltage across the inductive load (i.e., drain-source voltage of the upper switch v_{ds_H}) measured by double pulse test is added at terminals of the inductive load's circuit model. And then the current flowing through the inductive load in the simulation is monitored and compared with the tested inductive current. The switching waveform comparison in Figure 10.24(b) shows that the simulated inductive current I_L is almost identical to that based on test results during both turn-on and turn-off transients. In total, the derived high-frequency circuit model is accurate in both frequency and time domains and can be used for the following investigation.

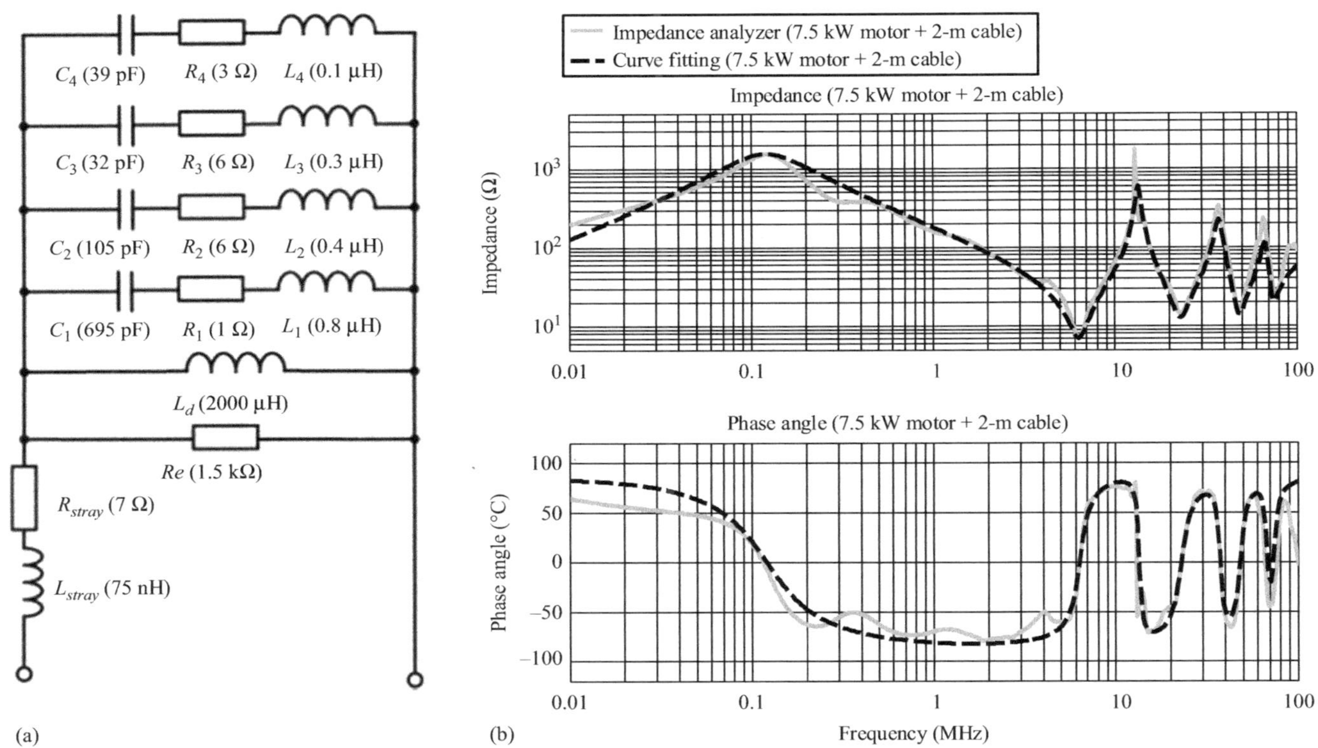

Figure 10.23 High-frequency impedance and circuit model of motor plus cable-based inductive load: (a) high-frequency behavior model; (b) impedance comparison

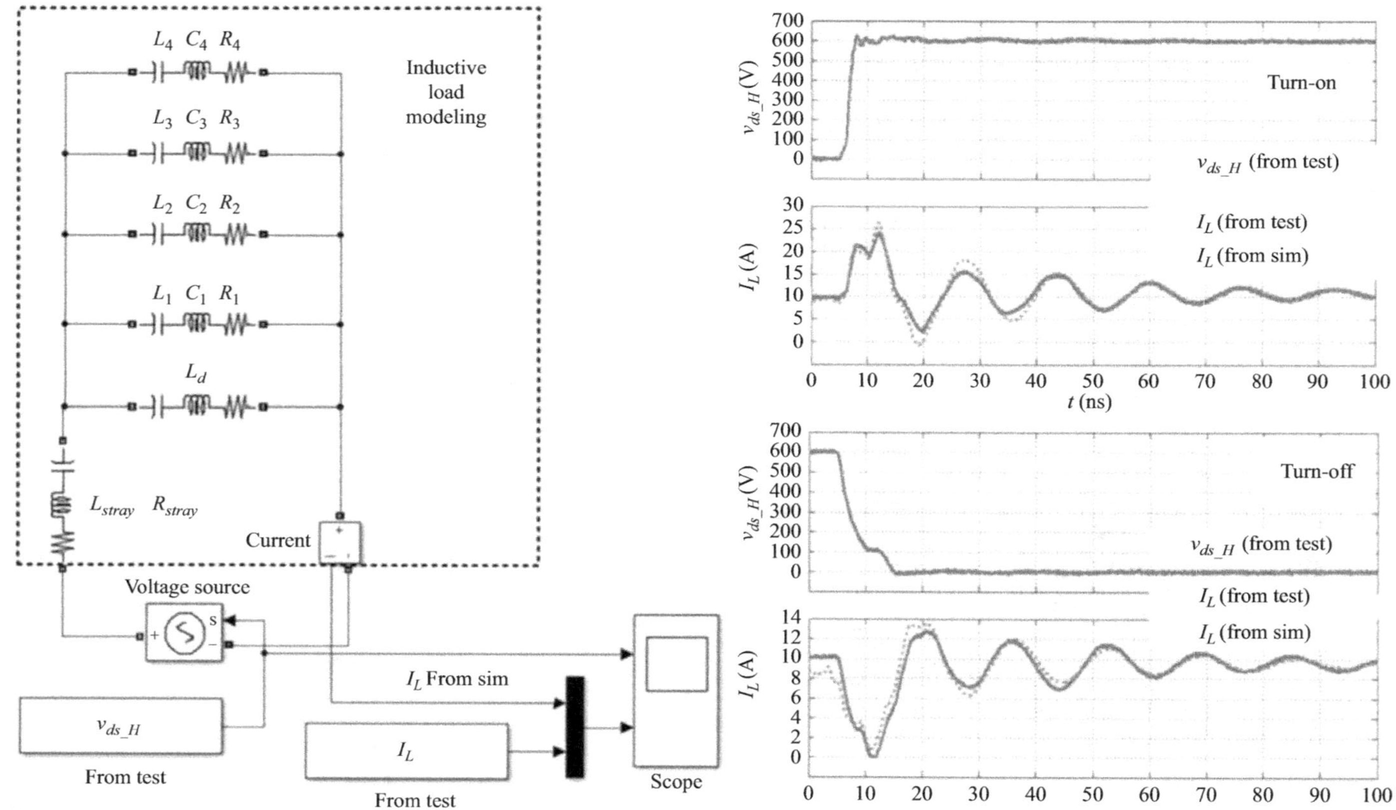

(a) (b)

Figure 10.24 Accuracy verification of the high-frequency modeling: (a) evaluation platform by MATLAB®/Simulink®; (b) test versus model-based simulation

10.2.2 Basic idea for mitigation of adverse effect from inductive load

According to the circuit model in Figure 10.23(a), the large inductance L_d can be considered as a current source during the switching transient. It is the *LRC* series resonant networks that affect the switching behavior. Taking turn-on transient as an example, as can be observed in Figure 10.25(b), during *dv/dt* transient, there will be resonant current excited per each *LRC* series resonant branch (i.e., I_1 to I_4). This additional current flows into the device and increases the channel current, resulting in the slower turn-on speed with larger switching losses. On the other hand, during the turn-off transient in Figure 10.25(c), the resonant current induced by *LRC* series resonant branch decreases the equivalent inductive current, leading to a longer turn-off time for charging/discharging of the device' output capacitance. Hence, the turn-off speed decreases as well.

To mitigate the impact of the inductive load's parasitics on the switching performance, resonant current suppression during the switching transient is critical. There are two basic ideas: (1) minimization of the resonant currents for the *LRC* branches with respect to the high resonant frequency (e.g., f_2 to f_4 in Figure 10.26(a)); and (2) regulation of the resonant period of the low resonant frequency-based *LRC* branch (e.g., f_1 Figure 10.26(a)) to be much longer than the switching commutation time, such that there is almost no response for this series resonant network to the switching voltage excitation during the switching transient.

To achieve these two ideas, an auxiliary inductor with small EPC is introduced in series with the existing inductive load to modify the high-frequency impedance of the inductive load. As can be found in Figure 10.26(a), first, this inductor has an excellent inductive characteristic at high frequency. Hence, the *LRC* branches with respect to the resonant frequency of f_2 to f_4 will disappear. Second, this inductor is capable of tuning the resonant frequency of the $L_1R_1C_1$ branch from f_1 to f_{ring}, enabling its corresponding resonant period T_{ring} to be much longer than the switching commutation time. And then the resonant current I_1 remains small during the switching transient (see Figure 10.26(b)).

10.2.3 Design criteria of the auxiliary filter

According to the basic ideas discussed earlier, the inductance for the auxiliary inductor L_{aux} is determined by f_{ring}, which is directly related to the switching voltage commutation time. However, the switching voltage commutation time changes as the operating condition varies. Therefore, it is critical to select a switching voltage commutation time that enables the switching performance not to be affected by the inductive load under most of operating conditions. As can be observed in Figure 10.27, the switching voltage commutation time tested by the double pulse tester greatly depends on the operating current, and the longest switching voltage commutation time occurs during the turn-off transient under the light inductive load. If f_{ring} is determined based on the turn-off switching voltage commutation time at the extremely light load, the required L_{aux} will be fairly large, but the benefit we can obtained is limited since the turn-off switching loss at the

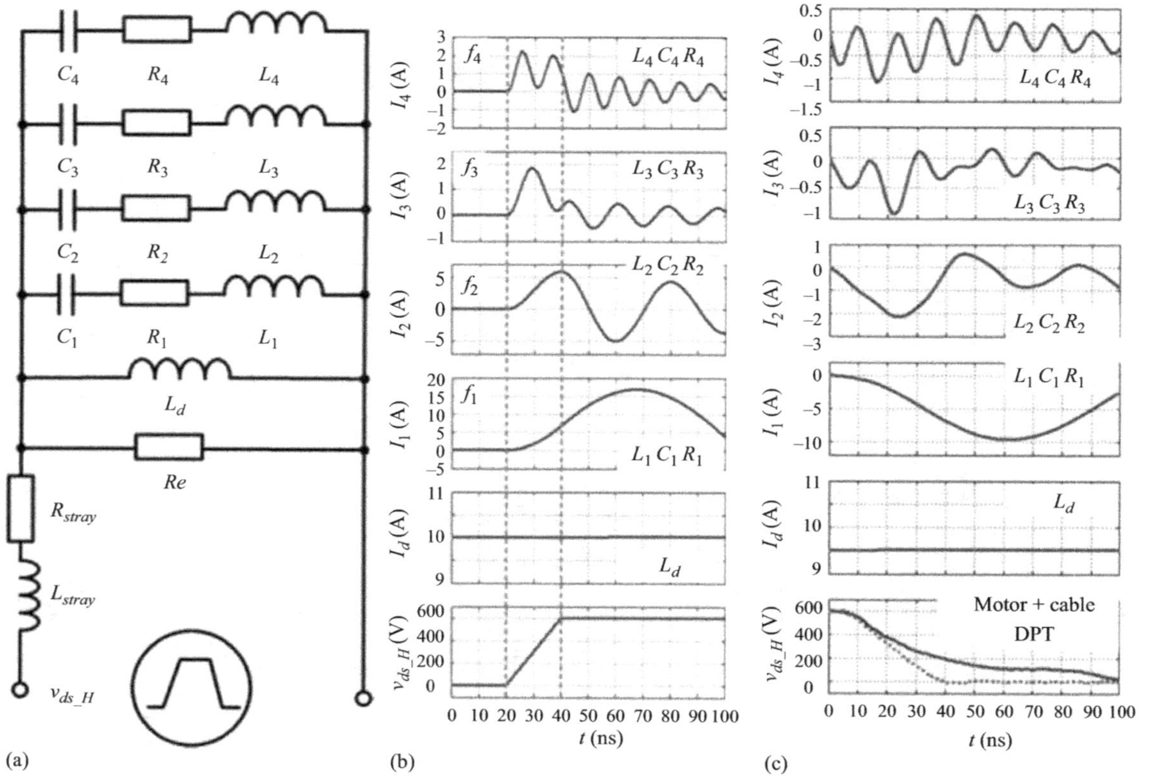

Figure 10.25 Impact of parasitics of inductive load on the switching behavior: (a) circuit model; (b) during turn-on transient; (c) during turn-off transient

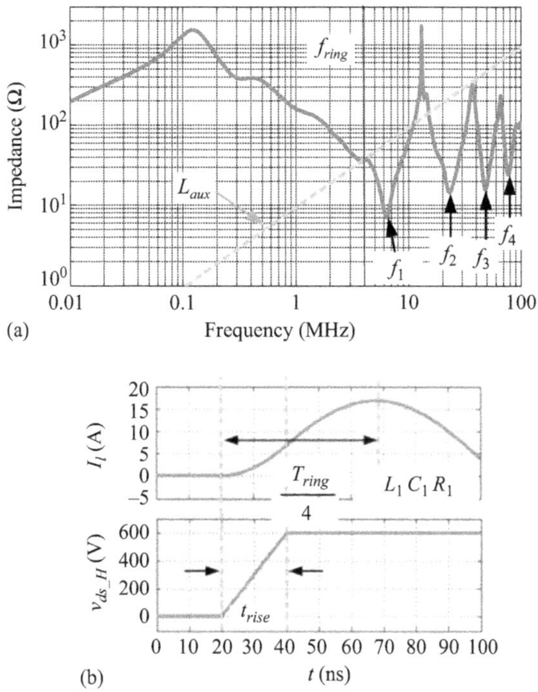

Figure 10.26 Basic ideas for mitigation of the adverse effect from inductive load: (a) high-frequency impedance comparison: inductive load versus auxiliary inductor; (b) resonant current I_l during the switching voltage commutation time

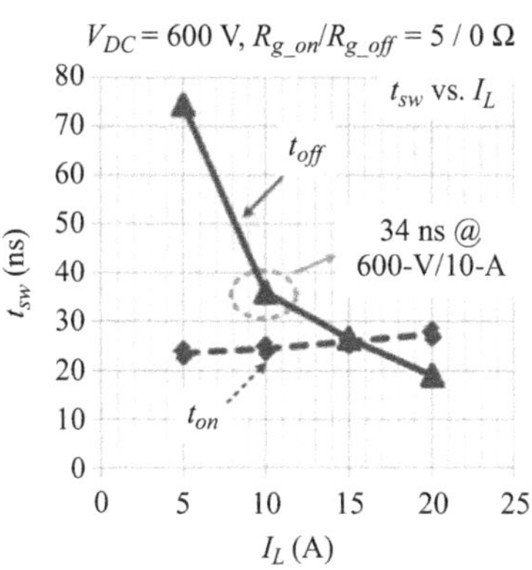

Figure 10.27 Voltage commutation time dependence on load current measured by double pulse test

light load is always almost zero. Thus, it is wise to select a switching voltage commutation time which is longer than all of the turn-on commutation time and part of the turn-off commutation time so that except the turn-off time at the light load, the switching time under the rest of the operating conditions together with the switching losses under all of the operating points will not be affected by the parasitics of the inductive load. In the following case study, f_{ring} is determined by the switching voltage commutation time of 34 ns during turn-off transient under the operating condition of 600 V/10 A (see Figure 10.27).

Based on the aforementioned design idea, f_{ring} should satisfy,

$$\frac{1}{f_{ring}} \gg t_{v,worst} \tag{10.1}$$

where $t_{v,worst}$ refers to the switching voltage commutation time at the selected worst operating conditions (e.g., 34 ns at 600 V/10 V in this case study). f_{ring} of 3 MHz is selected so that its corresponding resonant period T_{ring} is 10 times longer than $t_{v,worst}$. Combining with the high-frequency impedance of the inductive load, a 2 μH auxiliary inductor is needed. Figure 10.28 illustrates that after insertion of a 2 μH auxiliary inductor, high resonant frequencies, such as f_2, f_3, and f_4 existing in the original motor plus cable load is suppressed. Moreover, the resonant frequency of f_1 is shifted from 6 to 3 MHz. In total, a 2 μH auxiliary inductor achieves all the design requirements described earlier. Considering the three-phase conversion system, the auxiliary inductor per phase is 1.4 μH.

In addition to the inductance selection, physical design of the auxiliary inductor is important, because it affects the loss dissipated by this auxiliary inductor. The design objective is to minimize this extra loss so that combining with the switching loss reduction due to the auxiliary inductor, the total power loss of the converter can decrease. In this case study, considering the relatively small inductance, an air core inductor is employed. The design methodology is described as follows.

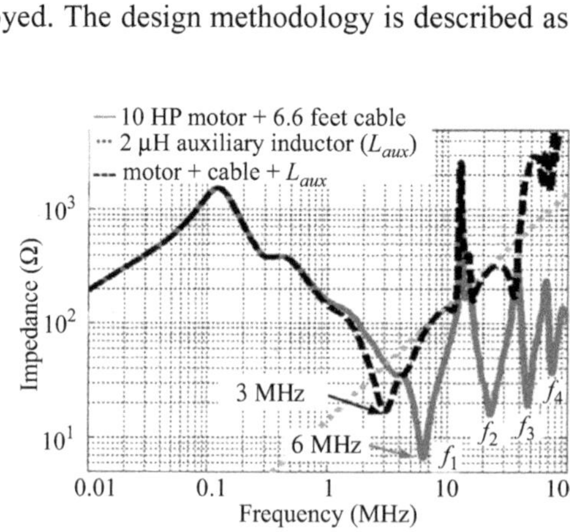

Figure 10.28 High-frequency impedance comparison: inductive load versus auxiliary inductor versus inductive load in series with the auxiliary inductor

First, AWG of the magnet wire is selected based on the maximum operating current. In this case study, based on 11 A RMS current per each phase in the three-phase voltage source inverter together with 4 A/mm^2 maximum current density for copper wires, 12 AWG magnet wire is selected.

Second, diameter d and number of turns N of the air core inductor are designed to achieve the required inductance with the shortest length of the copper wire (i.e., smallest $N \times MLT$ (mean-length-per turn)) for the minimization of power loss dissipated by this auxiliary inductor. Figure 10.29(a) shows a typical air core inductor, and its inductance is given by

$$L = (d^2 + N^2)/(18d + 40l) \tag{10.2}$$

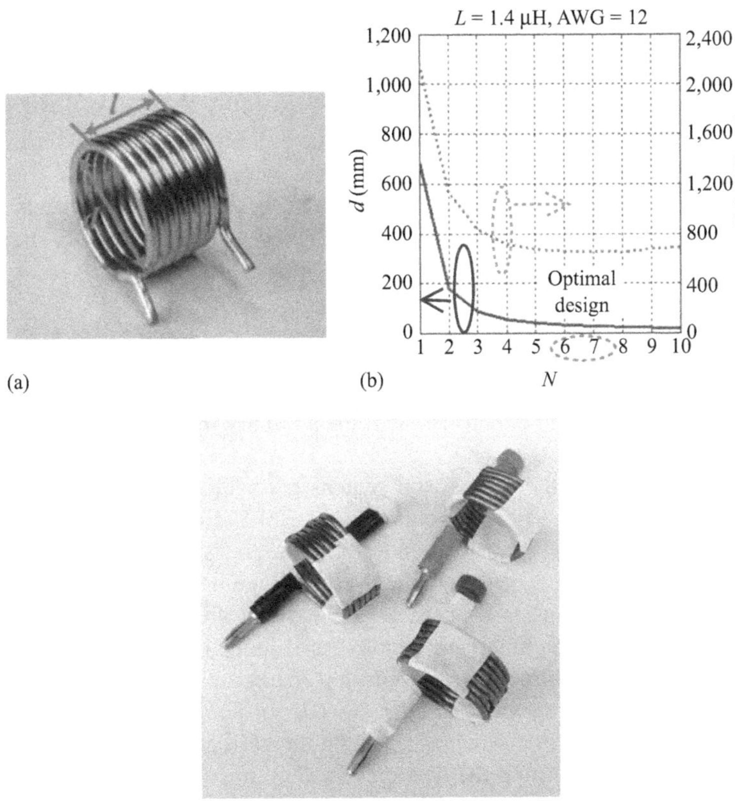

Figure 10.29 Design of auxiliary inductors: (a) typical air core inductor; (b) design optimization of air core inductor dependence on N and corresponding N × MLT; (c) three air core inductors designed as auxiliary inductors

Table 10.3 Parameters of the designed air core inductor

Max. current density	RMS current per phase	AWG	N	d
4.0 A/mm^2	11 A	12	7	30 m

Table 10.4 Three comparison groups with different inductive loads

First group	Double pulse test-based inductor (DPT)
Second group	7.5 kW induction motor plus 2-meter power cable (IM-PC)
Third group	Induction motor plus power cable with auxiliary inductor (IM-PC-L_{aux})

where L is inductance in μH, d is the coil diameter in inches, N is number of turns, l is coil length in inches, which can be determined by N and diameter of the copper wire (i.e., AWG). According to (10.2), the relationship between N and d to achieve a 1.4 μH air core inductor is illustrated in Figure 10.29(b). To minimize the length of the copper wire (i.e., $N \times MLT$), there will be an optimal design in terms of N and d. In this case study with 12 AWG copper wire, the smallest $N \times MLT$ is achieved when N of 6 or 7 is selected. In the end, based on the parameters listed in Table 10.3, three air core inductors are designed and fabricated, as shown in Figure 10.29(c).

10.2.4 Experimental verification

Comparison experiments are carried out under three different groups (see Table 10.4) to evaluate the effectiveness of the auxiliary inductor on the switching performance improvement.

Figure 10.30 shows the comparison switching waveforms with three different inductive loads under the operating condition of 600 V/10 A with 5-Ω gate resistance during the turn-on transient and 0-Ω gate resistance during the turn-off transient. As can be observed from the overall switching waveforms during both turn-on and turn-off transients, there is a 6.0 MHz frequency ringing in the switching current with the motor plus cable-based inductive load (see dashed red waveforms in Figure 10.30). After inserting 1.4 μH air core inductor per each phase, the ringing frequency is varied from 6.0 to 3.0 MHz (see dotted black waveforms in Figure 10.30), which agrees with the aforementioned design.

Also, as predicted, during the turn-on transient, the added auxiliary inductor improves the switching performance over the original motor plus cable (IM-PC)-based inductive load: *dv/dt* increases from 40 to 43 V/ns, and the turn-on switching loss reduces from 166 to 158 μJ. Additionally, thanks to the auxiliary inductor, the corresponding switching behavior of the converter with the motor plus cable load is almost identical to that by using the optimally designed DPT-based inductor load.

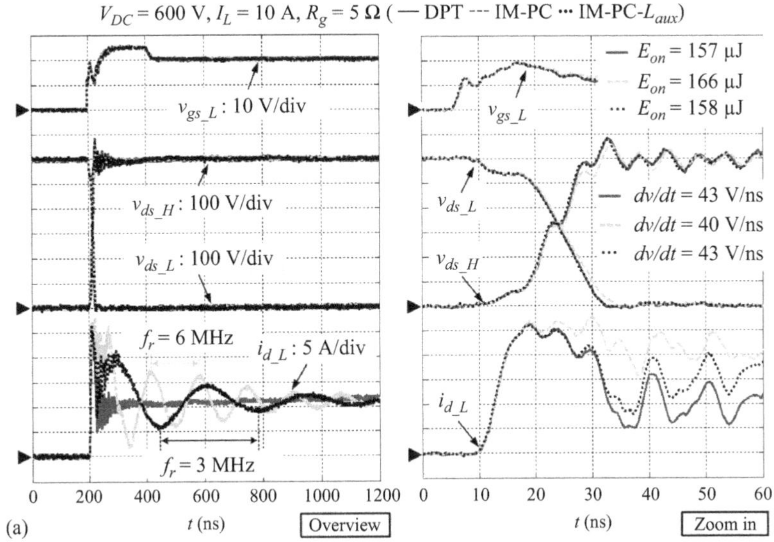

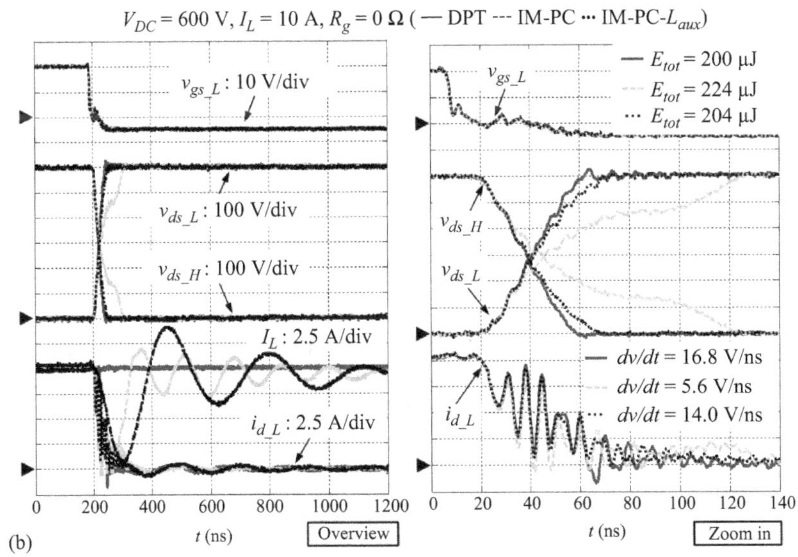

Figure 10.30 Switching waveform comparisons among different inductive load: (a) turn-on transient; (b) turn-off transient

Similarly, during the turn-off transient, auxiliary inductor allows *dv/dt* with motor plus cable-based inductive load increases from 5.6 to 14.0 V/ns, which almost approaches to the *dv/dt* of 16.8 V/ns by using the DPT inductor. Furthermore, the total switching losses of motor plus cable inductive load with the

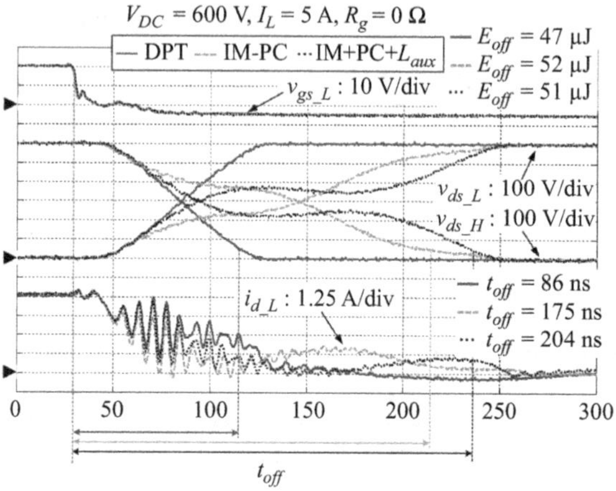

Figure 10.31 Comparison of turn-on switching performance among different inductive load

auxiliary inductor decreases from 224 to 204 μJ, which is nearly the same as 200 μJ, the total switching losses achieved by DPT inductor.

In summary, during the switching transient, adding the auxiliary inductor as described earlier can successfully decouple the interaction between the inductive load and power device, enabling the switching performance with the practical inductive load (e.g., motor plus cable-based inductive load in this case study) to mostly achieve the excellent behavior when the optimally designed DPT inductor is employed.

Additionally, based on the design consideration of the auxiliary inductor, to minimize the required inductance, there will be penalty for the turn-off time at the light load (e.g., <10 A in this case study). As can be observed in Figure 10.31, under the operating current of 5 A, the turn-off time with the motor plus cable-based inductive load becomes even longer when the auxiliary inductor is applied. This is because that the relatively long turn-off time at the light operating current cannot avoid the impact of the resonant current induced by the parasitics of the inductive load. Also, the auxiliary inductor cause the resonant frequency to decrease from 6.0 to 3.0 MHz, which means that the duration of the resonant current with lower resonant frequency becomes longer, and then the turn-off time by using the auxiliary inductor turns to be longer as well. However, the tested turn-off switching loss at the light load current stays nearly constant, and theoretically equals to the energy stored in the junction capacitance of the power device and its antiparallel diode. Figure 10.31 illustrates that although the turn-off time difference with different inductive loads are significant, their turn-off switching losses are almost identical.

Figure 10.32 shows more switching data comparison among different inductive loads under different operating currents. Figure 10.32(b) shows that, thanks to the auxiliary inductor, the total switching loss with motor plus cable becomes lower,

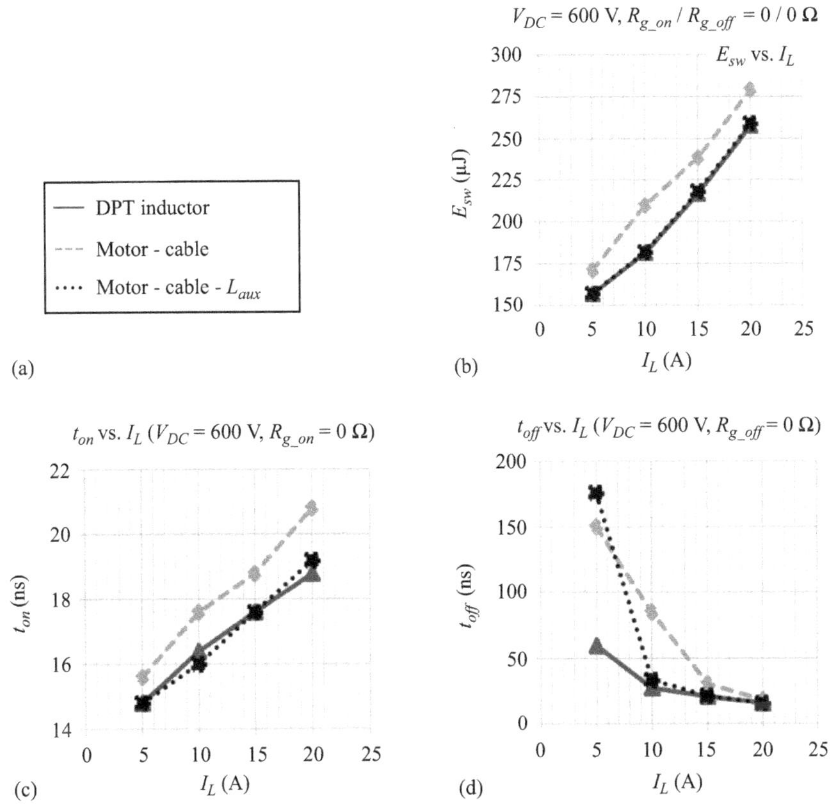

Figure 10.32 Comparison test results dependence on load inductor with 600-V DC bus voltage and 0-Ω gate resistance: (a) labels of comparison groups; (b) E_{sw} dependence on I_L; (c) t_{on} dependence on I_L; (d) t_{off} dependence on I_L

and is almost identical to the switching losses tested by using DPT inductor. Similar behavior can be found in terms of turn-on switching time in Figure 10.32(c). In Figure 10.32(d), except the light load (i.e., <10 A in this case study), the turn-off time with motor plus cable-based inductive load after the insertion of the auxiliary inductor decreases, and becomes identical to the turn-off time when the DPT inductor is employed.

Figure 10.33 illustrates the comparison test results for different junction temperatures. As predicted, the experimental data demonstrate the effectiveness of the auxiliary inductor on the switching performance improvement.

In summary, the auxiliary inductor described is capable of mitigating the interaction between the inductive load and power device among a wide operating range. Except the turn-off time at the light load, the switching times under the rest of the operating conditions together with the switching losses under all of the

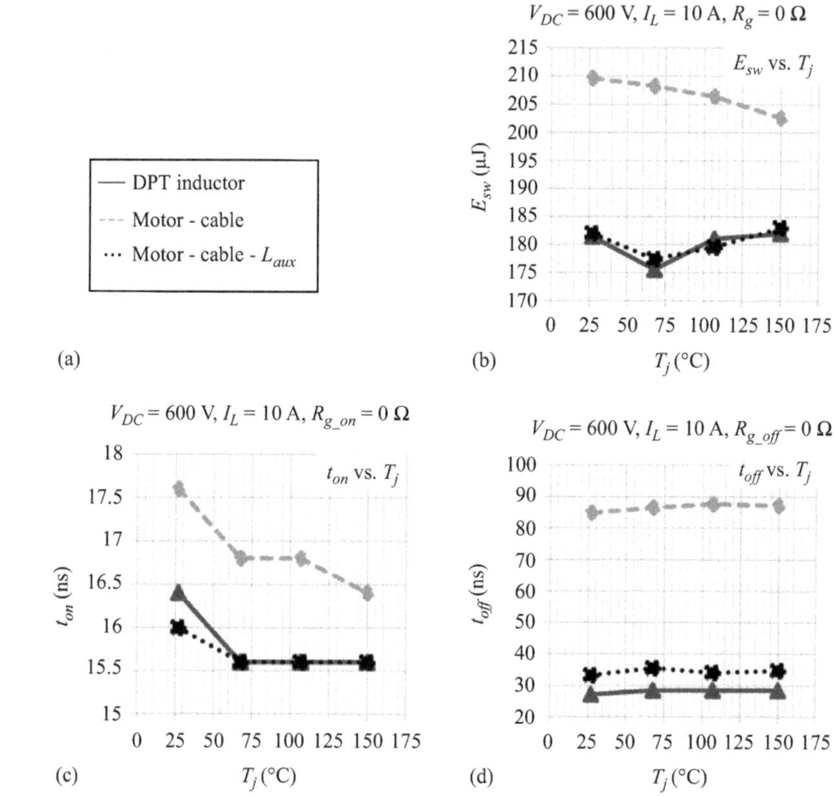

Figure 10.33 Comparison test results dependence on junction temperatures under 600 V/10 A with 0 Ω gate resistance: (a) labels of comparison groups; (b) E_{sw} dependence on T_j; (c) t_{on} dependence on T_j; (d) t_{off} dependence on T_j

operating points will no longer be affected by the parasitics of the motor plus cable-based inductive load.

10.3 Case study: validation of the DPT results vs. switching performance in an actual system

This section presents a case study based on SiC devices in a three-phase inverter-fed motor drive. The ultimate objective is to verify that when the actual inverter is operating, the tested switching performance of SiC devices can repeat the excellent switching behavior obtained by using the optimally designed DPT based on the techniques described in the previous sections.

A three-phase voltage source inverter with Wolfspeed 1200 V/20 A SiC MOSFETs is built, as shown in Figure 10.34. Based on this three-phase voltage

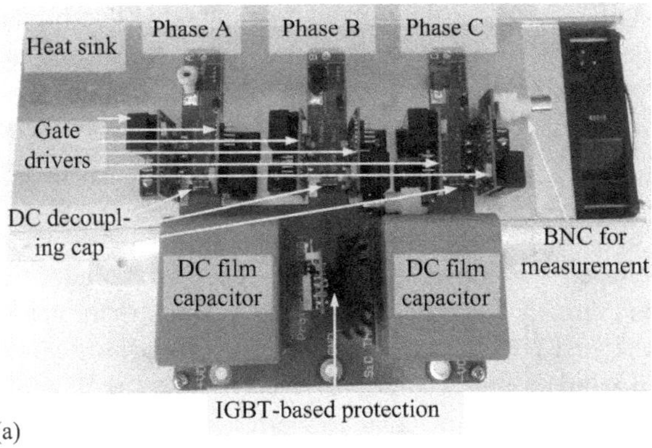

(a)

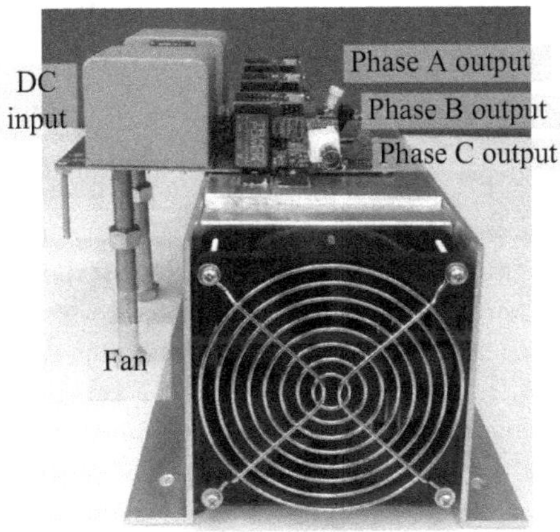

(b)

Figure 10.34 SiC-based three-phase voltage source converter: (a) top view; (b) lateral view

source inverter, a set of experiments has been devised to make a comprehensive comparison of switching performance. The methodology for experimental verification is described as follows.

1. Evaluation of the switching performance difference between the DPT board with the dynamic current measurement included and the phase-leg in a three-phase VSC. It is important to demonstrate that power devices in a three-phase VSC circuit have similar switching behaviors to those by using DPT board.
2. Evaluation of the interaction of phase-legs and impact of heat sink on the switching performance. DPT circuit merely depends on a phase-leg configuration. However in an actual three-phase VSC, there will be two more phase-legs as well as a heat sink. Based on the investigation and experimental

verification in Section 10.1, the influence of the other two phase-legs and heat sink on the switching performance is slight. But previous conclusion was drawn based on Wolfspeed first-generation SiC MOSFETs. Therefore, it is worthwhile to re-evaluate the influence of phase-legs and heat sink on the switching behavior according to the latest version of three-phase VSC, even though the conclusion is not expected to change as will be verified later in this section.

3. Evaluation of ultimate switching performance of SiC devices in the three-phase VSC when the motor is rotating. This switching behavior will be monitored and directly compared with the switching performance tested by the optimally designed DPT. Once they have the similar switching characterization, we can validate the accuracy of the DPT results to represent the true switching performance in an actual converter.

10.3.1 Switching performance comparison between DPT board and three-phase VSC

Figure 10.35 shows the comparison of switching loop interconnection design between one phase-leg of three-phase VSC and DPT board. Even if the same placement and PCB layout design together with the gate driver board are applied for both the DPT board and VSC, the power loop inductance of the DPT board would be inevitably larger due to the insertion of the coaxial shunt for the switching current measurement. Therefore, switching performance tested in DPT board and three-phase VSC will generally be different. However, as discussed in Chapter 6, thanks to the optimal PCB layout design of the DPT board, the extra power loop inductance induced by coaxial shunt is fairly small, resulting in only slight difference in the switching behaviors of the three-phase VSC and DPT board.

Figure 10.36 illustrates the switching waveform comparison between the DPT board and three-phase VSC when the DPT-based load inductor is employed under the operating condition of 600 V/10 A with 0 Ω gate resistance. Although the switching current is not directly monitored in the three-phase VSC, the comparison of gate voltage and drain-source voltages indicates that the switching performance in the three-phase VSC is almost identical to that tested by the DPT board. Also, based on different voltage dips due to the snubber effect of power loop inductance during turn-on *di/dt* transient together with voltage spikes, we can identify that the power loop inductance in the three-phase VSC is smaller than that in the DPT board.

Considering the practical inductive load consisting of a 7.5 kW induction motor plus 2-m power cable with 2 μH auxiliary inductor (MCL), the switching waveform comparison between DPT board and three-phase VSC are displayed in Figure 10.37. Similar to the observation in Figure 10.36, the difference of the switching waveform between DPT board and three-phase VSC is slight.

More comparison test results dependence on load currents for both the DPT load and MCL load are shown in Figures 10.38 and 10.39. It demonstrates that the switching time during both turn-on and turn-off transients with different test boards are almost the same. Even better, due to the smaller power loop parasitic inductance for the three-phase VSC, its corresponding over-voltages are

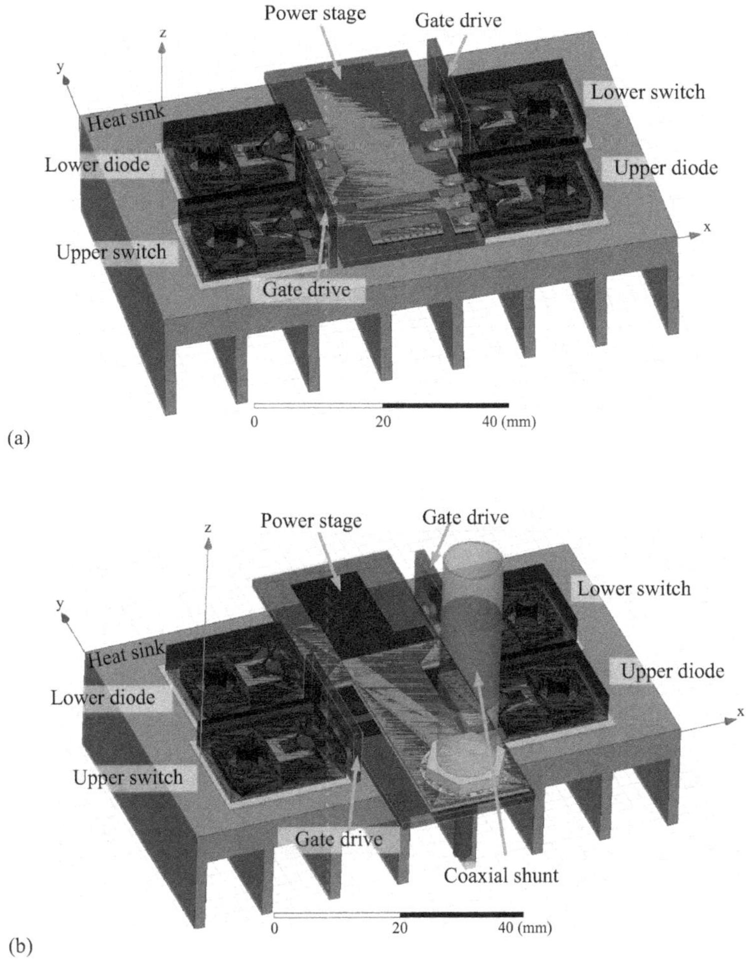

Figure 10.35 Switching loop interconnection design comparison: one phase-leg of three-phase VSC versus DPT board: (a) one phase-leg of three-phase VSC; (b) phase-leg configuration of DPT

always lower than those tested by the DPT board. The definition of $V_{os}(\%)$ in Figures 10.38(c) and 10.39(c) is given by

$$V_{os}(\%) = \frac{(V_{os} - V_{DC})}{V_{DC}} \times 100\% \qquad (10.3)$$

where V_{os} is the peak value of the drain-source voltage; V_{DC} refers to the DC bus voltage.

In total, the excellent switching waveforms based on DPT board can be achieved in the phase-leg of three-phase VSC.

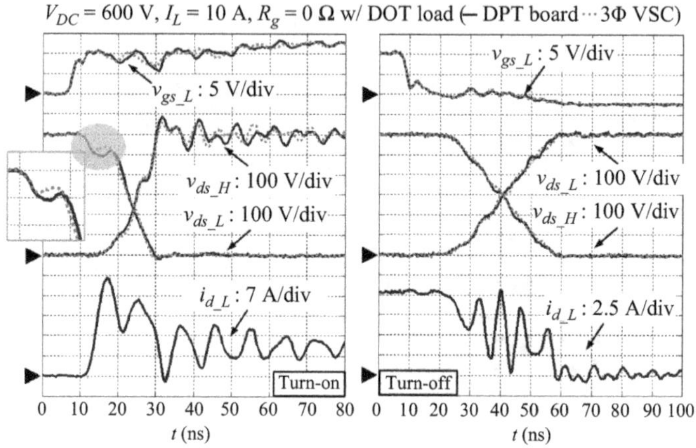

Figure 10.36 Switching waveform comparison with DPT load inductor between DPT board and three-phase VSC

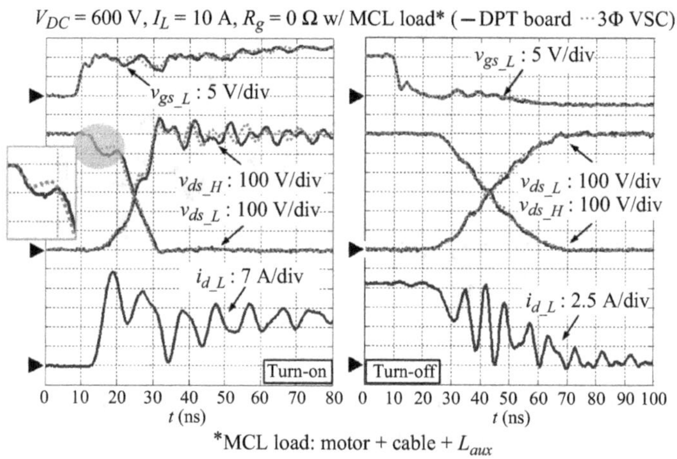

Figure 10.37 Switching waveform comparison with MCL load between DPT board and three-phase VSC

10.3.2 Interaction of phase-legs and impact of heat sink on the switching performance

The DPT circuit merely depends on a phase-leg configuration, but in actual three-phase VSC, there will be two more phase-legs as well as a heat sink. The impact of phase-legs and heat sink on the switching performance is experimentally shown here based on a three-phase VSC hardware.

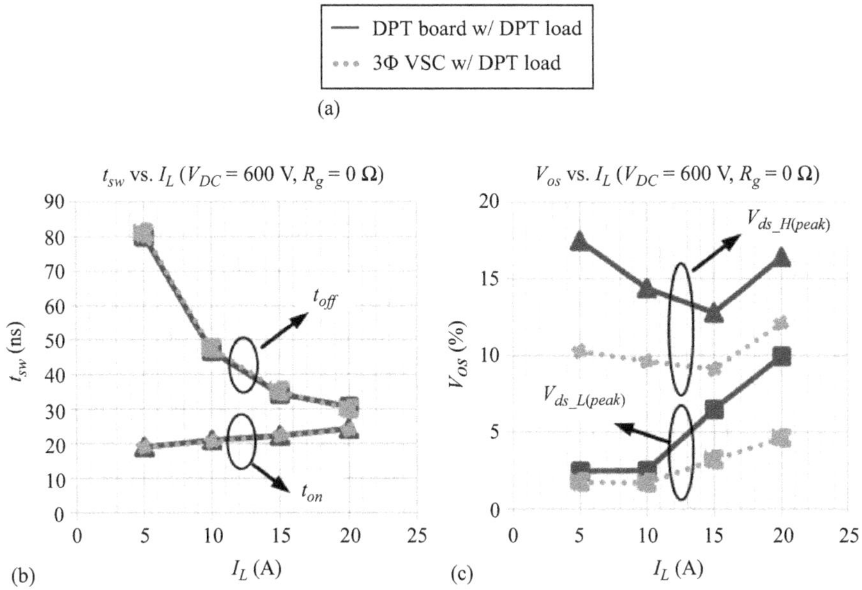

Figure 10.38 Comparison test results dependence on load currents with DPT load inductor: (a) labels of comparison groups; (b) switching time dependence on load currents; (c) voltage spike dependence on load currents

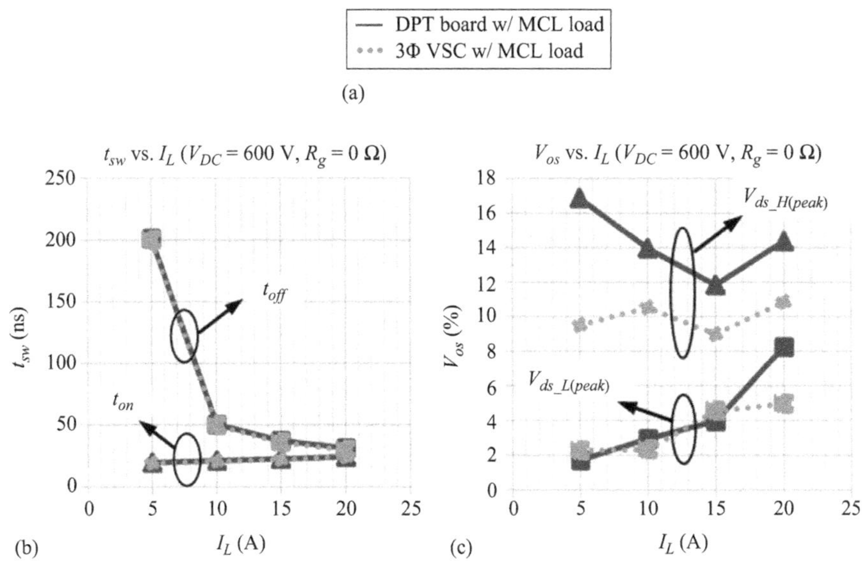

Figure 10.39 Comparison test results dependence on load currents with MCL load: (a) labels of comparison groups; (b) switching time dependence on load currents; (c) voltage spike dependence on load currents

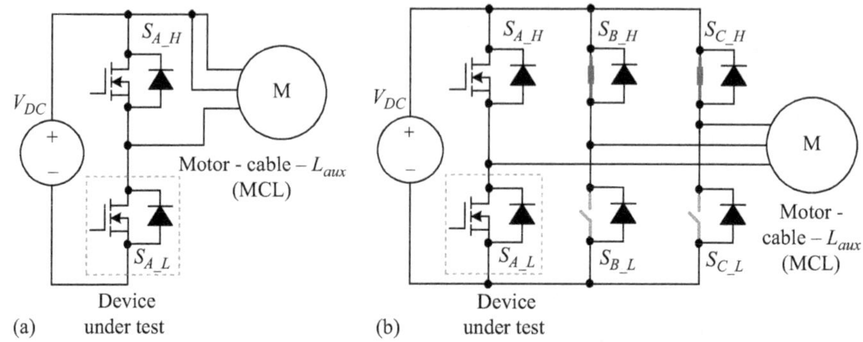

Figure 10.40 Test circuits for evaluation of the impact of two more phase-legs' parasitics on the switching behavior: (a) PL-MCL; (b) 3Φ-MCL

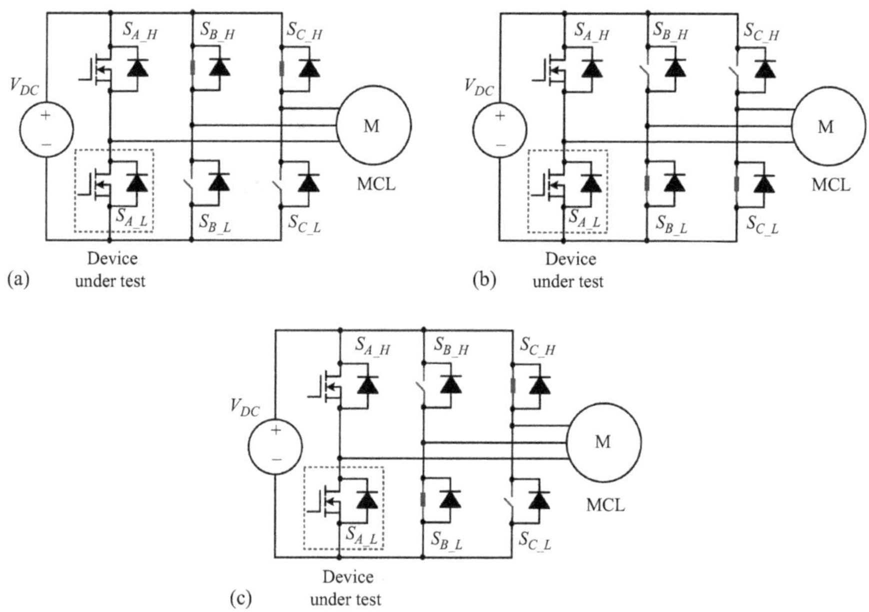

Figure 10.41 Test circuits for evaluation of the impact of switch states of two more phase-legs on the switching behavior: (a) 3Φ-MCL-1; (b) 3Φ-MCL-2; (c) 3Φ-MCL-3

According to test circuits for evaluation of the impact of two more phase-legs on the switching behavior, as shown in Figures 10.40 and 10.41, the switching waveforms in Figures 10.42 and 10.43 illustrate that neither the parasitics induced by phases B and C nor the different inductive load connection structures due to the

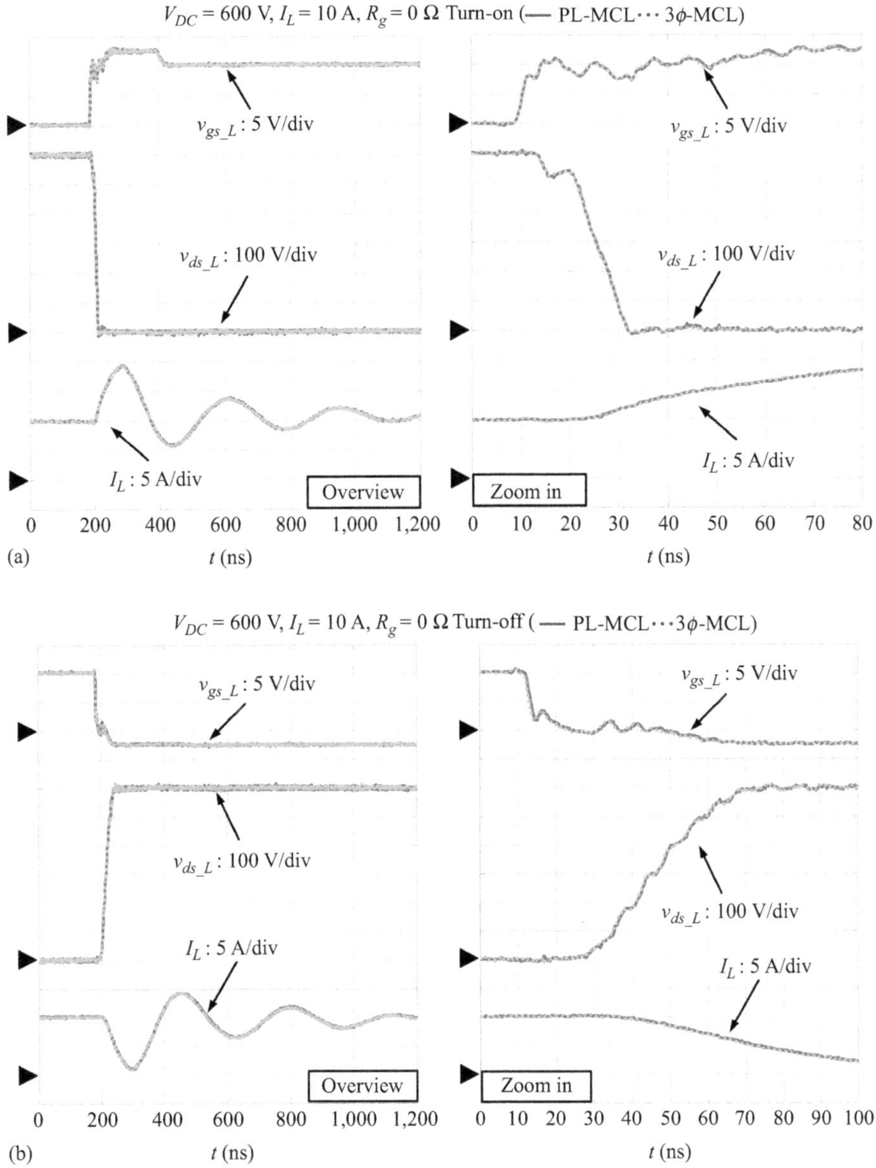

Figure 10.42 Impact of parasitics induced by two more phase-legs on the switching performance: (a) turn-on transient; (b) turn-off transient

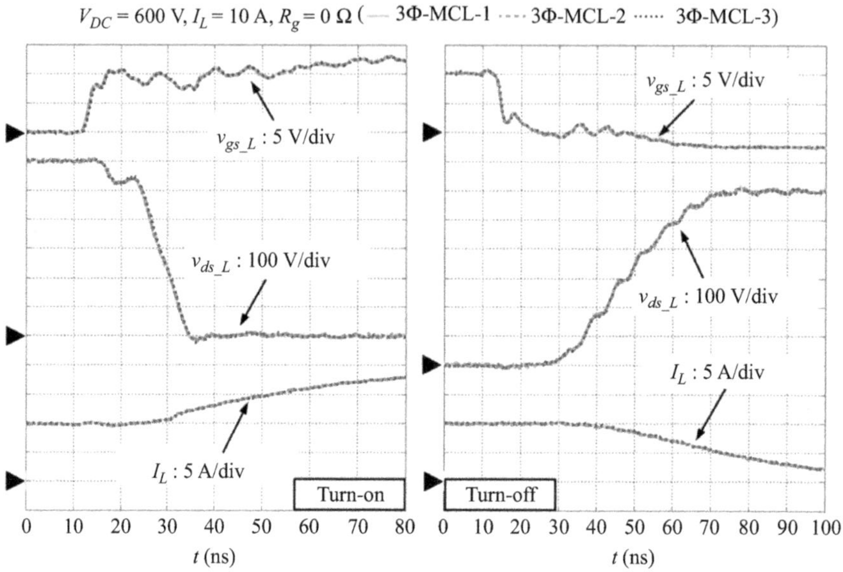

Figure 10.43 Impact of switch states of two more phase-legs on the switching behavior

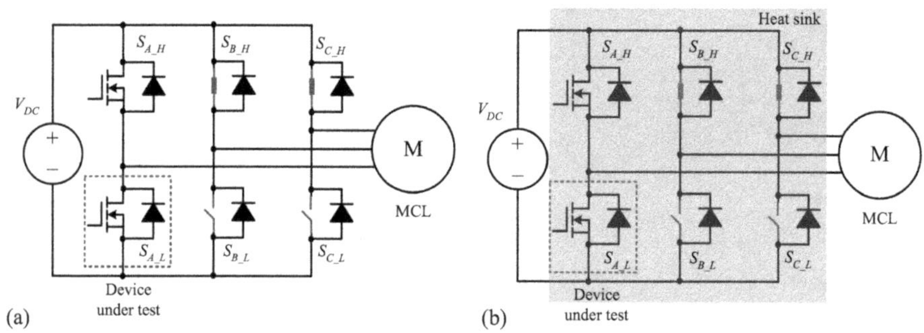

Figure 10.44 Test circuits for evaluation of the impact of heat sink on the switching behavior: (a) 3Φ-MCL-1; (b) 3Φ-MCL-HS-1

switches' on–off states have significant effect on the switching performance since the gate voltage and drain-source voltage under different test configurations are almost identical. Similarly, based on test circuits in Figure 10.44, the switching waveform comparison in Figure 10.45 shows that the capacitive coupling effect

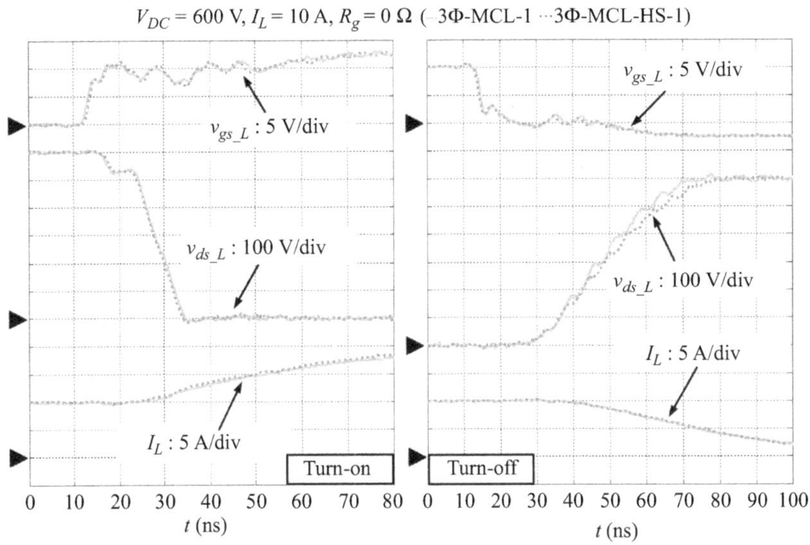

Figure 10.45 Impact of heat sink on the switching behavior

induced by heat sink on the switching performance is insignificant. More comparison test results in Figure 10.46 indicate that under different operating conditions, the influence of two more phase-legs and heat sink on the switching time of SiC devices in the three-phase VSC is negligible, which agrees with the conclusion drawn in Section 10.1.

10.3.3 Validation of the accuracy of DPT results in comparison with switching performance in an actual converter

Since the aforementioned test results are based on the pulse test, it is necessary to evaluate the switching performance in a converter when it operates and compare with the DPT results for the ultimate validation. Figure 10.47 shows the gate signal, line-to-line voltage and phase current of the three-phase inverter when driving the induction motor. And then, we pick one switching cycle when the current direction enables the device under test to be the operating switch, and compare the switching waveforms between the running test and previous double pulse test. The switching waveforms in Figure 10.48 shows that under the operating condition of 600 V/10 A with 0 Ω gate resistance, when the induction motor rotates, the switching performance is almost identical to that tested with MCL load by using the DPT board (DPT-MCL) during both turn-on and turn-off transients. Additionally,

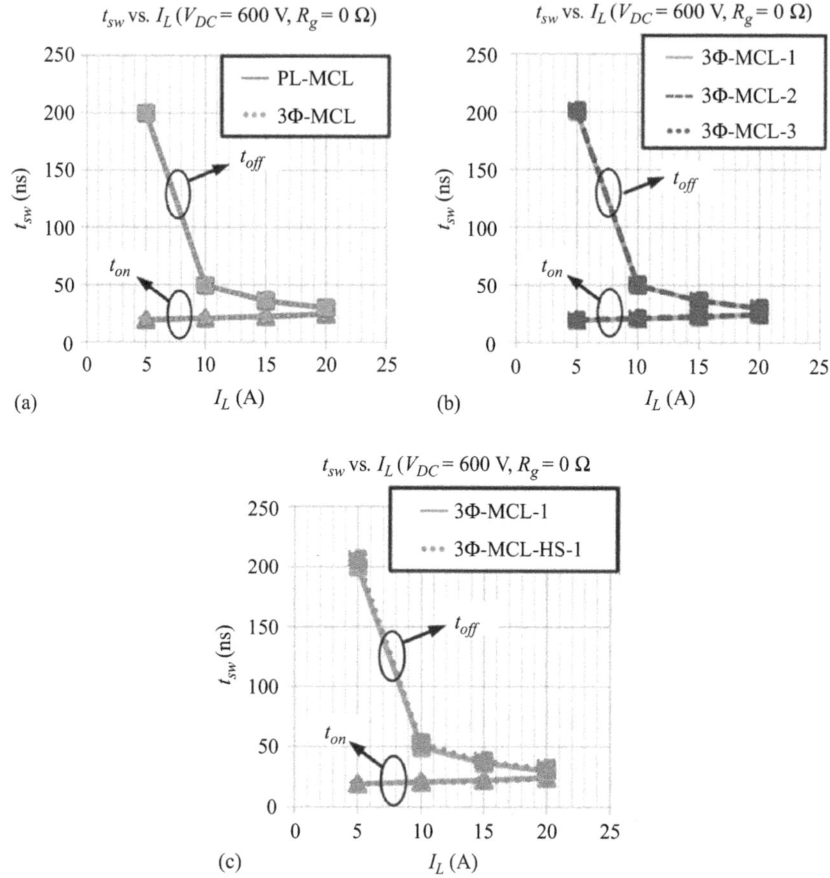

Figure 10.46 Comparison test results dependence on load currents with MCL load for the evaluation of the impact of two more phase-legs and heat sink on the switching performance: (a) impact of phase-legs' parasitics; (b) impact of phase-legs' switch states; (c) impact of heat sink

the switching performance of SiC devices in the actual three-phase VSC can mostly repeat the excellent switching behavior measured by the DPT board with the optimally designed DPT load (DPT-DPT).

More comparison test results in Figure 10.49 further verify that as compared to the switching performance tested by the DPT board, the similar high switching-speed performance of SiC devices can be realized in the three-phase VSC.

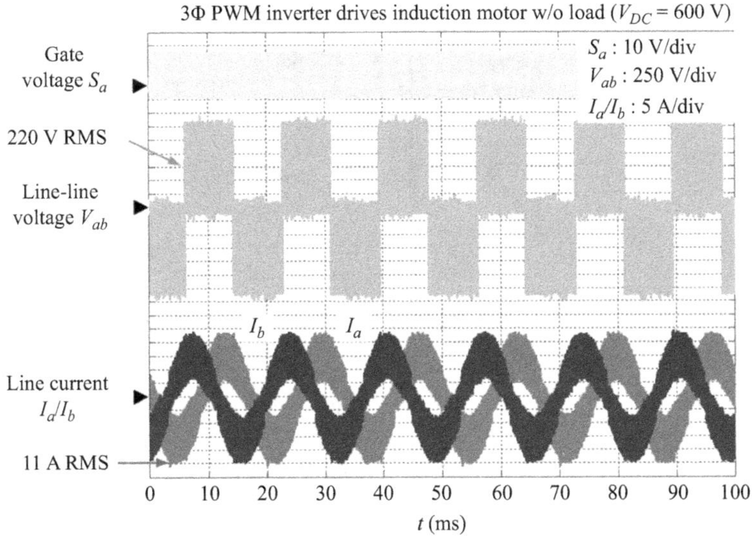

Figure 10.47 Typical waveforms of three-phase voltage source inverter-fed motor drives under continuous operating condition

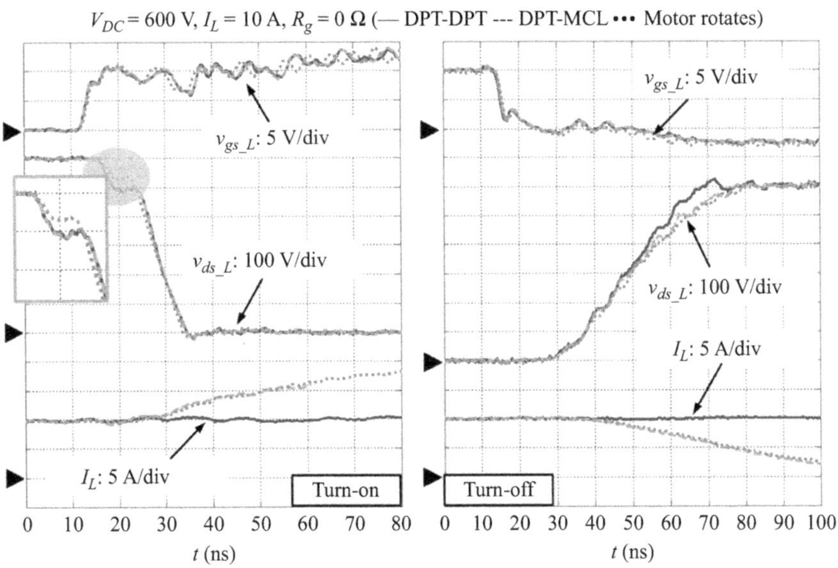

Figure 10.48 Ultimate switching waveform comparisons: three-phase voltage source inverter-fed motor drives versus double pulse tester

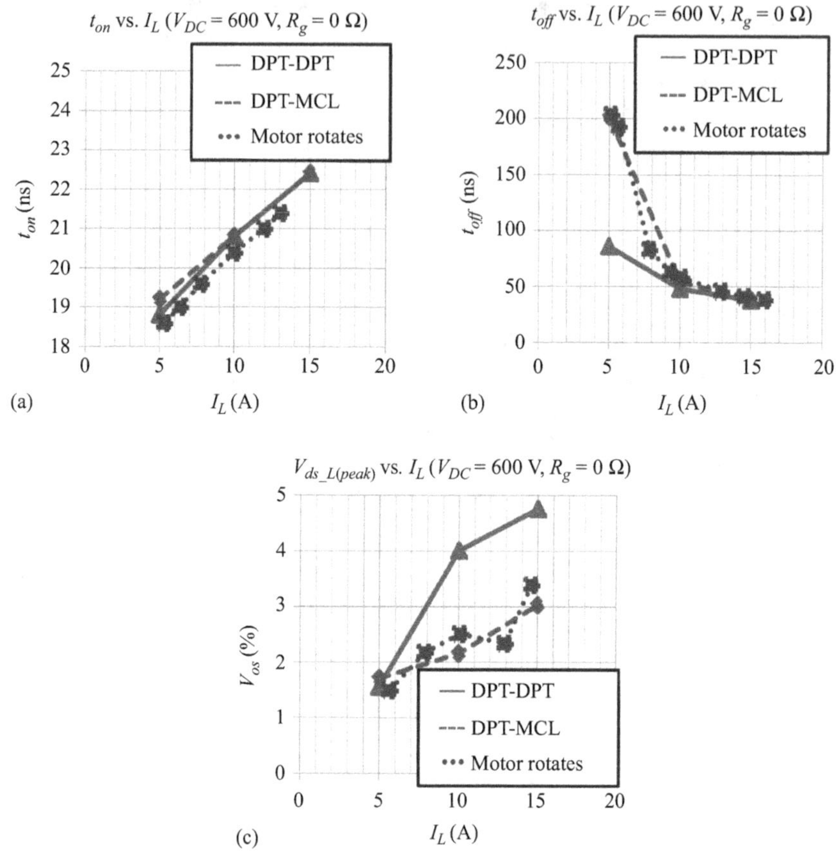

Figure 10.49 Ultimate comparison test results dependence on load currents: three-phase voltage source converter versus double pulse tester: (a) turn-on time dependence on load current; (b) turn-off time dependence on load current; (c) voltage spike dependence on load current

10.4 Summary

1. Due to intrinsic characteristics of WBG devices, such as small junction capacitance, small specific on-resistance, and high *di/dt* and *dv/dt* during fast switching transients, their switching behavior becomes more susceptible to parasitics and noise of the application circuit, compared with slower Si devices. Thus, the simple and layout-optimized DPT with an optimally designed load inductor in the phase-leg configuration may not sufficiently represent a more complex configuration of actual converters and loads.
2. In the three-phase converter, three potentially impacting factors of switching performance were evaluated and discussed, including motor loads, phase-legs,

and heat sink. It shows that a motor, with an induction motor as the example, especially with a relatively long power cable, will significantly impact the switching performance. In addition, the interactions among three phase-legs cannot be ignored unless decoupling capacitors are mounted close to each phase-leg to support the DC bus voltage during the switching transient. Also, the coupling capacitance due to the heat sink equivalently increases the junction capacitance of power devices; however, its influence on the switching behavior in motor drives is small considering the relatively large capacitive parasitics of the motor load. Therefore, nonideal high-frequency characteristics of inductive load is identified as the most critical impacting factor of the switching performance.

3. Two approaches can be employed to improve the accuracy of the dynamic characterization considering the impact of practical loads: one is to conduct DPT with the practical loads connected instead of the optimally designed DPT inductor, which makes the dynamic characterization application dependent and less general. The other approach is to develop solution to suppress the adverse effects of the impact of practical loads, allowing the conventional DPT results with DPT load inductor to represent the true switching behavior of WBG devices in practice.
4. A dedicated auxiliary inductor can be designed and inserted between the converter and inductive load so that the parasitics of the load will not be "seen" from the converter side during the switching transient. The auxiliary inductor has been experimentally demonstrated capable of mitigating the interaction between the inductive load and power device under different operating conditions. Except the turn-off time at the light load, the switching times under the rest of the operating conditions together with the switching losses under all of the operating points will not be affected by the parasitics of the motor plus cable-based inductive load.
5. A case study based on a three-phase VSC with Wolfspeed 1200 V/20 A SiC MOSFETs is presented to validate that the switching behavior in the actual three-phase VSC-fed motor drive can mostly repeat the excellent switching performance tested by the optimally designed DPT board with DPT inductor based on the techniques described earlier.

References

[1] B. Mirafzal, G. L. Skibinski, R. M. Tallam, D. W. Schlegel, and R. A. Lukaszewski, "Universal induction motor model with low-to-high frequency-response characteristics," *IEEE Transactions on Industry Applications*, vol. 43, pp. 1233–1246, 2007.

[2] Z. Zhang, G. Ben, F. Wang, *et al.*, "Methodology for switching characterization evaluation of wide band-gap devices in a phase-leg configuration," in *the Proceedings of 2014 Twenty-Ninth Annual IEEE Applied Power Electronics Conference and Exposition (APEC)*, Fort Worth, TX, USA, 2014, pp. 2534–2541.

[3] Z. Chen, "Characterization and modeling of high-switching-speed behavior of SiC active devices," MS Thesis, Electrical Engineering, Virginia Polytechnic Institute and State University, Blacksburg, 2009.

[4] Cree CMF20120D Datasheet, Rev A. [Online]. Available: http://www.cree.com/~/media/Files/Cree/Power/Data%20Sheets/CMF20120D.pdf

[5] A. Boglietti and E. Carpaneto, "Induction motor high frequency model," in *Conference Record of the 1999 IEEE Industry Applications Conference, Thirty-Fourth IAS Annual Meeting*, Phoenix, AZ, USA, 1999, pp. 1551–1558, vol. 3.

[6] M. Schinkel, S. Weber, S. Guttowski, W. John, and H. Reichl, "Efficient HF modeling and model parameterization of induction machines for time and frequency domain simulations," *Twenty-First Annual IEEE Applied Power Electronics Conference and Exposition (APEC)*, Dallas, TX, USA, 2006, pp. 1181–1186.

[7] W. Liwei, C. N. m. Ho, F. Canales, and J. Jatskevich, "High-frequency modeling of the long-cable-fed induction motor drive system using TLM approach for predicting overvoltage transients," *IEEE Transactions on Power Electronics*, vol. 25, pp. 2653–2664, 2010.

[8] I. Josifovic, J. Popovic-Gerber, and J. A. Ferreira, "Improving SiC JFET switching behavior under influence of circuit parasitics," *IEEE Transactions on Power Electronics*, vol. 27, pp. 3843–3854, 2012.

[9] Z. Zhang, F. Wang, L. M. Tolbert, B. J. Blalock, and D. Costinett, "Evaluation of switching performance of SiC devices in PWM inverter fed induction motor drives," *IEEE Transactions on Power Electronics*, vol. 30, no. 10, pp. 5701–5711, 2015.

Chapter 11
Topology consideration

In addition to the two-level phase-leg based voltage source converter that has been the focus in prior chapters, there are many other widely used converter types in practical applications, such as current source converters and multilevel voltage source converters. For the fast WBG devices, their switching behavior becomes more susceptible to parasitics and noise of the application circuit. Therefore, the dynamic characterization results based on the conventional two-level phase-leg DPT, even considering the influence of three-phase system as discussed in Chapter 10, usually cannot match well with the actual switching performance in other non-two-level-phase-leg based topologies.

This chapter aims at investigating the impacting factors on the switching performance originated from practical topologies that are not based on the simple two-level phase-leg configuration. The main focus here is the hard switching-based topology where the dynamic characteristics of devices would significantly affect the power converter performance. Two typical representatives, including current source converter and three-level active neutral-point clamped (ANPC) converter, are selected to understand the difference of switching commutation loop and the influence on the switching performance between these non-two-level-phase-leg based converters and the conventional two-level phase-leg based DPT.

First, based on the typical modulation scheme, the switching commutation of each topology is analyzed. Accordingly, the switching loops are derived and compared with those in the conventional DPT. Considering more complicated topologies versus the conventional DPT circuit, the critical non-operating devices that are beyond the main switching commutation loop but will affect the switching performance of the operating switch are identified; and then their influence on the dynamic characterization is described. Finally, based on these case studies, a general discussion is given to summarize the impact of other non-two-level-phase-leg based topologies on the dynamic characterization.

11.1 Current source converter

The current source converter features a voltage step-down function, smaller filter size, inrush current limiting capability, and the potential to achieve high

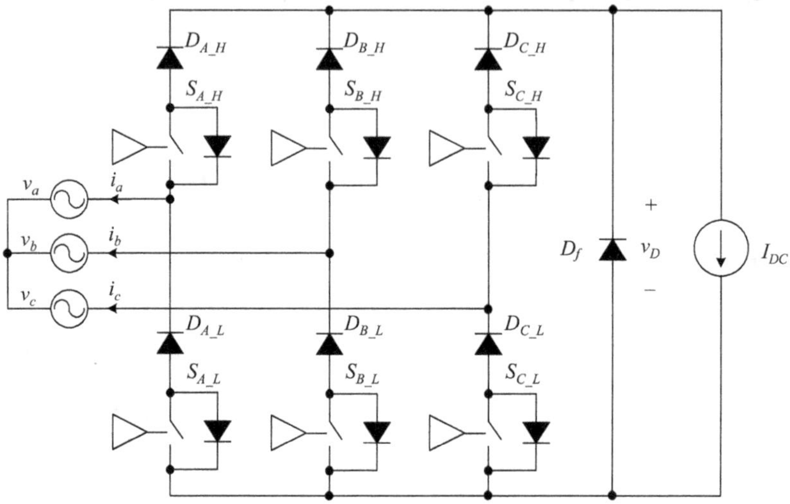

Figure 11.1 Current source converter

efficiency [1–5]. It can be used as active front-end in high-efficiency power supplies for telecommunication and data centers with DC distribution [1, 2], where the 480 or 380 V AC voltage can be stepped down to 380 or 400 V DC in a single conversion stage. It can be used as the off-board fast charger for electric vehicles as well [3]. In high-power applications such as arc furnace and induction melting, the current source converter-based rectifier can provide high current under comparatively low output voltage [4, 5]. For variable speed motor drive application, it can reduce the input filter size and increase power density [6, 7].

Unlike the voltage source converter with unidirectional DC voltage link, current source converter features a unidirectional DC current link. As illustrated in Figure 11.1, current source converter consists of six current-unidirectional voltage-bidirectional switch cells, which are physically implemented by the main active switches in anti-series connection with diodes.

11.1.1 Switching commutation analysis

The modulation scheme utilized in the current source converter determines the switching commutation loop and is described here first for convenience for readers. The input voltages v_a, v_b, and v_c and the fundamental components i_a, i_b, and i_c of the input currents are defined in (11.1), where V_m and I_m are the peak values of the sinusoidal voltage and current respectively; ω is the AC angular frequency, and φ is the phase angle difference between v_a and i_a.

$$\begin{cases} v_a = V_m\cos(\omega t + \varphi) \\ v_b = V_m\cos\left(\omega t - \dfrac{2\pi}{3} + \varphi\right) \\ v_c = V_m\cos\left(\omega t + \dfrac{2\pi}{3} + \varphi\right) \end{cases}$$

$$\begin{cases} i_a = I_m\cos(\omega t) \\ i_b = I_m\cos\left(\omega t - \dfrac{2\pi}{3}\right) \\ i_c = I_m\cos\left(\omega t + \dfrac{2\pi}{3}\right) \end{cases} \tag{11.1}$$

According to input voltages and input currents, 12 sectors are defined for the current source converter as shown in Figure 11.2(a) [8, 9]. On the space vector plane in Figure 11.2(b), the input current space vector $i^*_{abc} = \sqrt{\frac{3}{2}}I_m e^{j\omega t}$ can be synthesized by six active space vectors $\vec{I_x}, x = 1, \ldots, 6$ and a zero vector $\vec{I_0}$. The symbol $[S_x, S_y], x, y = 1, \ldots, 6, x \neq y$ denotes the on-state of the switches S_x and S_y in a space vector. Considering both switching loss and modulation index range, i^*_{abc} is usually composed by two consecutive active vectors and the zero vector in each sector [10].

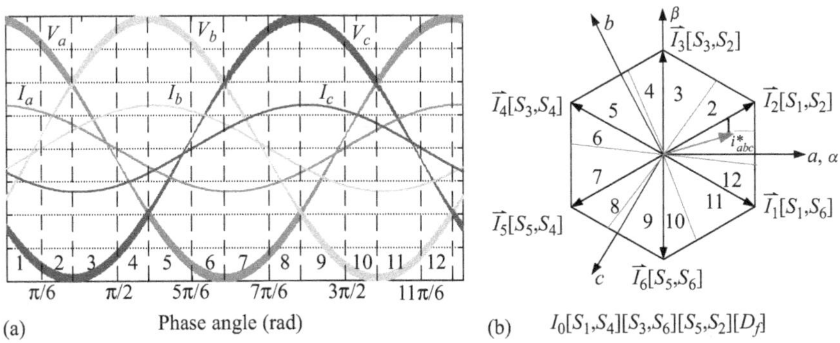

Figure 11.2 Sector division in space vector PWM for current source converter: (a) fundamental components of input current and voltage; (b) space vector plane

The "three-switch" and "four-switch" space vector modulation schemes, which contain three and four transitions in a switching period respectively, are most widely used in the current source converter, because they achieve a good compromise between switching loss and harmonic current level [1–16]. Four modulation schemes are widely used to reduce the switching loss [8, 9, 17–19]. Both symmetric schemes were introduced in [8], where Symmetric Scheme I (SS-I) is named "half-wave symmetrical modulation (HSM)" and Symmetric Scheme II (SS-II) "modified full-wave symmetrical modulation (MFSM)." The zero vector is placed between two active vectors in SS-I and after the two active vectors in SS-II. The space vectors are arranged to make the switching voltage the lowest in SS-II so that the switching loss is assumed to be minimum [8, 9]. Asymmetric Scheme III (US-III) in [17] is able to realize soft switching and reduce the switching loss. Asymmetric Scheme IV (US-IV) is applied in charge control of current source converter in [18, 19].

The commutations in four modulation schemes are different, as shown in Table 11.1. SS-I has only switch-diode commutations, while SS-II includes both types of commutations. US-III has no positive turn-off transition, while US-IV has no positive turn-on transition in the switch-switch commutation. To make the analysis completed, in the following discussion, SS-II is selected since it covers all the possible switching commutations.

Figure 11.3(a)–(d) shows the transition process in Sector 1 (0–30°) (see Table 11.1 modulation scheme SS-II), based on the chosen modulation method. In Sector 1, commutation happens between S_{A_H} and freewheeling diode D_f, or between S_{B_L} and S_{C_L}.

Table 11.1 Comparison of switching commutation between four different modulation schemes

	Output voltage v_D		**Number of transitions**			
	Sector 12 $V_a > V_c \geq V_b$	***Sector 1*** $V_a > V_b > V_c$	**Switch-switch commutation**		**Switch-diode commutation**	
			Off	***On***	***Off***	***On***
SS-I	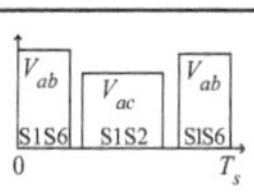	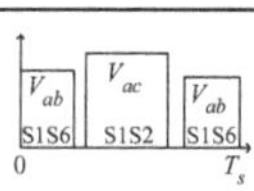	–	–	2	2
SS-II	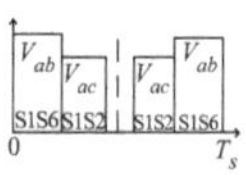	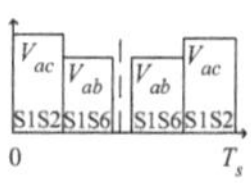	1	1	1	1
US-III	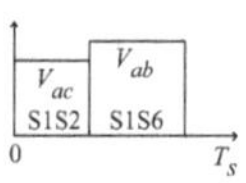	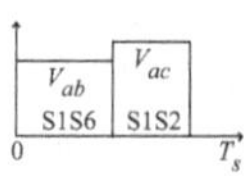	–	1	1	1
US-IV	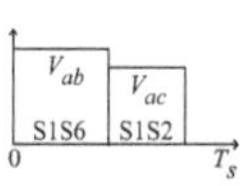	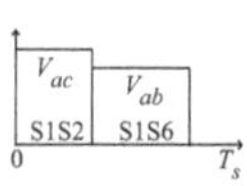	1	–	1	1

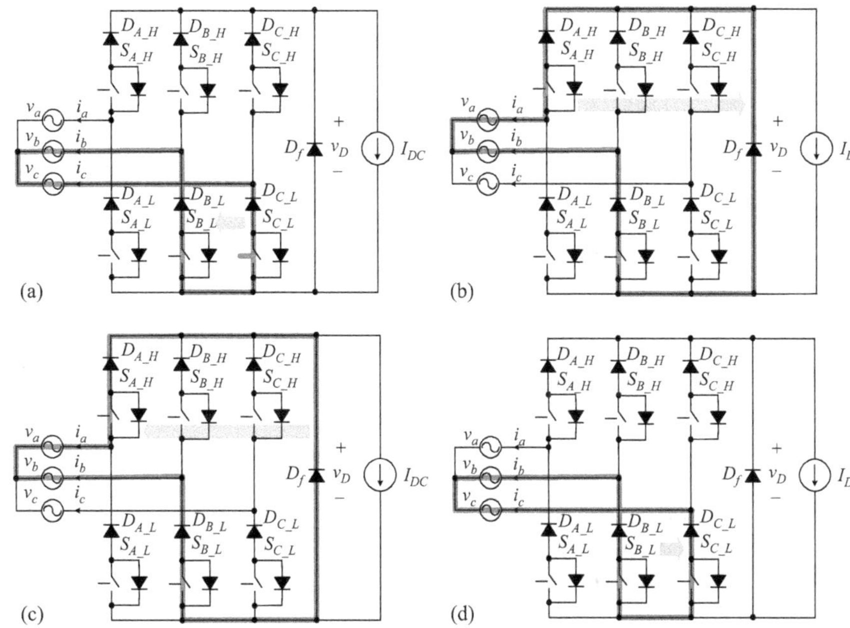

Figure 11.3 Equivalent circuit of the switching commutation in Sector 1: (a) current commutates from S_{C_L} to S_{B_L}; (b) current commutates from S_{A_H} to D_f; (c) current commutates from D_f to S_{A_H}; (d) current commutates from S_{B_L} to S_{C_L}

11.1.2 Comparison of switching commutation loop with conventional DPT

According to the switching commutation analysis discussed earlier, Figure 11.4 illustrates the two switching commutation loops in the current source converter versus commutation loop in the conventional DPT. Generally speaking, more switch elements in the current source converter are involved, making the switching commutation complicated as compared to that in the conventional DPT.

To make the comparison clearer, based on the circuit theory, the circuit in Figure 11.4 can be further simplified as shown in Figure 11.5, by setting all voltages sources to short circuits and all current sources to open circuits for dynamic analysis. It can be observed that (1) the switching commutation in the current source converter still occurs between the active switch and the diode (e.g., S_{B_L} and D_{C_L} in Figure 11.5(a) and S_{A_H} and D_f in Figure 11.5(b)), which is exactly the same as that in the conventional DPT (S_L and D_H in Figure 11.5(c)); and (2) more switch elements in the on-state are in the switching loop (e.g., S_{C_L} and D_{B_L} in Figure 11.5(a) and S_{B_L}, D_{B_L}, and D_{A_H} in Figure 11.5(b)), leading to more insertion inductance. As a result, in comparison with the switching loop that merely consists of two devices in the conventional DPT, the switching performance in the

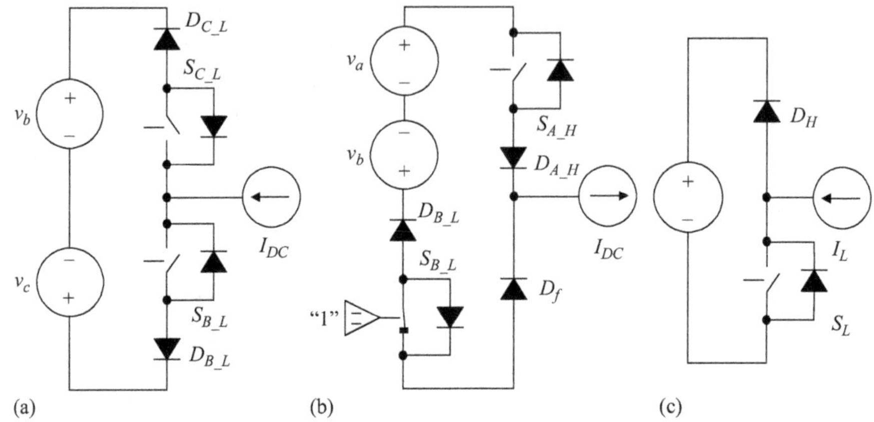

Figure 11.4 Switching commutation circuit comparison between current source converter and conventional DPT: (a) switching commutation circuit between S_{C_L} and S_{B_L}; (b) switching commutation circuit between D_f and S_{A_H}; (c) switching commutation circuit in the conventional DPT

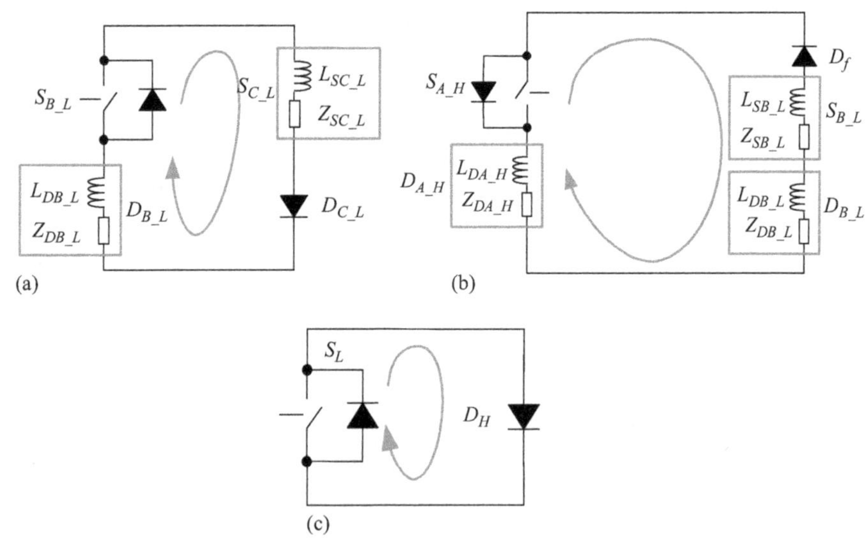

Figure 11.5 Simplified switching commutation loop comparison between current source converter and conventional DPT: (a) simplified switching commutation loop between S_{C_L} and S_{B_L}; (b) simplified switching commutation loop between D_f and S_{A_H}; (c) simplified switching commutation loop in the conventional DPT

Table 11.2 Voltage variation across nonactive switch cells during the switching commutation

Nonactive switch cells	Switching commutation between S_{C_L} and S_{B_L}	Switching commutation loop between D_f and S_{A_H}
S_{A_L}/D_{A_L}	$\boldsymbol{v_{ab} \leftrightarrow v_{ac}}$	$v_{ab} \leftrightarrow v_{ab}$
S_{A_H}/D_{A_H}	$0 \leftrightarrow 0$	Active switch
S_{B_L}/D_{B_L}	Active switch	$0 \leftrightarrow 0$
S_{B_H}/D_{B_H}	$v_{ab} \leftrightarrow v_{ab}$	$\boldsymbol{v_{ab} \leftrightarrow 0}$
S_{C_L}/D_{C_L}	Synchronous switch	$V_{cb} \leftrightarrow v_{cb}$
S_{C_H}/D_{C_H}	$v_{ac} \leftrightarrow v_{ac}$	$\boldsymbol{v_{bc} \leftrightarrow v_{ac}}$
D_f	$\boldsymbol{v_{ab} \leftrightarrow v_{ac}}$	Synchronous diode

current source converter is expected to be worse with more parasitic ringing and dynamic voltage stress.

11.1.3 Consideration of other non-active devices

In addition to the switches associated with the switching loop, it is worth noting the other non-active devices in the current source converter and their effects on the switching performance of the active switch. Table 11.2 summarizes the potential changes across the non-active switch cells during switching commutations discussed earlier. It is found that during the switching commutation between S_{C_L} and S_{B_L}, voltages across S_{A_L}/D_{A_L} switch cell and D_f change with the variation of v_{bc}; while during the switching commutation between D_f and S_{A_H}, voltage variation of v_{ab} incurs across S_{B_H}/D_{B_H} and S_{C_H}/D_{C_H} switch cells. Considering these switch elements are all in the off-state, the added capacitance is equivalently paralleled with the active switch, as illustrated in Figure 11.6.

During the switching transient with *dv/dt* across nonactive switches, their resultant displacement current will adversely affect the switching performance of the active switch, leading to slower switching speed and increased switching loss. In the meantime, due to the inevitable physical distance between the actual switching loop and critical non-active switches, a relatively large stray inductance L_{stray} in Figure 11.6 is added in series with the capacitance of the non-active switches, which will exacerbate the voltage spike across these nonactive switches.

11.2 Three-level ANPC converter

Compared with two-level phase-leg configuration-based topologies, multilevel voltage source converters have several advantages to achieve higher performance power conversion, such as lower voltage stress per device, reduced *dv/dt*, better power quality and lower EMI noise [20, 21]. The three-level active neutral point

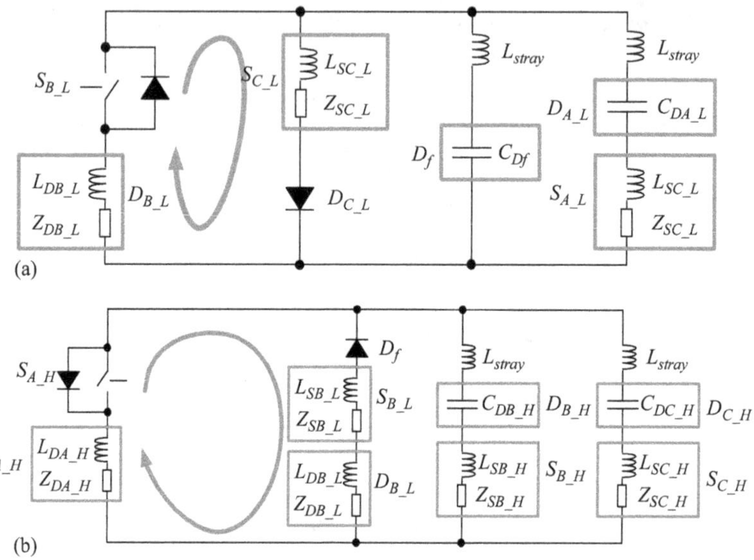

Figure 11.6 *Switching commutation in current source converter with the consideration of critical nonactive switches: (a) switching commutation between* S_{C_L} *and* S_{B_L} *with the consideration of critical nonactive switches; (c) switching commutation between* D_f *and* S_{A_H} *with the consideration of critical nonactive switches*

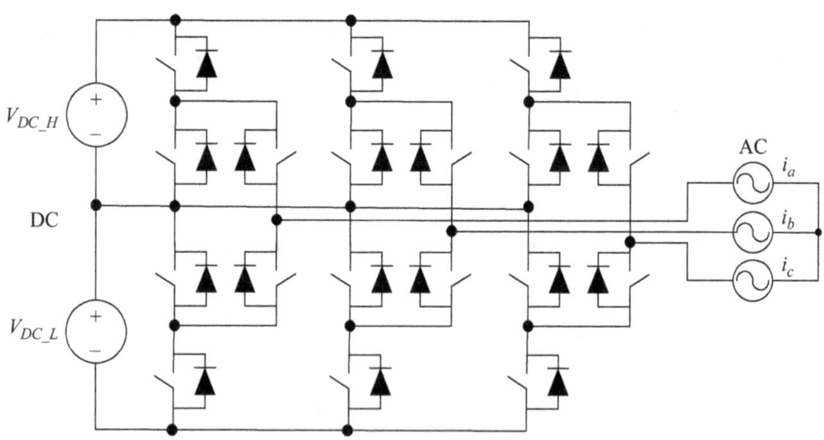

Figure 11.7 *Three-level active neutral-point clamped converter*

clamped (ANPC) converter is one of the typical representatives of multilevel converters that are widely used in many industry applications [20–22].

Figure 11.7 illustrates the circuit of a three-level ANPC converter. Since it belongs to the voltage source converter family, DC voltage and AC current are the source and load of the ANPC. Similar to the discussion on the impact of other phase-legs on the three-phase system in Chapter 10, considering the decoupling capacitor applied to each phase, the impact of the other two phases of three-level ANPC on the switching commutation is negligible. Accordingly, the following description focuses on a single phase-leg of three-level ANPC converter for the switching commutation analysis.

11.2.1 Switching commutation analysis

As shown in Figure 11.8, there are six switches for a three-level ANPC-based single phase-leg. In general, two modulation schemes are adopted in this switching cell, which are summarized as outer loop and inner loop. Details are discussed in [22]. Figures 11.9 and 11.10 show two modulation schemes and their corresponding commutation loops, respectively. In the following description, two assumptions are made: one is the AC current flowing out of the midpoint of the single phase, and the other is the switching commutation taking place during the positive voltage line cycle. Considering the symmetry of the circuit and operation principle, the other cases are similar and will not be repeated.

Specifically, for outer loop mode in Figure 11.9, S_2 and S_3 switch once per line cycle and are called line cycle switches, while S_1, S_p, S_n, and S_4 work as high-frequency switches. During the positive line cycle, S_2 and S_n are constantly on while S_3 and S_4 stay off. Almost all the switching commutations occur between S_1 and S_p in this case. As highlighted in Figure 11.9(a), the switching commutation loop consists of S_1 and S_p.

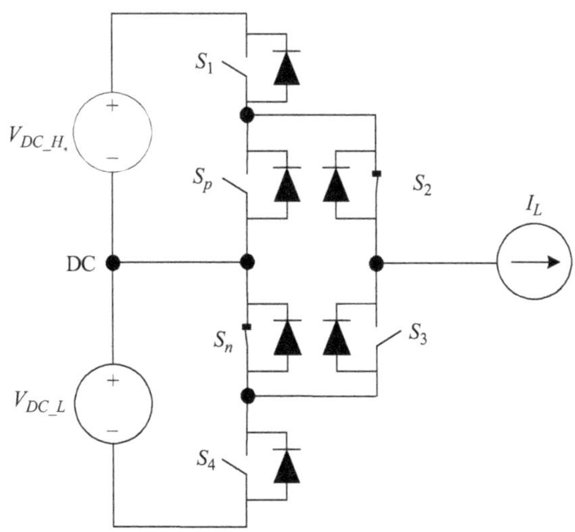

Figure 11.8 Single phase-leg of the three-level ANPC converter for switching commutation analysis

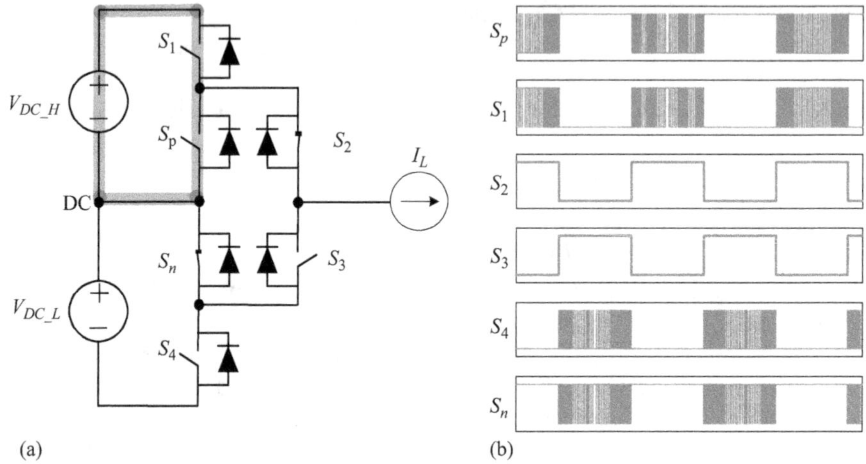

Figure 11.9 Switching loop and control signals for the outer loop mode: (a) switching loop in positive line cycle; (b) modulation scheme and gate signals

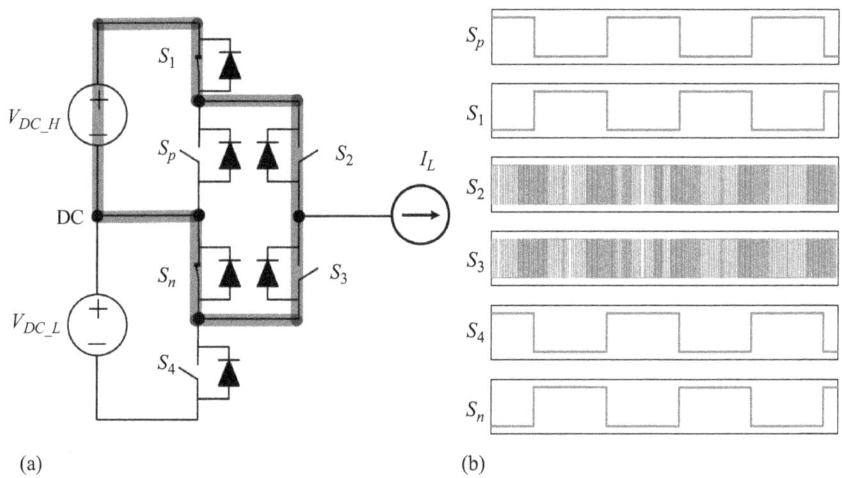

Figure 11.10 Switching loop and control signals for the inner loop mode: (a) switching loop in positive line cycle; (b) modulation scheme and gate signals

In the inner loop mode as shown in Figure 11.10, S_p, S_1, S_4, and S_n become the line frequency switches while S_2 and S_3 are high-frequency switches. Among them, S_1 and S_n are constantly on; S_p and S_4 remain off; and S_2 and S_3 are the main active/synchronous switches. As a result, the main commutation loop of the inner loop mode is formed by switches S_1, S_2, S_3, and S_n.

11.2.2 Comparison of switching loop with conventional DPT

According to the switching commutation analysis discussed earlier along with the simplification based on circuit theory, Figure 11.11 illustrates the two simplified switching commutation loops in three-level ANPC converter. It shows that the main switching commutation loop for the outer loop mode is similar to that in the conventional DPT; while for the inner loop mode, two more switch elements in the on-state are involved in the switching loop, leading to more parasitic inductance and worse switching performance. In addition, the actual switching commutation always occurs between one active switch and the freewheeling diode of the complementary device.

11.2.3 Consideration of other non-active devices

Based on the main switching commutation loop, the switching performance for the outer loop mode should be the same as that in the conventional DPT. However, considering the influence of the non-active switches in the single phase-leg of the three-level ANPC, different conclusion will be drawn. Table 11.3 summarizes the

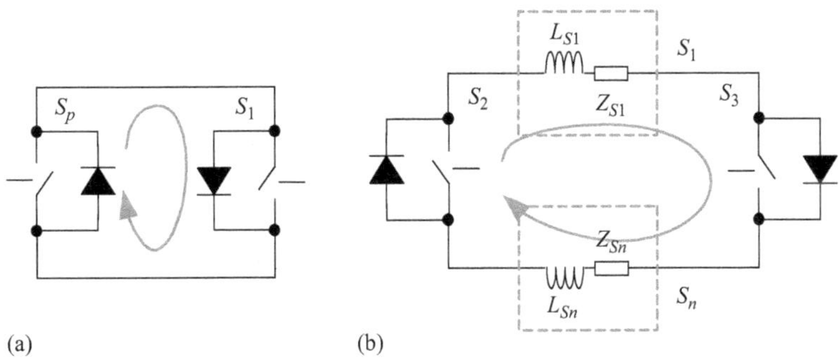

Figure 11.11 Simplified switching commutation loop in three-level ANPC converter: (a) simplified switching commutation loop for the outer loop mode; (b) simplified switching commutation loop for the inner loop mode

Table 11.3 Voltage variation across non-active switch during the switching commutation

Non-active switch	Switching commutation for the outer loop mode	Switching commutation for the inner loop mode
S_1	Synchronous switch	$0 \leftrightarrow 0$
S_p	Active switch	$V_{DC\_H} \leftrightarrow V_{DC\_H}$
S_2	$0 \leftrightarrow 0$	Synchronous switch
S_3	$V_{DC\_H} \leftrightarrow 0$	Active switch
S_n	$0 \leftrightarrow 0$	$0 \leftrightarrow 0$
S_4	$V_{DC\_L} \leftrightarrow V_{DC\_L}$	$V_{DC\_L} \leftrightarrow V_{DC\_L}$

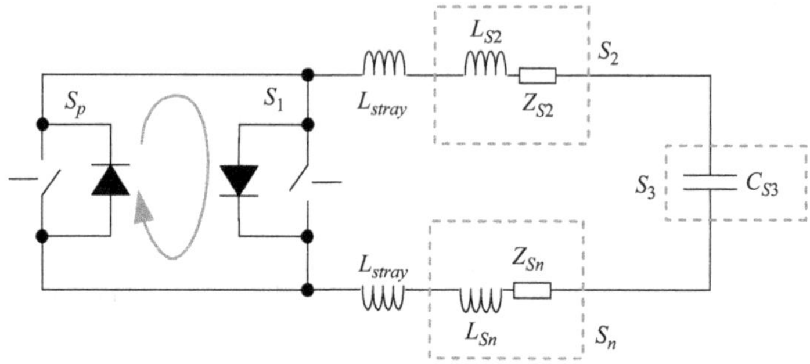

Figure 11.12 Simplified switching commutation for the outer loop mode with the consideration of critical non-active switches

voltage variation of the non-active switch during the switching commutation for both outer and inner loop modes. For the outer loop mode, S_3 in the off-state is paralleled with S_p via S_2 and S_n with the voltage variation of V_{DC_H} during the switching transition. In the end, the actual switching commutation for the outer loop mode is derived as shown in Figure 11.12. Additional output capacitance of S_3 reduces the switching speed and increases the switching loss. Also, excessive voltage spikes across S_3 will be generated due to the un-damped LRC series resonance network consisting of S_2, S_n, and stray inductance introduced by inter-connecting traces.

On the other hand, with respect to the inner loop mode, the voltages across all the non-active switches stay constant. No additional capacitance originates from the three-level circuit. In the end, the equivalent circuit illustrated in Figure 11.11(b) is able to illustrate the actual switching commutation for the inner loop mode.

11.3 Discussion and summary

1. Considering non-two-level-phase-leg based topologies, the switching performance becomes worse as compared to that characterized based on the conventional DPT that is based on the two-level phase-leg configuration. This is more critical for the fast switching WBG devices, of which the switching behavior becomes more susceptible to parasitics and noise of the application circuit.
2. According to two case studies of current source converter and three-level ANPC converter, several general findings are observed and summarized as follows:
   (a) No matter how complicated the topology is, the switching commutation always occurs between an active switch and a freewheeling diode, which is fundamentally the same as that in the conventional DPT.
   (b) The impacting factors of the switching performance in non-two-level-phase-leg based topologies come from two aspects: (1) some of the non-active switches in the on-state are involved in the main switching loop, that is, they are in series with the switching commutating switches. As a result, more insertion inductance is added, leading to more severe parasitic ringing

and voltage overshoot; and (2) some of the non-active switches in the off-state are paralleled with the switching commutating switches, equivalently adding additional junction capacitance, and therefore slowing down the switching speed and increasing the switching loss. Furthermore, special attention should be paid on the over voltage of the non-active switches since an un-damped LRC series resonance network can be easily formed due to the relatively large stray inductance introduced by the interconnection of the actual switching commutation loop with the non-active switches.

References

[1] A. Stupar, T. Friedli, J. Minibock, and J. W. Kolar, "Towards a 99% efficient three-phase buck-type PFC rectifier for 400-V DC distribution systems," *IEEE Transactions on Power Electronics*, vol. 27, no. 4, pp. 1732–1744, 2012.

[2] F. Xu, B. Guo, L. M. Tolbert, F. Wang, and B. J. Blalock, "Design and performance of an all-SiC three-phase buck rectifier for high efficiency data center power supplies," in *Proceedings of the IEEE Energy Conversion Congress & Exposition (ECCE)*, Raleigh, NC, USA, September 2012, pp. 2927–2933.

[3] T. B. Soeiro, T. Friedli, and J. W. Kolar, "Design and implementation of a three-phase buck-type third harmonic current injection PFC rectifier SR," *IEEE Transactions on Power Electronics*, vol. 28, no. 4, pp. 1608–1621, 2013.

[4] A. Siebert, A. Troedson, and S. Ebner, "AC to DC power conversion now and in the future," *IEEE Transactions on Industry Applications*, vol. 38, no. 4, pp. 934–940, 2002.

[5] S. A. Richter, P. Gaertner, D. Hirschmann, and R. W. De Doncker, "Design of a PWM current source rectifier for high power induction melting applications," in *Proceedings of the 13th European Conference on Power Electronics and Applications (EPE)*, Barcelona, Spain, September 2009, pp. 1–9.

[6] B. Wu, *High-Power Converters and AC Drives*, John Wiley & Sons. Inc., Hoboken, New Jersey, 2006.

[7] R. Cuzner, D. Drews, W. Kranz, A. Bendre, and G. Venkataramanan, "Power-dense shipboard-compatible low-horsepower variable-frequency drives," *IEEE Transactions on Industry Applications*, vol. 48, no. 6, pp. 2121–2128, 2012.

[8] T. Halkosaari and H. Tuusa, "Optimal vector modulation of a PWM current source converter according to minimal switching losses," in *Proceedings of the IEEE Power Electronics Specialists Conference (PESC)*, Galway, Ireland, June 2000, vol. 1, pp. 127–132.

[9] M. H. Bierhoff and F. W. Fuchs, "Loss minimized pulse width modulation of IGBT current source converters," in *Proceedings of the IEEE Industrial Electronics Annual Conference (IECON)*, Paris, France, November 2006, pp. 1739–1744.

[10] M. Baumann, T. Nussbaumer, and J. W. Kolar, "Comparative evaluation of modulation methods of a three-phase buck + boost PWM rectifier. Part I: theoretical analysis," *IET Power Electronics*, vol. 1, no. 2, pp. 255–267, 2008.

[11] M. H. Bierhoff and F. W. Fuchs, "Semiconductor losses in voltage source and current source IGBT converters based on analytical derivation," in *Proceedings of the IEEE Power Electronics Specialists Conference (PESC)*, Aachen, Germany, June 2004, vol. 4, pp. 2836–2842.

[12] C. J. Cass, R. Burgos, F. Wang, and D. Boroyevich, "Three-phase ac buck rectifier using normally-on SiC JFETs at 150 kHz switching frequency," in *Proceedings of the IEEE Power Electronics Specialists Conference (PESC)*, Orlando, FL, USA, June 2007, pp. 2162–2167.

[13] M. Bierhoff and F. W. Fuchs, "Analytical evaluation of the total harmonic current in three phase voltage and current source converters," in *Proceedings of the European Conference on Power Electronics and Applications (EPE)*, Dresden, Germany, September 2005, pp. 10.

[14] J. Espinoza, G. Joos, and K. Jin, "DSP based space vector PWM pattern generators for current source rectifiers and inverters," in *Proceedings of the Canadian Conference on Electrical and Computer Engineering*, September 1995, vol. 2, pp. 979–982.

[15] Y. Li, B. Wu, D. Xu, and N. R. Zargari, "Space vector sequence investigation and synchronization methods for active front-end rectifiers in high-power current-source drives," *IEEE Transactions on Industrial Electronics*, vol. 55, no. 3, pp. 1022–1034, 2008.

[16] Y. Nishida, T. Kondoh, M. Ishikawa, and K. Yasui, "Three-phase PWM-current-source type PFC rectifier (theory and practical evaluation of 12 kW real product)," in *Proceedings of the Power Conversion Conference*, Osaka, Japan, 2002, vol. 3, pp. 1217–1222.

[17] D. J. Tooth, N. McNeill, S. J. Finney, and B. W. Williams, "A new soft-switching technique using a modified PWM scheme requiring no extra hardware," *IEEE Transactions on Power Electronics*, vol. 16, no. 5, pp. 686–693, 2001.

[18] K. Wang, D. Boroyevich, and F. C. Lee, "Charge control of three-phase buck PWM rectifiers," in *Proceedings of the IEEE Applied Power Electronics Conference and Exposition (APEC)*, New Orleans, LA, USA, February 2000, vol. 2, pp. 824–831.

[19] C. J. Cass, R. Burgos, F. Wang, and D. Boroyevich, "Improved charge control with adjustable input power factor and optimized switching pattern for a 150 kHz three-phase buck rectifier," in *Proceedings of the IEEE Applied Power Electronics Conference and Exposition (APEC)*, Austin, TX, USA, February 2008, pp. 1200–1206.

[20] S. Kouro, M. Malinowski, K. Gopakumar, *et al.*, "Recent advances and industrial applications of multilevel converters," *IEEE Transactions on Industrial Electronics*, vol. 57, no. 8, pp. 2553–2580, 2010.

[21] J. Rodriguez, J.-S. Lai, and F. Z. Peng, "Multilevel inverters: a survey of topologies, controls, and applications," *IEEE Transactions on Industrial Electronics*, vol. 49, no. 4, pp. 724–738, 2002.

[22] Y. Jiao, S. Lu, and F. C. Lee, "Switching performance optimization of a high power high frequency three-level active neutral point clamped phase leg," *IEEE Transactions on Power Electronics*, vol. 29, no. 7, pp. 3255–3266, 2014.

Appendix A

Recommended equipment and components list for DPT setup

Category	No.	Part number	Description
DPT main components	1	Sorensen, XG 850 W	DC power supply – 600 V, 1.4 A
	2	Tektronix, 2230-30-1	Power supply, triple output, 30 V, 5 A, 120 W
	3	Tektronix, AFG1062	ARB function generator, 2 channel, 60 MHz
	4	Micrometals, FS-300060-2	Powder core for DPT load inductor
	5	MWS Wire Industries, HPN, 12 Gauge, MW80-C	Magnet wire for DPT inductor
Measurement	6	T & M research products, SSDN-10	0.1 Ω 2 GHz shunt resistor
	7	Tektronix, TPP1000	1 GHz, 10× for gate-source voltage measurement
	8	Tektronix, TPP0850	800 MHz, 50×, 2.5 kV, single-ended for drain-source voltage measurement
	9	Tektronix, DPO4104B-L	Oscilloscope, 4 analog channels, 1 GHz BW, 5 GS/s, 5M points
Tape and wire	10	3M Scotch 35 Handy Pack	Electrical tape for insulation
	11	3M Electronic Specialty, Adhesive Tapes 1/2X36YD SOLDER TAPE	Kapton tape for insulation with better temperature performance
	12	Mil. Spec. Wire, 1000 V AC, 22 Gauge	Auxiliary power supply connecting wire
	13	Mil. Spec. Wire, 1000 V AC, 12 Gauge	DC source connecting wire
	14	Solid wire, 300V AC Voltage, 22 Gauge	Solid wire for measurement (tip adapter)
	15	Jonard Industries, Wire Wrap, Hook-up Wire	Ultra-thin wire for measurement and debugging
Soldering	16	Weller Wtcpt	60 W variable-temperature soldering station
	17	Weller No. Eto, Conical, 1/32" Point Width	Soldering tip – thin
	18	Weller No. Etr, Screwdriver, 1/16" Point Width	Soldering tip – fat

(Continues)

(*Continued*)

Category	No.	Part number	Description
	19	Kester Solder 63/37 .015" DIA. 1LB SPOOL	Soldering wire – small
	20	Kester Solder RA FLUX 16AWG 60/40	Soldering wire – large
	21	Chemtronics 2-5L	Solder removal braid
	22	Jonard Tools DP-100	Desoldering pump
	23	Chemtronics 2.75" Polypropylene 500 PER BAG	Swab for soldering and PCB board cleaning
	24	Chemicals SUPER DUSTER 152 400G (14 OZ) AEROSOL	Duster for board cleaning
	25	Chemicals E-LINE FLUX REMOVER 10 oz aerosol	Flux remover
	26	Metcal Solder Removal Lead Free Brass Pad	Soldering tool cleaners
	27	Chip Quik, Solder SODR PASTE W/LD(T5) 63n/37 10cc SYRNG NC	Solder paste
	28	Chip Quik, Soldering Flux Tack Flux No-clean 10cc w/Plunger & Tip	Flux
	29	Apex Tool Group, Wire Stripping & Cutting Tools Xcelite 5-13/16 Knife Light Duty Precision	Precision cutter
	30	Metcal, Liquid Dispensers & Bottles Needle Dispense Kit	Dispensing tip kit
Fixture, connector, kits	31	TE Connectivity AMP Connectors 5227222-3	BNC jack
	32	Amphenol RF Cable Assemblies BNC St Plug-BNC St Plug RG-58 24 in.	50 Ω BNC cable, 24 inch
	33	Amphenol RF, RF Cable Assemblies BNC St Plug-BNC St Plug RG-58 36in	50 Ω BNC cable, 36 inch
	34	Johnson/Cinch Connectivity Solutions RF Cable Assemblies SMA TO BNC PLGS 36"	SMA to BNC cable for PWM signal transmission
	35	Test Connectors PK 25 EA 136-0962-00 AND 131-4209-00	Gate-source voltage measurement tip adapter
	36	Tektronix 013-0291-02	Drain-source voltage measurement tip adapter

(Continues)

(*Continued*)

Category	No.	Part number	Description
	37	Tektronix 13036700	BNC to probe tip adapter
	38	TDK Corporation ZCAT3035-1330-BK	Ferrite core (choke) for noise immunity enhancement of DPT logic circuit and measurement
	39	Yageo, Resistor Kits, 0603 E24 5% 0R, 1R-22M	SMD Resistor kit for DPT circuit debugging
	40	Wurth Electronics, 0603, 1 pF- 10uF	Capacitor kit for DPT circuit debugging
Mechanical	41	Stanley 60-002	Phillips screwdriver
	42	Stanley 60-004	Flat blade screwdriver
	43	Stanley 84-096	Long nose pliers
	44	Proto J206G	Diagonal cut pliers
	45	Stanley 84-199	Wire stripper/crimper
	46	Aven Tools 18032ARS, OO	OO style tweezers
	47	Aven Tools 18072EZ, 7	7 style tweezers
	48	1/4" Hex Size, 1/2" Length, 4-40 Thread Size	Male–female threaded hex standoff
	49	1/4" Hex Size, 1/2" Length, 6-32 Thread Size	Male–female threaded hex standoff
	50	1/4" Hex Size, 1/2" Length, 8-32 Thread Size	Male–female threaded hex standoff
	51	6/6 Nylon, 1/4" Hex Size, 1/4" Length, 4-40 Thread Size	Female threaded hex standoff
	52	6/6 Nylon, 1/4" Hex Size, 1/4" Length, 6-32 Thread Size	Female threaded hex standoff
	53	6/6 Nylon, 1/4" Hex Size, 1/4" Length, 8-32 Thread Size	Female threaded hex standoff
Personal protection equipment (PPE)	54	Eyeglass-Fit Panoramic Safety Glasses 55035T301	Eyeglass-fit panoramic safety glasses with clear scratch-resistant and fog-free lens, black frame

Appendix B

Data processing code for dynamic characterization

```
clear all;
clc;

%Specify the folder where the data is, if different from current working
     directory
testpoint_filepath = '400V_15A\';
%If using Tek5000 series, this should be file name except _Chx.csv at end
%If using Tek4000 series, this should be complete file name
%On and off file names can be the same, if you only save one waveform
filename_on = 0;%'tek0000ALL.csv'; %Set to 0 if you do not want to process
     turn-on waveform
filename_off = 'tek0005ALL.csv'; %Set to 0 if you do not want to process
     turn-off waveform

%Settings for data importing
TekDPO5xxx = false; %If using Tek 5000-series scope
TekDPO4xxx = true; %If using Tek 4000-series scope
plotlines = true; %Plot vertical lines showing transient subinterval
     triggers
biasadjust = false; %Adjust dc bias for vds and id
plot_channel_current = true; %Calculate and plot Coss and channel current
     (requires separate Coss import)
alignment_check = false; %Check id-vds alignment (deskew) graphically
compensate_Esw = true; %Compensate switching loss for conduction loss that
     would otherwise occur during transient interval
id_channel_number = 4;
vds_channel_number = 1;
vgs_channel_number = 2;
i_adj = 1; %Vertical scale adjustment for current measurement, if not scaled
     properly in scope
id_alignment_ns = 0; %Number of ns to move id earlier in time for deskew
     correction (shifts id to the left in plots)

%Transient waveform window areas, for analyzing and plotting
turnon_window_stepback = 30; %How long before transient midpoint to start
     waveform window, in ns
turnon_window_time = 80; %Total window time of desired turn-on transition
     view, in ns
turnoff_window_stepback = 30; %How long before transient midpoint to start
     waveform window, in ns
turnoff_window_time = 80; %Total window time of desired turn-on transition
     view, in ns
filter_samples = 1; %Number of samples to use for differentiation (minimum
     1, higher filters noise better but loses accuracy)
```

```
%Parameters for plotting time-base waveforms
vds_scale = 100; %V/div
vgs_scale = 2; %V/div
id_on_scale = 10; %A/div
id_off_scale = 5; %A/div
Pon_scale = 5e3; %W/div
Poff_scale = 1e3; %W/div
time_grid_increment = 10; %ns/div
linewidth = 3;
vdsColor = [0 0 1];
vgsColor = [0 1 0];
idColor = [1 0 0];
powerColor = [0 0 0];
ichanColor = [0 1 1];
icossColor = [1 0.5 0.2];
fontsize = 16;

%Device/PCB parameters
Lloop = 4.7e-9; %Estimated power loop inductance
Cstray = 10e-12;
Rdson = 52e-3; %On-resistance of device under test
Rshunt = 0.100; %Resistance of coaxial current shunt
Vbd = 2; %Forward voltage of body diode
Rbd = Rdson*2; %Series resistance of body diode. This can be set to 0 if not
     known.
% Deskew plot settings. Set them all to -1 if you are not sure yet.
deskew_start = 20; %Start time in ns for x scale
deskew_stop = 30; %Stop time in ns for x scale
deskewVmin = 300; %Min voltage for y scale
deskewVmax = 450; %Max voltage for y scale

if (filename_on > 0)
%%%%%%%%%%%%%%%%%%%%%%%%%%%%%%%%%%%%%%%%%
%%%%%%%%%%%%% Turn-on waveform analysis %%%%%%%%%%%%%
%%%%%%%%%%%%%%%%%%%%%%%%%%%%%%%%%%%%%%%%%

%This section of the code processes a full data set that has been captured
%for analyzing turn-on waveform data (all variables ending with "_onwf").
%This data set likely has turn-off data as well, so that data must be
%processed partially, but it is not used for turn-off transient analysis.

%Import data
if (TekDPO5xxx == true) && (TekDPO4xxx == false) %Import 5000 series data
     format
  M = csvread([testpoint_filepath filename_on '_Ch' num2str(vgs_
       channel_number) '.csv'],0,3);
  time_onwf = (M(:,1) - M(1,1));
  deltat_onwf = (M(2,1)-M(1,1));
  vgs_onwf = M(:,2);
  M = csvread([testpoint_filepath filename_on '_Ch' num2str(vds_
       channel_number) '.csv'],0,3);
  vds_onwf = M(:,2);
  M = csvread([testpoint_filepath filename_on '_Ch' num2str(id_channel_
       number) '.csv'],0,3);
  id_onwf = -M(:,2);
  vector_length = length(time_onwf);
elseif (TekDPO4xxx == true) && (TekDPO5xxx == false) %Import 4000 series
     data format
```

```
  M = csvread([testpoint_filepath filename_on],21,0);
  time_onwf = (M(:,1)-M(1,1));
  vgs_onwf = M(:,vgs_channel_number+1);
  vds_onwf = M(:,vds_channel_number+1);
  id_onwf = M(:,id_channel_number+1);
  deltat_onwf = (M(2,1)-M(1,1));
  vector_length = length(time_onwf);
else
  return; %Not a compatible data format
end

% Check for inf and replace with closest non-inf data point
for m=1:vector_length
  if((vds_onwf(m)==Inf)|(vds_onwf(m)==-Inf))
    vds_onwf(m) = vds_onwf(m-1);
  end
  if((id_onwf(m)==Inf)|(id_onwf(m)==-Inf))
    id_onwf(m) = id_onwf(m-1);
  end
  if((vgs_onwf(m)==Inf)|(vgs_onwf(m)==-Inf))
    vgs_onwf(m) = vgs_onwf(m-1);
  end
end

%Adjust id if not scaled properly in scope
id_onwf = id_onwf*i_adj;

% Time shift id to correct deskew (I-V misalignment)
id_alignment_pts = round(id_alignment_ns * 1e-9 / deltat_onwf);
id_length = length(id_onwf);
if (id_alignment_pts > 0)
  id_onwf(1:id_length-id_alignment_pts) = id_onwf(id_alignment_pts+1:
       id_length);
else
  id_onwf(-id_alignment_pts+1:id_length) = id_onwf(1:id_length+id_
       alignment_pts);
end

%%%%%%%%%%%Full window view of turn-on data set%%%%%%%%

%Set up plot window
fig = figure(1); clf; hold on;
fig.Name = 'Turn-On Waveform: Full Capture';
fullPlot = gca;
fullPlot.XGrid = 'on';
fullPlot.YGrid = 'on';
fullPlot.GridLineStyle = '--';
fullPlot.GridColor = [0.6 0.6 0.6];
fullPlot.GridAlpha = 0.5;
fullPlot.FontSize = fontsize;
fullPlot.XAxis.Visible = 'on';
fullPlot.YAxis.Visible = 'off';
fullPlotYAxis = fullPlot.YAxis;

%Plot lines
xlabel(['Time (\mus)']);
fullPlotXAxisLength = time_onwf(end)*1e6;
```

```
ymax = 0; ymin = 0;
% V_DS
vdsOnLine = line((time_onwf)*1e6, vds_onwf/vds_scale);
vdsOnLine.Color = vdsColor;
vdsOnLine.LineWidth = linewidth;
% V_GS
vgsOnLine = line(time_onwf*1e6, vgs_onwf/vgs_scale);
vgsOnLine.Color = vgsColor;
vgsOnLine.LineWidth = linewidth;
% I_D
idOnLine = line(time_onwf*1e6, id_onwf/id_on_scale);
idOnLine.Color = idColor;
idOnLine.LineWidth = linewidth;

% Add triangles to show y=0 point in scope-view plots
fullPlotYAxisLength = fullPlotYAxis.Limits(2)-fullPlotYAxis.Limits(1);
hold on;
fill([0 0 fullPlotXAxisLength/30]+fullPlotXAxisLength/50, [-1 1 0]*
     fullPlotYAxisLength/20, vgsColor);
fill([0 0 fullPlotXAxisLength/30]+fullPlotXAxisLength/100, [-1 1 0]
     *fullPlotYAxisLength/20, idColor);
fill([0 0 fullPlotXAxisLength/30], [-1 1 0]*fullPlotYAxisLength/20,
     vdsColor);
line([0 0], [fullPlotYAxis.Limits], 'color','black','linewidth',2)
line([1 1]*fullPlotXAxisLength, [fullPlotYAxis.Limits], 'color',
     'black','linewidth',2)
line([0 fullPlotXAxisLength], [1 1]*fullPlotYAxis.Limits(1), 'color',
     'black','linewidth',2)
line([0 fullPlotXAxisLength], [1 1]*fullPlotYAxis.Limits(2), 'color',
     'black','linewidth',2)

%%%%%%%%%%%%%%%% Find turn-off time indices %%%%%%%%%%%%%%%%
j = 1;
while (vds_onwf(j) > 0.5*max(vds_onwf)) %Find and skip first turn-on, if it
     exists
  j = j + 1;
end
j = j + ceil(turnon_window_time*1e-9/deltat_onwf); %Skip ahead
while (vds_onwf(j) < 0.5*max(vds_onwf)) %Now find turn-off transient of
     interest
  j = j + 1;
end
%Set beginning and end of turn-off transient (but this data will not be used
     for
%turn-off transient analysis, since it comes from the turn-on data set)
j = j - floor(turnoff_window_stepback*1e-9/deltat_onwf);
turnoff_start_onwf = j;
j = j + floor(turnoff_window_time*1e-9/deltat_onwf);
turnoff_end_onwf = j;

%%%%%%%% Find turn-on time indices %%%%%%%%%%
while (vds_onwf(j) > 0.5*max(vds_onwf))
  j = j + 1;
end
%Set beginning and end of turn-on transient (this is the data that will be
%used for analysis of the turn-on transient)
j = j - floor(turnon_window_stepback*1e-9/deltat_onwf);
turnon_start_onwf = j;
```

```
j = j + floor(turnon_window_time*1e-9/deltat_onwf);
turnon_end_onwf = j;

while (vds_onwf(j) < 0.5*max(vds_onwf))
  j = j + 1;
end
%If it is captured in data set, this will find the end of the second pulse
second_turnoff_start_onwf = j - floor(turnoff_window_stepback*1e-9/
     deltat_onwf);

%Plot dashed vertical lines indicating the turn-on and turn-off transient
%locations on the full turn-on waveform capture
if (plotlines == true)
     line([turnoff_start_onwf turnoff_start_onwf]*deltat_onwf*1e6,
          fullPlotYAxis.Limits,'Color',[1 0.2 0.7],'LineStyle','--',
          'LineWidth',2);
     line([turnoff_end_onwf turnoff_end_onwf]*deltat_onwf*1e6,
          fullPlotYAxis.Limits,'Color',[1 0.2 0.7],'LineStyle','--',
          'LineWidth',2);
     line([turnon_start_onwf turnon_start_onwf]*deltat_onwf*1e6,
          fullPlotYAxis.Limits,'Color',[1 0.2 0.7],'LineStyle','--',
          'LineWidth',2);
     line([turnon_end_onwf turnon_end_onwf]*deltat_onwf*1e6,
          fullPlotYAxis.Limits,'Color',[1 0.2 0.7],'LineStyle','--',
          'LineWidth',2);
end

% Estimate load current based on average value 20 ns prior to start of turn-
     off waveform
IL_onwf = mean(id_onwf(turnoff_start_onwf-floor(20e-9/deltat_onwf):
     turnoff_start_onwf));

%Adjust for any dc bias error (not generally recommended, but optional)
if (biasadjust == true)
  id_dcbias_onwf = mean(id_onwf(turnon_start_onwf-floor(20e-9/
       deltat_onwf):turnon_start_onwf));
  id_onwf = id_onwf - id_dcbias_onwf;
  vds_dcbias_onwf = mean(vds_onwf(turnon_end_onwf:second_turnoff_
       start_onwf)) - IL_onwf*Rdson;
  vds_onwf = vds_onwf - vds_dcbias_onwf;
  % Recalculate IL with bias adjusted
  IL_onwf = mean(id_onwf(turnoff_start_onwf-floor(20e-9/deltat_onwf):
       turnoff_start_onwf));
end

% Find steady-state on values for vgs based on average 20 ns prior to start of
     turn-off waveform
% (this should be the positive driving voltage)
Vdrv_onwf = mean(vgs_onwf(turnoff_start_onwf-floor(20e-9/deltat_onwf):
     turnoff_start_onwf));
% Find average vds between 20 ns after turn-off and beginning of turn-on
Vdc_onwf = mean(vds_onwf(turnoff_end_onwf+floor(20e-9/deltat_onwf):
     turnon_start_onwf));

%%%%%%%% Find turn-on trigger points for subinterval analysis %%%%%%%%%%
j = turnon_start_onwf;
while (vgs_onwf(j-1) < 0.2*Vdrv_onwf)
```

```
  j = j + 1;
end
while (mean(vgs_onwf(j-2:j-1)) < vgs_onwf(j))
  j = j - 1;
end
%t1 is point where vgs starts to rise
t1_onwf = j;
t1_time_onwf = (time_onwf(t1_onwf)-time_onwf(turnon_start_onwf))*1e9;
while (id_onwf(j) < IL_onwf)
  j = j + 1;
end
%t3 is point where id = IL
t3_onwf = j;
t3_time_onwf = (time_onwf(t3_onwf)-time_onwf(turnon_start_onwf))*1e9;
while (id_onwf(j-1) < id_onwf(j))
  j = j - 1;
end
%t2 is point where id starts to rise
t2_onwf = j;
if (t2_onwf < turnon_start_onwf)
  t2_onwf = turnon_start_onwf + 1;
end
t2_time_onwf = (time_onwf(t2_onwf)-time_onwf(turnon_start_onwf))*1e9;
while (vds_onwf(j) > 0)
  j = j + 1;
end
%t4 is point where vds crosses 0 V, or in some cases where it reaches its
%first trough point then starts increasing again (only if it never crosses
%zero)
t4_onwf = j;
if (t4_onwf > turnon_end_onwf)
  t4_onwf = turnon_end_onwf - 1;
end
t4_time_onwf = (time_onwf(t4_onwf)-time_onwf(turnon_start_onwf))*1e9;

%%%%%%%% Calculate switching times %%%%%%%%%
td_on = t2_time_onwf - t1_time_onwf; %turn-on delay time
tcr = t3_time_onwf - t2_time_onwf; %current rise time
tvf = t4_time_onwf - t3_time_onwf; %voltage fall time
t_on = t4_time_onwf - t1_time_onwf; %total turn-on time


%%%%%%%%%%%%%%% Calculate switching energy %%%%%%%%%%%%
Pon = vds_onwf.*id_onwf;
% Eonwf_cum((turnon_start_onwf:turnon_end_onwf)-turnon_start_onwf+1) =
     cumtrapz(time_onwf(turnon_start_onwf:turnon_end_onwf),
     Pon((turnon_start_onwf:turnon_end_onwf)-turnon_start_onwf+1));
if (compensate_Esw == true)
  %Optionally, compensate for conduction loss that would have occurred
  %during the time interval of switching, to avoid double-counting (Ideal
  %duty cycles D and 1-D usually used for conduction loss calculation,
  %neglecting switching transient time)
  Eon_comp = (IL_onwf * Rdson^2) * (t4_onwf-t2_onwf)*deltat_onwf;
else
  Eon_comp = 0;
end
% Turn-on loss is integral of I*V during tcr and tvf only
Eon = trapz(time_onwf(t2_onwf:t4_onwf),Pon(t2_onwf:t4_onwf)) - Eon_
     comp;
```

```
%%%%%%%%%%%%%%%%%%%% Calculate dv/dt %%%%%%%%%%%%%%%%%%%%
j = t3_onwf;
while (vds_onwf(j) > Vdc_onwf*0.9)
   j = j + 1;
end
% Find adjusted t3 where vds = 0.9Vdc
t3_90percent = j;
while (vds_onwf(j) > Vdc_onwf*0.1)
   j = j + 1;
end
% Find adjusted t4 where vds = 0.1Vdc
t4_10percent = j;
tvf_9010 = (t4_10percent - t3_90percent)*deltat_onwf*1e9;
% Using 10 & 90% points neglects the exponential fall at start and end, and
% instead focuses on nearly-linear fall of vds
dvdt_on(1:t4_10percent-t3_90percent+1) = (vds_onwf(t3_90percent+1:
      t4_10percent+1) - vds_onwf(t3_90percent:t4_10percent))./
      (deltat_onwf);
dvdt_on_pk = max(-dvdt_on); %Peak dv/dt
dvdt_on_avg = mean(-dvdt_on); %Average dv/dt, which is typically used to
      describe switching speed

%Display results in command window
disp('Turn-on waveform analysis:');
fprintf(' V_dc %4.0f V',Vdc_onwf);
fprintf('\n I_L %4.1f A',IL_onwf);
fprintf('\n Eon %4.1f uJ',Eon*1e6);
fprintf('\n td_on %4.1f ns',td_on);
fprintf('\n tcr %4.1f ns',tcr);
fprintf('\n tvf %4.1f ns',tvf);
fprintf('\n Avg dv/dt %4.1f V/ns',dvdt_on_avg*1e-9);
fprintf('\n Peak dv/dt %4.1f V/ns \n\n',dvdt_on_pk*1e-9);



%%%%%%%%%%%%%%%%%% Plot turn-on transient %%%%%%%%%%%%%%%%%%%%

% Set up plot window
ta = turnon_start_onwf;
tb = turnon_end_onwf;
fig = figure(2); clf; hold on;
fig.Name = 'Turn-On Transient';

% Setup Power Subplot and plot power line
powerSubPlot = subplot(2, 1, 1);
powerSubPlot.XGrid = 'on';
powerSubPlot.YGrid = 'on';
powerSubPlot.Position = [0.01 0.77 0.98 0.22];
powerSubPlot.XAxis.Visible = 'off';
powerSubPlot.YAxis.Visible = 'off';
powerSubPlot.FontSize = fontsize;
powerSubPlot.XTick = [0:time_grid_increment:turnon_window_time-mod
      (turnon_window_time,time_grid_increment)];
powerSubPlot.YTick = [floor(min(Pon(ta:tb)/Pon_scale)):1:ceil(max(Pon
      (ta:tb)/Pon_scale))];
powerSubPlot.GridLineStyle = '--';
powerSubPlot.GridColor = [0.6 0.6 0.6];
powerSubPlot.GridAlpha = 0.5;
```

```
powerLine = line((time_onwf(ta:tb)-time_onwf(ta))*1e9, Pon(ta:tb)/
     (Pon_scale));
powerLine.Color = powerColor;
powerLine.LineWidth = linewidth;
powerYAxis = powerSubPlot.YAxis;
powerYAxis.Limits = [floor(min(Pon/Pon_scale)) ceil(max(Pon/Pon_
     scale))];
powerYAxisLength = powerYAxis.Limits(2)-powerYAxis.Limits(1);
axis([0 turnon_window_time powerYAxis.Limits]);
% Add triangle to show y=0 point in scope-view plot
hold on;
fill([0 0 turnon_window_time/30], [-powerYAxisLength/8 powerYAxisLength/
     8 0],'black');
line([0 0], [powerYAxis.Limits], 'color','black','linewidth',2)
line([1 1]*turnon_window_time, [powerYAxis.Limits], 'color','black',
     'linewidth',2)
line([0 turnon_window_time], [1 1]*powerYAxis.Limits(1), 'color',
     'black','linewidth',2)
line([0 turnon_window_time], [1 1]*powerYAxis.Limits(2), 'color',
     'black','linewidth',2)

if (plotlines == true)
%   line([t1_time_onwf t1_time_onwf],[powerYAxis.Limits],'Color',[1 0.2
     0.7],'LineStyle','--','LineWidth',2);
   line([t2_time_onwf t2_time_onwf],[powerYAxis.Limits],'Color',[1 0.2
     0.7],'LineStyle','--','LineWidth',2);
%   line([t3_time_onwf t3_time_onwf],[powerYAxis.Limits],'Color',[1 0.2
     0.7],'LineStyle','--','LineWidth',2);
   line([t4_time_onwf t4_time_onwf],[powerYAxis.Limits],'Color',[1 0.2
     0.7],'LineStyle','--','LineWidth',2);
end

% Setup id/vds/vgs subplot and plot these lines
measSubP = subplot(2, 1, 2);
measSubP.Position = [0.01 0.16 0.98 0.6];
measSubP.XGrid = 'on';
measSubP.YGrid = 'on';
measSubP.GridLineStyle = '--';
measSubP.GridColor = [0.6 0.6 0.6];
measSubP.GridAlpha = 0.5;
measSubP.FontSize = fontsize;
measSubP.XAxis.Visible = 'on';
measSubP.YAxis.Visible = 'off';
measSubP.XTick = [0:time_grid_increment:turnon_window_time-mod
     (turnon_window_time,time_grid_increment)];
measSubP.YTick = [-100:1:100];
measYAxis = measSubP.YAxis;
xlabel(['Time (ns)']);
line([0 turnon_window_time],[0 0],'Color',[0.4 0.4 0.4],'LineStyle',
     '-.','LineWidth',3);
ymax = 0; ymin = 0;
% V_DS
vdsOnLine = line((time_onwf(ta:tb)-time_onwf(ta))*1e9, vds_onwf(ta:tb)/
     vds_scale);
vdsOnLine.Color = vdsColor;
vdsOnLine.LineWidth = linewidth;
ymax = max([ymax ceil(max(vds_onwf(ta:tb)/vds_scale))]);
ymin = min([ymin floor(min(vds_onwf(ta:tb)/vds_scale))]);
```

```
% V_GS
vgsOnLine = line((time_onwf(ta:tb)-time_onwf(ta))*1e9, vgs_onwf(ta:tb)/
     vgs_scale);
vgsOnLine.Color = vgsColor;
vgsOnLine.LineWidth = linewidth;
ymax = max([ymax ceil(max(vgs_onwf(ta:tb)/vgs_scale))]);
ymin = min([ymin floor(min(vgs_onwf(ta:tb)/vgs_scale))]);
% I_D
idOnLine = line((time_onwf(ta:tb)-time_onwf(ta))*1e9, id_onwf(ta:tb)/
     id_on_scale);
idOnLine.Color = idColor;
idOnLine.LineWidth = linewidth;
ymax = max([ymax ceil(max(id_onwf(ta:tb)/id_on_scale))]);
ymin = min([ymin floor(min(id_onwf(ta:tb)/id_on_scale))]);

% Calculate Coss and channel currents if option is selected
% To use this function, you must generate a file Coss.mat with a variable
% Coss stored in it, which can be created using file
% "Process_capacitance_curves.m". This Coss vector must be scaled such
% that the voltage is equal to the index of each point (i.e., Coss(12) is
% the Coss of your device when Vds = 12 V).
% You must also select "filter_samples" wisely, based on how noisy your
% data set is.
if (plot_channel_current == true)
  load Coss.mat;
  % Calculate dvdt using a centered moving average, which is accomplished
  % by looking forward and backward in time equally (filter_samples) then
  % dividing the difference in vds by 2*filter_samples*deltat
  dvdt_calc = [(vds_onwf(ta+filter_samples:tb+filter_samples) - vds_onwf
       (ta-filter_samples:tb-filter_samples))/(deltat_onwf * 2 *
       filter_samples)];
  % Use imported Coss.mat data to find Coss at each sampled time.
  % Note that if vds < 1, Coss(1) is used, and if vds exceeds the Coss
  % vector that was imported, it uses the highest vds point on the curve.
  for mmm = 1:tb-ta + 1
    if (vds_onwf(mmm + ta - 1) < 1);
      Coss_calc(mmm) = Coss(1) + Cstray;
    elseif (vds_onwf(mmm + ta - 1) > length(Coss));
      Coss_calc(mmm) = Coss(end) + Cstray;
    else
      Coss_calc(mmm) = Coss(floor(vds_onwf(mmm + ta - 1))) + Cstray;
    end
    icoss_onwf(mmm) = -dvdt_calc(mmm) * Coss_calc(mmm);
    ichan_onwf(mmm) = icoss_onwf(mmm) + id_onwf(mmm + ta - 1);
  end
  % Plot icoss line
  icossOnLine = line((time_onwf(ta:tb)-time_onwf(ta))*1e9,icoss_onwf/
       id_on_scale);
  icossOnLine.Color = icossColor;
  icossOnLine.LineWidth = linewidth;
  ymax = max([ymax ceil(max(icoss_onwf/id_on_scale))]);
  ymin = min([ymin floor(min(icoss_onwf/id_on_scale))]);
  % Plot ichan line
  ichanOnLine = line((time_onwf(ta:tb)-time_onwf(ta))*1e9,ichan_onwf/
       id_on_scale);
  ichanOnLine.Color = ichanColor;
  ichanOnLine.LineWidth = linewidth;
  ymax = max([ymax ceil(max(ichan_onwf/id_on_scale))]);
  ymin = min([ymin floor(min(ichan_onwf/id_on_scale))]);
end
```

```
% Add triangles to show y=0 point in scope-view plots
axis([0 turnon_window_time ymin ymax]);
measYAxisLength = measYAxis.Limits(2)-measYAxis.Limits(1);
xTick = get(measSubP,'XTick');
measSubP.XTick = xTick(1:end-1);
hold on;
fill([0 0 turnon_window_time/30]+turnon_window_time/50, [-1 1 0]*
     measYAxisLength/20, vgsColor);
fill([0 0 turnon_window_time/30]+turnon_window_time/100, [-1 1 0]*
     measYAxisLength/20, idColor);
fill([0 0 turnon_window_time/30], [-1 1 0]*measYAxisLength/20, vdsColor);
% Add black border around plot
line([0 0], [measYAxis.Limits], 'color','black','linewidth',2)
line([1 1]*turnon_window_time, [measYAxis.Limits], 'color','black',
     'linewidth',2)
line([0 turnon_window_time], [1 1]*measYAxis.Limits(1), 'color',
     'black','linewidth',2)
line([0 turnon_window_time], [1 1]*measYAxis.Limits(2), 'color',
     'black','linewidth',2)


% Optionally, plot the subinterval trigger points t1-t4 to help verify
% calculations. If lines do not appear to be where they should be, then
% algorithm for detecting trigger points may need to be modified for your
% data set.
if (plotlines == true)
     line([t1_time_onwf t1_time_onwf],[ymin ymax],'Color',[1 0.2
     0.7],'LineStyle','--','LineWidth',2);
     line([t2_time_onwf t2_time_onwf],[ymin ymax],'Color',[1 0.2
     0.7],'LineStyle','--','LineWidth',2);
     line([t3_time_onwf t3_time_onwf],[ymin ymax],'Color',[1 0.2
     0.7],'LineStyle','--','LineWidth',2);
     line([t4_time_onwf t4_time_onwf],[ymin ymax],'Color',[1 0.2
     0.7],'LineStyle','--','LineWidth',2);
end


% Optionally, calculate di/dt and graphically examine I-V alignment to
% improve/correct deskew setting
if (alignment_check == true)
   % Calculate didt using a centered moving average, which is accomplished
   % by looking forward and backward in time equally (filter_samples) then
   % dividing the difference in id by 2*filter_samples*deltat
   didt = [(id_onwf(ta+filter_samples:tb+filter_samples) - id_onwf(ta-
        filter_samples:tb-filter_samples))/(deltat_onwf * 2 * filter_
        samples)];
   % This equation is based on a KVL of power loop, using settings
   % selected at beginning of this file.
   vds_calc = Vdc_onwf + Vbd + (IL_onwf-id_onwf(ta:tb))*Rbd - id_onwf(ta:
        tb)*Rshunt - Lloop*didt;
   % Adjust dc bias to make graphical alignment easier
   vds_calc_adjust = mean(vds_onwf(ta:t2_onwf)-vds_calc(1:t2_onwf-
        ta+1));
   vds_calc = vds_calc + vds_calc_adjust;
   % Plot vds and this new vds calculated using id waveform, and compare
   % the two graphically. If not well aligned, try adjusting
   % filter_samples or id_alignment_ns until the two match well.
   % IMPORTANT! - The deskew setting should not change unless your test
   % setup changes (board, cables, scope, shunt, etc.). Do not adjust
   % deskew or id_alignment_ns differently for different data points in
```

```
  % the same test data set.
  fig = figure(5); clf; hold on; hax = gca; grid on;
  set(fig, 'units', 'inches', 'position', [2 2 7.4 4.5])
  set(hax,'FontSize',fontsize);
  plot((time_onwf(ta:tb)-time_onwf(ta))*1e9,vds_onwf(ta:tb),
       '-','lineWidth',linewidth,'color',[0 0 0.8])
  plot((time_onwf(ta:tb)-time_onwf(ta))*1e9,vds_calc,'-','lineWidth',
       linewidth,'color',[0.5 0.1 0.5])
  line([t2_time_onwf t2_time_onwf],get(hax,'YLim'),'Color',[1 0.2
       0.7],'LineStyle','--','LineWidth',2);
  line([t3_time_onwf t3_time_onwf],get(hax,'YLim'),'Color',[1 0.2
       0.7],'LineStyle','--','LineWidth',2);

  xlabel('Time (ns)');
  ylabel('Voltage (V)');
  legend('v_{ds,meas}','v{ds,calc}','location','Southwest')
  %This scaling must be selected at the beginning of the file.
  if deskew_start>=0
    axis([deskew_start deskew_stop deskewVmin deskewVmax]);
  end
end

end


if (filename_off > 0)
%%%%%%%%%%%%%%%%%%%%%%%%%%%%%%%%%%%
%%%%%%%%%% Turn-off waveform analysis %%%%%%%%%%%%%
%%%%%%%%%%%%%%%%%%%%%%%%%%%%%%%%%%%

%This section of the code processes a full data set that has been captured
%for analyzing turn-off waveform data (all variables ending with "_offwf").
%This data set likely has turn-on data as well, so that data must be
%processed partially, but it is not used for turn-on transient analysis.

%Import data
if (TekDPO5xxx == true) && (TekDPO4xxx == false) %Import 5000 series data
     format
  M = csvread([testpoint_filepath filename_off '_Ch' num2str(vgs_
       channel_number) '.csv'],0,3);
  time_offwf = (M(:,1) - M(1,1));
  deltat_offwf = (M(2,1)-M(1,1));
  vgs_offwf = M(:,2);
  M = csvread([testpoint_filepath filename_off '_Ch' num2str(vds_
       channel_number) '.csv'],0,3);
  vds_offwf = M(:,2);
  M = csvread([testpoint_filepath filename_off '_Ch' num2str(id_
       channel_number) '.csv'],0,3);
  id_offwf = -M(:,2);
  vector_length = length(time_offwf);
elseif (TekDPO4xxx == true) && (TekDPO5xxx == false) %Import 4000 series
     data format
  M = csvread([testpoint_filepath filename_off],21,0);
  time_offwf = (M(:,1)-M(1,1));
  vgs_offwf = M(:,vgs_channel_number+1);
  vds_offwf = M(:,vds_channel_number+1);
  id_offwf = M(:,id_channel_number+1);
```

```
  deltat_offwf = (M(2,1)-M(1,1));
  vector_length = length(time_offwf);
else
  return; %Not a compatible data format
end

% Check for inf and replace with closest non-inf data point
for m=1:vector_length
  if((vds_offwf(m)==Inf)|(vds_offwf(m)==-Inf))
    vds_offwf(m) = vds_offwf(m-1);
  end
  if((id_offwf(m)==Inf)|(id_offwf(m)==-Inf))
    id_offwf(m) = id_offwf(m-1);
  end
  if((vgs_offwf(m)==Inf)|(vgs_offwf(m)==-Inf))
    vgs_offwf(m) = vgs_offwf(m-1);
  end
end

%Adjust id if not scaled properly in scope
id_offwf = id_offwf*i_adj;

% Time shift id to correct deskew (I-V misalignment)
id_alignment_pts = round(id_alignment_ns * 1e-9 / deltat_offwf);
id_length = length(id_offwf);
if (id_alignment_pts > 0)
  id_offwf(1:id_length-id_alignment_pts) = id_offwf(id_alignment_pts+
       1:id_length);
else
  id_offwf(-id_alignment_pts+1:id_length) = id_offwf(1:id_length+id_
       alignment_pts);
end

%%%%%%%%%%%Full window view of turn-off data set%%%%%%%%%

%Set up plot window
fig = figure(3); clf; hold on;
fig.Name = 'Turn-Off Waveform: Full Capture';
fullPlot = gca;
fullPlot.XGrid = 'on';
fullPlot.YGrid = 'on';
fullPlot.GridLineStyle = '--';
fullPlot.GridColor = [0.6 0.6 0.6];
fullPlot.GridAlpha = 0.5;
fullPlot.FontSize = fontsize;
fullPlot.XAxis.Visible = 'on';
fullPlot.YAxis.Visible = 'off';
fullPlotYAxis = fullPlot.YAxis;

%Plot lines
xlabel(['Time (\mus)']);
fullPlotXAxisLength = time_offwf(end)*1e6;
ymax = 0; ymin = 0;
% V_DS
vdsOffLine = line((time_offwf)*1e6, vds_offwf/vds_scale);
vdsOffLine.Color = vdsColor;
vdsOffLine.LineWidth = linewidth;
```

```
% V_GS
vgsOffLine = line(time_offwf*1e6, vgs_offwf/vgs_scale);
vgsOffLine.Color = vgsColor;
vgsOffLine.LineWidth = linewidth;
% I_D
idOffLine = line(time_offwf*1e6, id_offwf/id_on_scale);
idOffLine.Color = idColor;
idOffLine.LineWidth = linewidth;

% Add triangles to show y=0 point in scope-view plots
fullPlotYAxisLength = fullPlotYAxis.Limits(2)-fullPlotYAxis.Limits(1);
hold on;
fill([0 0 fullPlotXAxisLength/30]+fullPlotXAxisLength/50, [-1 1 0]
     *fullPlotYAxisLength/20, vgsColor);
fill([0 0 fullPlotXAxisLength/30]+fullPlotXAxisLength/100, [-1 1 0]
     *fullPlotYAxisLength/20, idColor);
fill([0 0 fullPlotXAxisLength/30], [-1 1 0]*fullPlotYAxisLength/20,
     vdsColor);
line([0 0], [fullPlotYAxis.Limits], 'color','black','linewidth',2)
line([1 1]*fullPlotXAxisLength, [fullPlotYAxis.Limits], 'color',
     'black','linewidth',2)
line([0 fullPlotXAxisLength], [1 1]*fullPlotYAxis.Limits(1), 'color',
     'black','linewidth',2)
line([0 fullPlotXAxisLength], [1 1]*fullPlotYAxis.Limits(2), 'color',
     'black','linewidth',2)

%%%%%%%%%%%%%%%% Find turn-off time indices %%%%%%%%%%%%%%%%%
j = 1;
while (vds_offwf(j) > 0.5*max(vds_offwf)) %Find and skip first turn-on, if
     it exists
  j = j + 1;
end
j = j + ceil(turnon_window_time*1e-9/deltat_offwf); %Skip ahead
while (vds_offwf(j) < 0.5*max(vds_offwf)) %Now find turn-off transient of
     interest
  j = j + 1;
end
%Set beginning and end of turn-off transient (this is the data that will be
%used for analysis of the turn-off transient)
j = j - floor(turnoff_window_stepback*1e-9/deltat_offwf);
turnoff_start_offwf = j;
j = j + floor(turnoff_window_time*1e-9/deltat_offwf);
turnoff_end_offwf = j;

%%%%%%%% Find turn-on time indices %%%%%%%%%%
while (vds_offwf(j) > 0.5*max(vds_offwf))
  j = j + 1;
end
%Set beginning and end of turn-on transient (but this data will not be used
     for
%turn-on transient analysis, since it comes from the turn-off data set)
j = j - floor(turnon_window_stepback*1e-9/deltat_offwf);
turnon_start_offwf = j;
j = j + floor(turnon_window_time*1e-9/deltat_offwf);
turnon_end_offwf = j;

while (vds_offwf(j) < 0.5*max(vds_offwf))
  j = j + 1;
end
```

```
%If it is captured in data set, this will find the end of the second pulse
second_turnoff_start_offwf = j - floor(turnoff_window_stepback*1e-9/
     deltat_offwf);

%Plot dashed vertical lines indicating the turn-on and turn-off transient
%locations on the full turn-on waveform capture
if (plotlines == true)
   line([turnoff_start_offwf turnoff_start_offwf]*deltat_offwf*
        1e6,fullPlotYAxis.Limits,'Color',[1 0.2 0.7],'LineStyle','--',
        'LineWidth',2);
   line([turnoff_end_offwf turnoff_end_offwf]*deltat_offwf*1e6,
        fullPlotYAxis.Limits,'Color',[1 0.2 0.7],'LineStyle','--',
        'LineWidth',2);
   line([turnon_start_offwf turnon_start_offwf]*deltat_offwf*1e6,
        fullPlotYAxis.Limits,'Color',[1 0.2 0.7],'LineStyle','--',
        'LineWidth',2);
   line([turnon_end_offwf turnon_end_offwf]*deltat_offwf*1e6,
        fullPlotYAxis.Limits,'Color',[1 0.2 0.7],'LineStyle','--',
        'LineWidth',2);
end

% Estimate load current based on average value 20 ns prior to start of turn-
     off waveform
IL_offwf = mean(id_offwf(turnoff_start_offwf-floor(20e-9/deltat_
     offwf):turnoff_start_offwf));

%Adjust for any dc bias error (not generally recommended, but optional)
if (biasadjust == true)
   id_dcbias_offwf = mean(id_offwf(turnon_start_offwf-floor(20e-9/
        deltat_offwf):turnon_start_offwf));
   id_offwf = id_offwf - id_dcbias_offwf;
   vds_dcbias_offwf = mean(vds_offwf(turnon_end_offwf:second_turnoff_
        start_offwf)) - IL_offwf*Rdson;
   vds_offwf = vds_offwf - vds_dcbias_offwf;
   % Recalculate IL with bias adjusted
   IL_offwf = mean(id_offwf(turnoff_start_offwf-floor(20e-9/deltat_
        offwf):turnoff_start_offwf));
end

% Find steady-state on values for vgs based on average 20 ns prior to start of
     turn-off waveform
% (this should be the positive driving voltage)
Vdrv_offwf = mean(vgs_offwf(turnoff_start_offwf-floor(20e-9/delta-
     t_offwf):turnoff_start_offwf));
% Find average vds between 20 ns after turn-off and beginning of turn-on
Vdc_offwf = mean(vds_offwf(turnoff_end_offwf+floor(20e-9/delta-
     t_offwf):turnon_start_offwf));

%%%%%%%% Find turn-off trigger points for subinterval analysis %%%%%%%%%
j = turnoff_start_offwf;
while (vgs_offwf(j) > Vdrv_offwf*0.9)
   j = j + 1;
end
while (vgs_offwf(j-1) > vgs_offwf(j)) || (vgs_offwf(j-2) > vgs_offwf(j))
   j = j - 1;
end
%t5 is point where vgs starts to fall
t5_offwf = j;
```

```
t5_time_offwf = (time_offwf(t5_offwf)-time_offwf(turnoff_start_offwf))
      *1e9;
% Now, find time when vds has risen to 5% of Vdc
while (vds_offwf(j) < Vdc_offwf*0.05)
   j = j + 1;
end
% Then, go backward until vds stops decreasing significantly
while (vds_offwf(j) > mean(vds_offwf(j-15:j)))
   j = j - 1;
end
%t6 is point where vds starts to rise
t6_offwf = j;
t6_time_offwf = (time_offwf(t6_offwf)-time_offwf(turnoff_start_offwf))
      *1e9;
while (id_offwf(j) > 0) %Find first negative-sloped zero crossing of id
   j = j + 1;
end
% Check that vds > Vdc at this zero crossing. If not, find next
% negative-sloped zero crossing until this is true. (Usually it will be
% the first zero crossing, but this may get complicated in cases where the
% device package inductance is significant).
while (vds_offwf(j) < Vdc_offwf)
   while (id_offwf(j) < 0)
      j = j + 1;
   end
   while (id_offwf(j) > 0)
      j = j + 1;
   end
end
% t7 is first negative-sloped point where id crosses 0 A that occurs after
% vds exceeds Vdc. In some cases, this may be a trough point rather than a
% zero crossing then starts increasing again (only if it never crosses 0 A).
t7_offwf = j;
t7_time_offwf = (time_offwf(t7_offwf)-time_offwf(turnoff_start_offwf))
      *1e9;


% In very light-load cases, id may have a non-negligible dc bias error
% following the turn-off transient, which may prevent id from crossing
% zero. In these cases, it may work better to use the minimum id point,
% if it approaches 0 A but does not cross. Alternatively, you can use the
% peak vds point, which is nearby to the actual t7, but may not be exactly
% the same moment (very tough to know for sure in some waveforms).
[peak_vds maxvds_index] = max(vds_offwf(turnoff_start_offwf:turnoff_
      end_offwf));
[minid_off minid_index] = min(id_offwf(turnoff_start_offwf:turnoff_
      end_offwf));
if (t7_offwf > turnoff_end_offwf)
%   t7_offwf = maxvds_index(1) + turnoff_start_offwf - 1;
   t7_offwf = minid_index(1) + turnoff_start_offwf - 1;
   t7_time_offwf = (time_offwf(t7_offwf)-time_offwf(turnoff_start_
         offwf))*1e9;
end


%%%%%%%%%%%%%% Calculate switching times %%%%%%%%%%%
td_off = t6_time_offwf - t5_time_offwf; %turn-off delay time
tvr = t7_time_offwf - t6_time_offwf; %turn-off voltage rise time
t_off = t7_time_offwf - t5_time_offwf; %total turn-off time
```

```
%%%%%%%%%%%%%% Calculate switching energy %%%%%%%%%%%
Poff = vds_offwf.*id_offwf;
% Eoffwf_cum((turnoff_start_offwf:turnoff_end_offwf)-turnoff_start_
      offwf+1) = cumtrapz(time_offwf(turnoff_start_offwf:turnoff_end_
      offwf),Pon((turnoff_start_offwf:turnoff_end_offwf)-turnoff_
      start_offwf+1));
if (compensate_Esw == true)
   %Optionally, compensate for conduction loss that would have occurred
   %during the time interval of switching, to avoid double-counting (Ideal
   %duty cycles D and 1-D usually used for conduction loss calculation,
   %neglecting switching transient time)
   Eoff_comp = (IL_offwf * Rdson^2) * (t7_offwf-t6_offwf)*deltat_offwf;
else
   Eoff_comp = 0;
end
% Turn-on loss is integral of I*V during tvr only
Eoff = trapz(time_offwf(t6_offwf:t7_offwf),Poff(t6_offwf:t7_offwf)) -
      Eoff_comp;


%%%%%%%%%%%%%%%%%% Calculate dv/dt %%%%%%%%%%%%%%%%%%%
j = t6_offwf;
while (vds_offwf(j) < Vdc_offwf*0.1)
   j = j + 1;
end
% Find adjusted t6 where vds = 0.1Vdc
t6_10percent = j;
while (vds_offwf(j) < Vdc_offwf*0.9)
   j = j + 1;
end
% Find adjusted t7 where vds = 0.9Vdc
t7_90percent = j;
tvr_1090 = (t7_90percent - t6_10percent)*deltat_offwf*1e9;
% Using 10 & 90% points neglects the exponential rise at start and end, and
% instead focuses on nearly-linear rise of vds
dvdt_off(1:t7_90percent-t6_10percent+1) = (vds_offwf(t6_10percent+1:
      t7_90percent+1) - vds_offwf(t6_10percent:t7_90percent))./(deltat_
      offwf);
dvdt_off_pk = max(dvdt_off); %Peak dv/dt
dvdt_off_avg = mean(dvdt_off); %Average dv/dt, which is typically used to
      describe switching speed


% Calculate peak vds and vds overshoot, using the measured Vdc.
% Note that this assumes Vdc and vds both have some dc bias error
% immediately following the transient, since this Vdc is usually higher
% than the actual dc voltage coming from the benchtop supply.
vds_overshoot = peak_vds - Vdc_offwf;


%Display results in command window
disp('Turn-off waveform analysis:');
fprintf(' V_dc        %4.0f V',Vdc_offwf);
fprintf('\n I_L          %4.1f A',IL_offwf);
fprintf('\n Eoff         %4.1f uJ',Eoff*1e6);
fprintf('\n td_off       %4.1f ns',td_off);
fprintf('\n tvr         %4.1f ns',tvr);
fprintf('\n Avg dv/dt   %4.1f V/ns',dvdt_off_avg*1e-9);
fprintf('\n Peak dv/dt  %4.1f V/ns',dvdt_off_pk*1e-9);
fprintf('\n Vds overshoot  %4.1f V \n\n',vds_overshoot);
```

```
%%%%%%%%%%%%%%%% Plot turn-off transient %%%%%%%%%%%%%%%%%%%%

% Set up plot window
ta = turnoff_start_offwf;
tb = turnoff_end_offwf;
fig = figure(4); clf; hold on;
fig.Name = 'Turn-Off Transient';

% Setup Power Subplot and plot power line
powerSubPlot = subplot(2, 1, 1);
powerSubPlot.XGrid = 'on';
powerSubPlot.YGrid = 'on';
powerSubPlot.Position = [0.01 0.77 0.98 0.22];
powerSubPlot.XAxis.Visible = 'off';
powerSubPlot.YAxis.Visible = 'off';
powerSubPlot.FontSize = fontsize;
powerSubPlot.XTick = [0:time_grid_increment:turnoff_window_time-mod
     (turnoff_window_time,time_grid_increment)];
powerSubPlot.YTick = [floor(min(Poff(ta:tb)/Poff_scale)):1:ceil(max
     (Poff(ta:tb)/Poff_scale))];
powerSubPlot.GridLineStyle = '--';
powerSubPlot.GridColor = [0.6 0.6 0.6];
powerSubPlot.GridAlpha = 0.5;
powerLine = line((time_offwf(ta:tb)-time_offwf(ta))*1e9, Poff(ta:tb)/
     (Poff_scale));
powerLine.Color = powerColor;
powerLine.LineWidth = linewidth;
powerYAxis = powerSubPlot.YAxis;
powerYAxis.Limits = [floor(min(Poff(ta:tb)/Poff_scale)) ceil(max(Poff
     (ta:tb)/Poff_scale))];
powerYAxisLength = powerYAxis.Limits(2)-powerYAxis.Limits(1);
axis([0 turnoff_window_time powerYAxis.Limits]);
% Add triangle to show y=0 point in scope-view plot
hold on;
fill([0 0 turnoff_window_time/30], [-powerYAxisLength/8 powerYAxi-
     sLength/8 0],'black');
line([0 0], [powerYAxis.Limits], 'color','black','linewidth',2)
line([1 1]*turnoff_window_time, [powerYAxis.Limits], 'color','black',
     'linewidth',2)
line([0 turnoff_window_time], [1 1]*powerYAxis.Limits(1), 'color',
     'black','linewidth',2)
line([0 turnoff_window_time], [1 1]*powerYAxis.Limits(2), 'color',
     'black','linewidth',2)


if (plotlines == true)
%    line([t5_time_offwf t5_time_offwf],[powerYAxis.Limits],'Color',[1
     0.2 0.7],'LineStyle','--','LineWidth',2);
  line([t6_time_offwf t6_time_offwf],[powerYAxis.Limits],'Color',[1 0.2
       0.7],'LineStyle','--','LineWidth',2);
  line([t7_time_offwf t7_time_offwf],[powerYAxis.Limits],'Color',[1 0.2
       0.7],'LineStyle','--','LineWidth',2);
end

% Setup id/vds/vgs subplot and plot these lines
measSubP = subplot(2, 1, 2);
measSubP.Position = [0.01 0.16 0.98 0.6];
measSubP.XGrid = 'on';
measSubP.YGrid = 'on';
```

```
measSubP.GridLineStyle = '--';
measSubP.GridColor = [0.6 0.6 0.6];
measSubP.GridAlpha = 0.5;
measSubP.FontSize = fontsize;
measSubP.XAxis.Visible = 'on';
measSubP.YAxis.Visible = 'off';
measSubP.XTick = [0:time_grid_increment:turnoff_window_time-mod
     (turnoff_window_time,time_grid_increment)];
measSubP.YTick = [-100:1:100];
measYAxis = measSubP.YAxis;
xlabel(['Time (ns)']);
line([0 turnoff_window_time],[0 0],'Color',[0.4 0.4 0.4],'LineStyle',
     '-.','LineWidth',3);
ymax = 0; ymin = 0;
% V_DS
vdsOffLine = line((time_offwf(ta:tb)-time_offwf(ta))*1e9, vds_offwf(ta:
     tb)/vds_scale);
vdsOffLine.Color = vdsColor;
vdsOffLine.LineWidth = linewidth;
ymax = max([ymax ceil(max(vds_offwf(ta:tb)/vds_scale))]);
ymin = min([ymin floor(min(vds_offwf(ta:tb)/vds_scale))]);
% V_GS
vgsOffLine = line((time_offwf(ta:tb)-time_offwf(ta))*1e9, vgs_offwf(ta:
     tb)/vgs_scale);
vgsOffLine.Color = vgsColor;
vgsOffLine.LineWidth = linewidth;
ymax = max([ymax ceil(max(vgs_offwf(ta:tb)/vgs_scale))]);
ymin = min([ymin floor(min(vgs_offwf(ta:tb)/vgs_scale))]);
% I_D
idOffLine = line((time_offwf(ta:tb)-time_offwf(ta))*1e9, id_offwf(ta:
     tb)/id_off_scale);
idOffLine.Color = idColor;
idOffLine.LineWidth = linewidth;
ymax = max([ymax ceil(max(id_offwf(ta:tb)/id_off_scale))]);
ymin = min([ymin floor(min(id_offwf(ta:tb)/id_off_scale))]);


% Calculate Coss and channel currents if option is selected
% To use this function, you must generate a file Coss.mat with a variable
% Coss stored in it, which can be created using file
% "Process_capacitance_curves.m". This Coss vector must be scaled such
% that the voltage is equal to the index of each point (i.e., Coss(12) is
% the Coss of your device when Vds = 12 V).
% You must also select "filter_samples" wisely, based on how noisy your
% data set is.
if (plot_channel_current == true)
   load Coss.mat;
   % Calculate dvdt using a centered moving average, which is accomplished
   % by looking forward and backward in time equally (filter_samples) then
   % dividing the difference in vds by 2*filter_samples*deltat
   dvdt_calc = [(vds_offwf(ta+filter_samples:tb+filter_samples) - vds_
        offwf(ta-filter_samples:tb-filter_samples))/(deltat_offwf * 2 *
        filter_samples)];
   % Use imported Coss.mat data to find Coss at each sampled time.
   % Note that if vds < 1, Coss(1) is used, and if vds exceeds the Coss
   % vector that was imported, it uses the highest vds point on the curve.
   for mmm = 1:tb-ta + 1
      if (vds_offwf(mmm + ta - 1) < 1);
         Coss_calc(mmm) = Coss(1) + Cstray;
      elseif (vds_offwf(mmm + ta - 1) > length(Coss));
         Coss_calc(mmm) = Coss(end) + Cstray;
```

```
    else
      Coss_calc(mmm) = Coss(floor(vds_offwf(mmm + ta - 1))) + Cstray;
    end
    icoss_offwf(mmm) = dvdt_calc(mmm) * Coss_calc(mmm);
    ichan_offwf(mmm) = -icoss_offwf(mmm) + id_offwf(mmm + ta - 1);
  end
  % Plot icoss line
  icossOffLine = line((time_offwf(ta:tb)-time_offwf(ta))*1e9,
       icoss_offwf/id_off_scale);
  icossOffLine.Color = icossColor;
  icossOffLine.LineWidth = linewidth;
  ymax = max([ymax ceil(max(icoss_offwf/id_off_scale))]);
  ymin = min([ymin floor(min(icoss_offwf/id_off_scale))]);
  % Plot ichan line
  ichanOffLine = line((time_offwf(ta:tb)-time_offwf(ta))*1e9,ichan_
       offwf/id_off_scale);
  ichanOffLine.Color = ichanColor;
  ichanOffLine.LineWidth = linewidth;
  ymax = max([ymax ceil(max(ichan_offwf/id_off_scale))]);
  ymin = min([ymin floor(min(ichan_offwf/id_off_scale))]);
end

% Add triangles to show y=0 point in scope-view plots
axis([0 turnoff_window_time ymin ymax]);
measYAxisLength = measYAxis.Limits(2)-measYAxis.Limits(1);
xTick = get(measSubP,'XTick');
measSubP.XTick = xTick(1:end-1);
hold on;
fill([0 0 turnoff_window_time/30]+turnoff_window_time/50, [-1 1 0]*
     measYAxisLength/20, vgsColor);
fill([0 0 turnoff_window_time/30]+turnoff_window_time/100, [-1 1 0]*
     measYAxisLength/20, idColor);
fill([0 0 turnoff_window_time/30], [-1 1 0]*measYAxisLength/20, vdsColor);
% Add black border around plot
line([0 0], [measYAxis.Limits], 'color','black','linewidth',2)
line([1 1]*turnoff_window_time, [measYAxis.Limits], 'color','black',
     'linewidth',2)
line([0 turnoff_window_time], [1 1]*measYAxis.Limits(1), 'color',
     'black','linewidth',2)
line([0 turnoff_window_time], [1 1]*measYAxis.Limits(2), 'color',
     'black','linewidth',2)

% Optionally, plot the subinterval trigger points t5-t7 to help verify
% calculations. If lines do not appear to be where they should be, then
% algorithm for detecting trigger points may need to be modified for your
% data set.
if (plotlines == true)
  line([t5_time_offwf t5_time_offwf],[ymin ymax],'Color',[1 0.2 0.7],
       'LineStyle','--','LineWidth',2);
  line([t6_time_offwf t6_time_offwf],[ymin ymax],'Color',[1 0.2 0.7],
       'LineStyle','--','LineWidth',2);
  line([t7_time_offwf t7_time_offwf],[ymin ymax],'Color',[1 0.2 0.7],
       'LineStyle','--','LineWidth',2);
end

end
```

Index

www.ingramcontent.com/pod-product-compliance
Lightning Source LLC
LaVergne TN
LVHW011911200926
848484LV00002B/595
9781785614910